1 MONTH OF
FREE
READING

at

www.ForgottenBooks.com

By purchasing this book you are eligible for one month membership to ForgottenBooks.com, giving you unlimited access to our entire collection of over 1,000,000 titles via our web site and mobile apps.

To claim your free month visit: www.forgottenbooks.com/free533627

ISBN 978-0-666-00002-6
PIBN 10533627

BRITISH·BIRDS

AN ILLUSTRATED MAGAZINE DEVOTED
TO THE BIRDS ON THE BRITISH LIST

EDITED BY

H. F. WITHERBY F.Z.S. M.B.O.U.

ASSISTED BY

W. P. PYCRAFT A.L.S. M.B.O.U.

Volume II.

JUNE 1908—MAY 1909

WITHERBY & CO.
326 HIGH HOLBORN LONDON.

20

PREFACE.

THE papers which have appeared in the pages of the volume now completed do not lose, we venture to think, by comparison with the contents of Volume I.; while the increase in the number of contributors may be taken, not only as an indication that BRITISH BIRDS justifies its existence, but no less as a sign that the study of our native avifauna is being pursued in a wider, and yet more thorough fashion than heretofore.

In the excellent series of articles on " Early British Ornithologists and their Works," we have been given a glimpse into the past, for which our readers will, we feel sure, join us in thanking Mr. W. H. Mullens, who has taken infinite pains to give accurate, as well as interesting information.

The investigation into the causes and spread of Wood-Pigeon Diphtheria, which we commenced in our first volume, has been advanced a stage further, and though the subject bristles with difficulties, we intend, with Dr. C. B. Ticehurst's aid, to pursue the enquiry.

The valuable articles and notes on the habits, and especially the nesting habits of birds, have been a feature of the volume, and special mention may be made of Mr. Noble's paper on the Ducks, and the interesting correspondence arising therefrom. The study of nestling birds, hitherto so strangely neglected, has in this volume made material progress, thanks to the work of Dr. C. B. Ticehurst and Miss A. Jackson, and we hope for more contributions on this subject.

Of the manner of the distribution, and the nesting areas of our summer migrants, our knowledge is meagre, and the Messrs. Alexanders' valuable paper illustrating their novel method of mapping out the haunts of selected species was therefore a most welcome contribution, and will serve as an invaluable model for further work. Hand in hand with research of this kind is that of marking birds, and considerable progress in this direction has recently been made. There are great possibilities in this method of tracing the movements and so on of individual birds, and we hope to devote special attention to it in the future. The subject of geographical races is linked with these migratory movements, and we are glad to note that a more general recognition of such races is being made. Our heartiest support will always be accorded to all who are endeavouring to add to our knowledge in this direction.

Finally, we may refer to the articles on the additions to our knowledge of British Birds recorded since 1899, and now that these are complete, our readers will be enabled to acquire an up-to-date knowledge of the subject by consulting Howard Saunders' "Manual," and the indices of the volumes of this Magazine. To keep these records up-to-date month by month, with the help of our contributors, and by reference to every contribution of interest . published elsewhere, will be our constant endeavour.

THE EDITORS.

May 1st, 1909.

LIST OF ILLUSTRATIONS.

BIRDS

ILLUSTRATED·MAGAZINE
OTED·TO·THE·BIRDS·ON
THE·BRITISH·LIST

E 1,

8.

Vol. II.
No. I.

ONTHLY·ONE·SHILLING·NET
526 HIGH·HOLBORN·LONDON
WITHERBY & Cº

TYPICAL FEATHERS FROM DUCKS' NESTS.

Nat. size.

(Reproduced direct from the feathers.)

(For explanation see p. 23)

BRITISH·BIRDS

EDITED BY H· F. WITHERBY FZS., M.B.O.U.
ASSISTED BY W. P. PYCRAFT. A.L.S. M B.O.U.

CONTENTS OF NUMBER 1, VOL. II. JULY 1. 1908

EDITORIAL.

THANKS to the generous support accorded it during its
first year of life, and to that which has been promised
already for the future, BRITISH BIRDS enters upon its
second year with the prospect of a useful career before it

The programme, which we are already able to announce.
for the next twelve months is sufficient in itself to show
that there will be no falling off in the interest of our
pages ; but, on the contrary, as the year wears on, doubtless we shall receive many other articles in every way as

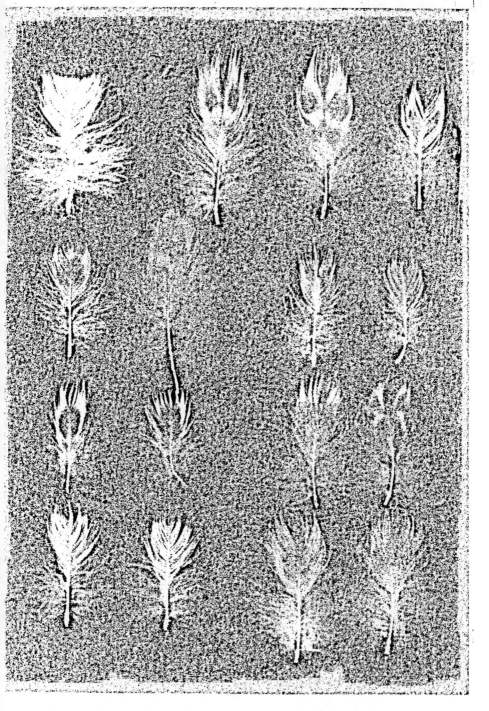

TYPICAL FEATHERS FROM DUCKS' NESTS.

Nat. size.

eproduced direct from the feathers.)

(For explanation see p. 23).

BRITISH·BIRDS

EDITED BY H. F. WITHERBY, F.Z.S., M.B.O.U.
ASSISTED BY W. P. PYCRAFT, A.L.S, M.B.O.U.

CONTENTS OF NUMBER 1, VOL. II. JUNE 1, 1908.

EDITORIAL.

THANKS to the generous support accorded it during its
first year of life, and to that which has been promised
already for the future, BRITISH BIRDS enters upon its
second year with the prospect of a useful career before it.

The programme, which we are already able to announce,
for the next twelve months is sufficient in itself to show
that there will be no falling off in the interest of our
pages ; but, on the contrary, as the year wears on, doubtless we shall receive many other articles in every way as

attractive and as valuable as those we have now the good fortune to announce. Of these, some, we hope, will bear on the themes to be presently suggested.

Not the least interesting matter in our new programme will be, we venture to think, the series of essays on Early British Ornithologists and their Work, by Mr. W. H. Mullens. As many of our readers doubtless know, there are few men so able to appraise the work of these old authors as Mr. Mullens, who has for some time been engaged in the study of these early authors, and during that time he has brought together an extensive collection of their books, many of which are quite inaccessible to the working ornithologist, and these are to be drawn upon for our benefit. There are some, indeed, who seem inclined to decry the labours of these pioneers—who mark only the inaccurate and, sometimes, absurd statements which passed with them for knowledge, and forget how difficult were the conditions under which they were compelled to labour. But the spirit of kindly appreciation shown by Mr. Mullens will enable us to realize that libraries in those days, even where they existed, were not easily accessible; and the dangers and difficulties of travel, even within the confines of Great Britain, were greater than we can readily imagine. We are, in short, inclined to forget that we have entered into their labours, and have built upon the foundations which they laid.

There are many aspects of the bionomical, or, as some prefer to call it, the œcological side of our study which demand more attention than they have generally met with among ornithologists of this country. And we hope that some of our readers may be induced to send us contributions on such subjects, for example, as bear upon the influence of climate on plumage, and on the inter-relations of species. On this last theme, there are several important cases awaiting *systematic* investigation: such, for instance, as the effect of the increasing numbers of Starlings on the Woodpecker; and of the decrease in the Swallow-tribe through the pugnacity of the House-Sparrow. How much of truth is there in the isolated statements which, of recent years, have been made on these subjects? Many other kindred problems will doubtless suggest themselves to our readers.

The subject of Economic Ornithology in this country has been scandalously neglected. So far, scraps of information, mostly incorrect and gathered at haphazard, generally by strongly biassed partizans, have been made to serve our needs. No attempt to remedy this state of affairs can possibly meet with success which is not made in all seriousness, and carried out on strictly scientific lines. One cannot " dabble " with a problem of this kind. We had hoped very much to be able to carry on a preliminary investigation of the kind we are so anxious to see carried out, but a careful calculation has convinced us that the cost of such an enterprise would be prohibitive. We must again express the hope that an investigation will be undertaken by the Board of Agriculture, as has long been done, both on the Continent and in America, and with magnificent results. To carry conviction such an investigation must be prosecuted by an impartial body, and one which can command the services of fully qualified experts, whose work must be carried out under conditions which leave no loophole for doubt.

In the present number will be found the first section of an article framed for the purpose of facilitating the identification of Ducks' eggs—a by no means easy matter. Read with the help of the coloured plates which the generosity of the author enables us to provide, we feel sure that this contribution will overcome the difficulty that has hitherto existed in the determination of doubtful cases.

Among other articles already promised we may mention the following: Mr. Boyd Alexander on the British migrants which he met with in his last great journey from the Niger to the Nile; Mr. E. Bidwell on Cuckoo fosterers; Mr. J. L. Bonhote on British birds which have bred in captivity; Mr. W. H. Kirkman on variations in the nest-building of the Common and Arctic Terns; Commander H. Lynes on the habits of our summer birds when on migration in the Mediterranean; Mr. M. J. Nicoll on the moult of the Swallow; and Prof. Lloyd Morgan on some aspects of the psychology of nest-building, or some kindred theme to be determined by him later. That this will prove a welcome and valuable contribution there is no need to doubt, for Prof. Morgan is the greatest authority on this difficult subject in this country.

As we have already announced, the articles on "Additions to our Knowledge of British Birds since 1899 " will be continued and completed in the present volume, and the results of the Wood-Pigeon enquiry will be given by Dr. C. B. Ticehurst in an early issue.   Other contributors to last year's volume will interest us again, and photography, as an aid to our science, will be to the fore.

We have given an outline of our programme ; but let no intending contributor think that our space is exhausted. We shall always find room for anything which we think should be put before the readers of BRITISH BIRDS.

Finally, we need hardly say that we shall continue to make a feature of " Notes," which, we hope, will increase in number and importance, while we shall make a point, as hitherto, of extracting from all sources information of importance to the student of British birds, and thus provide a current history of the subject ; and in this connection we must again ask our readers' help by drawing attention to papers and records which have escaped our notice.

THE EDITORS.

Pratincola rubetra

SOME EARLY BRITISH ORNITHOLOGISTS AND THEIR WORKS.

BY

W. H. MULLENS, M.A., LL.M., M.B.O.U.

I.—WILLIAM TURNER

(*circa* 1500—1568).

THE history of British ornithology may be said to commence from the time of William Turner, famous both as a naturalist and an author. Born just 400 years ago, this illustrious man, who is styled the "Father of English Botany," is perhaps best known for his researches in that department of natural history ; but he also excelled in several branches of zoology, and his claim to be considered the earliest responsible authority on the birds of this country is undeniable.

Before Turner's time, the available knowledge concerning British birds was small indeed. It is true that a quaint and very credulous writer, Giraldus Cambrensis (1146-1223) had in his *Topography of Ireland* (written in 1187, and first published in 1587) devoted ten chapters to a description of the birds of that country, but his observations, although made at first hand, are confused and unreliable, and more curious than instructive.* Passing mention of certain birds is also to be found in the itineraries of some of the earlier English writers, *e.g.*, William of Worcester (ob. 1480), and John Leland (ob. 1552), and some information concerning the Hawks and Game Birds can be obtained from the old books of the chase—the most famous example of which is *The Boke of St. Albans, containing the Treatises of Hawking, Hunting, and Coat-Armour*, printed at St. Albans, 1486, and attributed to Dame Juliana Barnes, or

---

* *cf.* also Forrest, " The Fauna of North Wales," p. xxv.

Berners. The early forest laws, and the different Acts
of Parliament enjoining the protection or destruction
of certain birds, may also be consulted with advantage,
but these necessarily include only a few species in their
enactments.

Small as the knowledge of birds was in this country, it
can hardly be said to have stood in better case in Continental
Europe. There the study of natural history had made
little or no advance since the days of Aristotle and Pliny.
It had been, in common with much else, enveloped and
obscured in the intellectual gloom of the Middle Ages.
Those few mediæval writers who concerned themselves
with the subject of natural history were content to derive
their information from the great Greek and Latin authors
of the classic age, and while attempting in no way to improve
or elaborate such information, they rather, in the spirit of
the age in which they wrote, disguised it with a mass of
superstition and ignorance. It must, of course, be
remembered that their books were chiefly written with a
medicinal purpose, and that their object was to set forth
the various strange curative properties which they ascribed
to the component parts of the birds and beasts they
mentioned, rather than to study or describe the animals
themselves. Among the more prominent mediæval authors
who treated of birds at any considerable length, it may
here suffice to mention the following :—Albertus Magnus
(ob. 1282), whose twenty-six books, *De Animalibus*, were
printed in 1478 ; Vicentius Belovacensis (ob. 1264),
whose *Speculum Nature* was published at Strasburg about
the same date ; and Bartholomew de Glanville, commonly
known as Bartholomæus Anglius (fl. 1230-1255), from
whose famous work, *De Proprietatibus Rerum*, first printed
at Basle, *circa* 1470, we can obtain a good idea of the
general state of knowledge concerning natural history in
the Middle Ages. Mention should also be made of a
work entitled (*H*)*Ortus Sanitatis*, commonly ascribed to
Johannes de Cuba, and published at Mainz in 1475. This,
though professedly a herbal, deals in its third *tractatus*

with " Birds and Flying-Things," and being the first printed book to contain illustrations of birds, must always be of interest to the student of early ornithology.

It was not until the middle of the sixteenth century that the general revival of learning throughout Europe touched the study of ornithology in particular, and that of natural history in general. This revival, as far as it affected ornithology, was largely due to the illustrious Conrad Gesner (1516-1565), and his able contemporaries, Pierre Belon (1517-1564), author of *L'Histoire de la Nature des Oyseaux* (Paris, 1555), Gybertus Longolius (1507-1543), who wrote the *Dialogus de Avibus*, and William Turner, the subject of this article. In no way inferior in ability to the authors mentioned, Turner was in point of publication their leader, his book *Avium . . . . historia*, appearing in 1544, eleven years before the ornithological works of Belon and Gesner were printed.

William Turner* was born at the beginning of the sixteenth century at Morpeth, in Northumberland, the exact date of his birth being unknown, as the registers of his native town date only from the year 1582. He is said to have been the son of a tanner, but of his childhood and early education we have no record. Through the influence of Thomas, Lord Wentworth, Turner in due course became a member of Pembroke Hall, Cambridge, where he graduated B.A. 1529-30. He became a Fellow of his College in 1531, and its Senior Treasurer in 1538. His M.A. degree he commenced in 1533. How long he retained his Fellowship is uncertain. Mr. Jackson thinks he may have held it until his marriage with Jane, daughter of George Ander, alderman of Cambridge.

At Cambridge, Turner was a contemporary of the famous John Caius, founder of the college which bears his name, and also one of our earliest writers on natural history (his *De rariorum animalium atque stirpium Historia* was

* The particulars of Turner's life are derived from those given in the facsimile reprint of Turner's "Libellus de re Herbaria," by Benjamin Daydon Jackson, F.L.S., privately printed, London, 1877. 1 vol., 4to.

published at London, 1570). It was probably during his
residence at Cambridge that Turner first directed his
attention to the study of birds, while there, no doubt the
fascination of the Fens fell upon him, as it has fallen on so
many since his time, and it was in the Fens that many of
his most valuable observations were made on birds which,
then resident, are now only known as rare stragglers to
this country. It seems unlikely that Turner could have
devoted much time to natural history before he went to the
University, as he himself informs us that he had never seen
the nest of the Water Ousel or Dipper, a somewhat curious
fact when we remember that he was a native of Northumberland.

It was in 1538, while still at Cambridge, that Turner
published his first work on natural history, entitled :—

> Libellus de / re Herbaria Novvs, / in quo her-
> barum aliquot no- / mina Græca, Latina &
> Anglica / habes, vna cum nomini- / bus officin-
> arum, in / gratiam stu- / diosæ iuuentutis nunc
> pri- / mum in lucem / æditus.

Such was the prevailing ignorance of those times
that, writing thirty years later, he bitterly complains
that he could get no assistance in his work from his
contemporaries :—

"Wher as I could learne never one Greke, neither
Latin, nor English name, even amongest the Phisicions
of any herbe or tre, suche was the ignorance in simples
at that tyme, and as yet there was no Englishe Herbal
but one, al full of unlearned Cacographees and falselye
naming of herbes."

During his stay at Cambridge, Turner became an
intimate friend of Nicholas Ridley (1500-1555), and of
Hugh Latimer, Ridley's fellow martyr at the stake.
From Ridley, Turner received his first instruction in
Greek, and, influenced by the teaching of the Reformers,
he now embraced those religious views for which he
laboured so zealously during the remainder of his life.
Leaving his University he travelled through a consider-

able part of England, preaching, and while at Oxford he
was imprisoned for preaching "without a call." When,
"At length being let loose, and banished, he travelled
into Italy."

In Italy Turner studied botany under Luca Ghini, at
Bologna, and took the degree of M.D. either at that
University or at Ferrara. Continuing his travels, he
visited the illustrious Conrad Gesner, at Zurich, and
became a firm friend and trusted correspondent of that
great naturalist.

Turner seems to have been at Basle in 1543, and the
following year at Cologne. From this latter place he
issued in 1544 his *Avium Præcipuarum* . . . . *historia*,
dedicated to Edward, Prince of Wales (afterwards
Edward VI.), and in the same year the posthumous work
of his friend, Gybertus Longolius, of Utrecht (1507-1543),
entitled *Dialogus de Avibus*. Turner's polemical works
now followed each other in quick succession, and were
prohibited by a proclamation of Henry VIII. On the
death of that monarch, Turner returned to England,
and whilst waiting for ecclesiastical preferment acted as
physician to the Lord Protector, Somerset.

At length, after several disappointments, Turner
obtained the Deanery of Wells in 1550.

The accession of Queen Mary saw Turner again a
fugitive, and his writings were once more prohibited in
England, and ordered to be destroyed wherever found.
He returned to his native country when Elizabeth
succeeded her sister, and was reinstated in his Deanery.
In 1564, however, he was again suspended for non-
conformity, and took up his abode in London. There
he died on the 7th July, 1568, and was buried in the
Church of St. Olave, Crutched Friars, where may be seen
a tablet to his memory erected by his widow.

The book on which Turner's fame as an ornithologist
rests has the following title :—

> "Avium / Præcipu / arum, quarum / apud
> Plinium et Ari- / stotelem mentio est, brevis

& / succincta historia. / Ex optimis quibusque
scripto- / ribus Contexta, Scholio illu / strata
& aucta. / Adjectis nominibus Græcis, Ger-
manicis & / Britannicis. / Per Dn. Guilielmum
Turnerum, artium & Me- / dicinæ doctorem /
Coloniæ excudebat Ioan. Gymnicus, / Anno
M.D.XLIIII."

1 Vol., 8vo., pages unnumbered, 157 (*cf. Ibis*, 1899, p. 153).

The above is the first edition. It was reprinted by
Dr. George Thackeray, Provost of King's College,
Cambridge, in 1823 ; the reprint is said to be as rare as
the original—and again by Mr. A. H. Evans, in 1903,
at the Cambridge University Press—Mr. Evans' edition
contains a full translation and many valuable notes.

Turner's object in writing this work is set out both
in the title and in the *Epistola Nuncupatoria* thereof.
This was to determine the principal kinds of birds named
by Aristotle and Pliny in their writings. In addition
to this, he also added copious notes on those species
which came under his own immediate observation,
" and in so doing he has produced the first book on birds
which treats them in anything like a scientific spirit,"
and not merely from a medical point of view. But the
great value of Turner's work consists in the fact that he
is always most careful to tell us whether he observed the
birds he describes in England or abroad, and it is for this
reason that his comments are of such importance to the
student of British ornithology. It must here suffice to
give a few short extracts.

Speaking of the Crane, he says :—" The smaller, that
is, younger, Cranes, are called by Pliny, Vipiones, as
young Doves are known as Pipiones. Cranes, moreover,
breed in England in marshy places ; I myself have often
seen their pipers [young Pigeons are still called pipers
in England], though some people born away from England
urge that this is false " (*cf.* Evans' Ed., p. 97).

And of the Kite, or " Kyte " :—" I know two sorts of
Kites, the greater and the less ; the greater is in colour

nearly rufous, and in England is abundant and remarkably rapacious. This kind is wont to snatch food out of children's hands, in our cities and towns. The other kind is smaller, blacker, and more rarely haunts cities. This I do not remember to have seen in England . . . ." (*cf.* Evans Ed., p. 117).

His remarks on the Black Tern, which ceased to breed in this country about the middle of the last century, the last recorded eggs having been taken in Norfolk in 1858, are also of considerable importance, especially when we consider that this bird still visits the British Isles with unfailing regularity :—" There is another small bird of this kind called Stern* in local dialect, which is so like the sea Lari that it seems to differ from them only in its size and colour ; for it is a Larus, though smaller than the sea Lari, and blacker. Throughout the whole of summer, at which time it breeds, it makes such an unconscionable noise that by its unrestrained clamour it almost deafens those who live near lakes and marshes. This, I certainly believe to be the bird whose vile garrulity gave rise to the old proverb ' Larus partavit.' It is almost always flying over lakes and swamps, never at rest, but always open-mouthed for prey. This bird nests in thick reed-beds " (*cf.* Evans' Ed., p. 79).

The care and trouble which Turner took in verifying the statements of Aristotle and Pliny is shown in the following passage :—" The Mergus [*i.e.*, Cormorant], a sad-coloured bird, is nearly equal to a goose in size, with the bill long and hooked at the end ; it is web-footed, heavy in the body, and the attitude is upright in the sitting bird. Pliny writes that it nests on trees, but Aristotle says on sea-rocks. What each man saw or learnt from the reports of bird-catchers, he has set down in writing. And I have observed both birds myself, for I have seen Mergi nesting on sea-cliffs about the mouth of the Tyne river, and on lofty trees in Norfolk with the Herons " (*cf.* Evans' Ed., p. 111).

* " The Black Tern (*Sterna nigra*)."

The following affords us some idea of the value in which
the Godwit was held as a table bird :—" Furthermore, the
bird (which ' the English call the Godwit, or Fedoa '*)
is so much like the Woodcock that, if it were not a little
larger, and did not the breast verge upon ash-colour,
the one of them could hardly be distinguished from the
other. It is found in marshy places and on river banks.
The beak is long ; but in captivity it feeds on wheat,
just as our Pigeons do. With us it sells for thrice as much
again as any Woodcock, so much does its flesh tickle the
palates of our magnates " (cf. Evans' Ed., p. 45).

Equally interesting are his observations of the Hobby,
Hen-Harrier, Water Ousel (or Dipper),† Bald-Buzzard
(or Marsh-Harrier), Osprey, Wheatear, Sandpiper,
Fieldfare, Cuckoo, Black-headed Gull, and many other
birds, and though he fell into the prevailing error
with regard to the generation of the Bernacle Goose,
the fault was hardly his own. Misled by the accounts
he had read and heard on this subject, he was by
no means convinced, and as he tells us :—" Inasmuch
as it seemed hardly safe . to trust the vulgar, and
by reason of the rarity of the thing I did not quite credit
Gyraldus [i.e., Giraldus Cambrensis], while I thought on
this, of which I now am writing, I took counsel of a
certain man, whose upright conduct, often proved by me,
had justified my trust, a theologian by profession and an
Irishman by birth, Octavian by name, whether he thought
Gyraldus worthy of belief in this affair." The said
Octavian, however, not only informed our author that
the popular fable was a fact, but, further, " taking oath
upon the very Gospel which he taught," stated that he
had seen and handled the young Bernacles as they emerged
from the fungi of wood rotted in the sea, and even promised
to forward Turner " some of these growing Chicks."

* *Vide* Newton " Dict. Birds," p. 248.

† The name " Dipper " was first applied to the Water Ouzel by
Marmaduke Tunstall in his " Ornithologia Britannica." London.
1771. 1 vol., folio.

Turner was held in great. estimation by Gesner, who quotes him freely in his writings under the title of Turnerus Anglus (*cf.* Evans' Ed., p. XI., etc.). There is no evidence that Turner studied mammals, but he certainly published one or more works on ichthyology, besides supplying Gesner with much information about the fishes of Great Britain.*

In taking leave of William Turner, it only remains to add that the authentic books of this remarkable man number no less than thirty-nine, and to quote the description of him given by John Ray :—" *Vir solidæ eruditionis et judicii.*"

* *cf.* Art. by Rev. H. A. Macpherson. "Zoologist," 1898, p. 337.

Parus ater

ON THE OCCURRENCE OF SCHLEGEL'S PETREL (*ŒSTRELATA NEGLECTA*) IN CHESHIRE.

A NEW BRITISH AND EUROPEAN BIRD.

BY

ROBERT NEWSTEAD, A.L.S., & T. A. COWARD, F.Z.S.

On April 1st, 1908, an example of one of the " Dove-like Fulmars " was found dead under a tree near Tarporley, Cheshire, by a man who attends the weekly market at Chester.   On the fourth day after its dis-covery the bird was offered to Mr. Arthur Newstead, who subsequently purchased it ; it is now in his posses-sion.   The bird, a male, was examined by one of us while it was still in the flesh, an outline drawing was made of it, and the colour of the soft parts, the weight, and other details, carefully noted.   The bird was in an excellent state of preservation, and, as might be expected with a Petrel, there was no indication that it had been in captivity.

The distinguishing characters of the bird are as follows :—Uniformly brown, paler beneath ; forehead and cheeks faintly mottled ; " exposed portion of the outer primary beneath—white towards the base of the inner web."* Tarsi, proximal third of the toes, and webs, bluish-grey ; the rest of the feet black.   Tail very slightly rounded.   Bill black.   Irides dark hazel.

The details are :—Upper-surface dark brown, head and neck decidedly greyer ; all the feathers edged with paler brown, with the exception of some of the scapulars, which are also decidedly darker (blackish-brown) than the feathers of the back ; forehead and cheeks mottled with brown.   Under-surface greyish-brown, in a strong

---

* Salvin, *Cat. Birds in coll. Brit. Museum*, XXV., p. 397.

light presenting a marked grey reflection ; traces of narrow, interrupted, obscure dark bands on the breast, which are evident *only when closely examined in a good light*.   Under tail-coverts dark-brown ; bases white. Under wing-coverts and axillaries brown with paler margins ; primaries blackish-brown, bases of inner webs and shafts white.   Concealed bases of all the feathers white, a character most strongly marked on the neck and

Male example of Schlegel's Petrel (*Œstrelata neglecta*) found dead at Tarporley, Cheshire, April 1st, 1908.
(Photographed by ALFRED NEWSTEAD.)

breast, where the grey-brown tips barely cover the underlying portions, so that on the slightest displacement of the feathers the white proximal portions show distinctly through.

Total length, 15 inches ; wing 11.1 ; tail, central and lateral rectrices, 4 ; bill, 1.7 ; tarsus, 1.5 ; middle and outer toe, 2.1 ; inner toe, 1.7.   Weight, 16 oz.

* Salvin, *op. cit.*, p. 412.

This example agrees best with the dark-breasted form of *Œ. neglecta* (Schlegel), but this species, according to Salvin,* has the tarsi and basal portions of the toes yellow. However this may be, we find on comparing our specimen with an example of the dark-breasted form of *Œ. neglecta*, in the collection of the Liverpool Museum, and with the specimens in the Natural History Museum, South Kensington, that they are, we think, specifically identical. The plumage agrees in almost every detail. Furthermore, Salvin* states that " great variation exists as to the colour of the under-surface, some birds being nearly uniform greyish-brown."

Œ. arminjoniana, Gigl. and Salvad., comes very near it, but this species is said to have the " under tail-coverts white,"† and there are other marked differences.

Both the species hitherto recorded for the British Isles [*Œ. hesitata*, Kuhl., and *Œ. brevipes* (Peale)], belong to that section of the genus *Œstrelata* in which the exposed portion of the outer primary beneath is dark not white, so that, apart from other differences, the Cheshire specimen cannot be either of these.

Regardless, therefore, of the difference in the colour of the feet and legs, we have come to the conclusion that our specimen is referable to *Œstrelata neglecta*, and that this species should be added to our fauna as a wanderer to the British Isles. Drs. Bowdler Sharpe and Du Cane Godman, to whom we showed the specimen, are of opinion that our conclusion is warranted.

Œ. neglecta is known only as a South Pacific species ; it has been obtained in the neighbourhood of the Kermadec Islands, but little is known about its range. Apparently it has never before been recorded as occurring in Europe. On March 25th the wind in Cheshire veered from the south-east to the west, and later to the N.N.W. On the 27th it backed to the S.S.W., rising in force, and remained westerly until the 31st, when, as registered at Manchester, it was blowing

---

* Salvin, *op. cit.*, p. 412.   † Salvin, *op. cit.*, p. 413.

with an average velocity of 21 miles an hour.   Tarporley
is about 11 miles S.S.E. of the Mersey Estuary, 16 miles
S.E. of the Dee Estuary, some 60 miles E. of Cardigan
Bay, and over 100 miles N. of the Bristol Channel.
From the condition of the bird when found we conclude
that it dropped towards the end of the month, probably
on or about March 31st, when the westerly winds were
at their strongest.

The bird was exhibited at the meeting of the
Zoological Society held on May 12th, and at the meeting
of the British Ornithologists' Club held on May 20th
last.

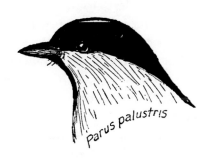

Parus palustris

ON THE IDENTIFICATION OF DUCKS' EGGS.

BY

HEATLEY NOBLE, M.B.O.U.

(PLATE I.)

THE eggs of various species of *Anatidæ* are so frequently sent to me for identification that I have gained some experience in the matter, and it has been thought that the results of that experience might be of some interest to readers of BRITISH BIRDS.

There are three means of identification—(1) the eggs themselves ; (2) the down found in the nests ; (3) the feathers which are generally mixed with the down. The last provides by far the most important and certain means of identification, although it is seldom mentioned by writers on this subject. The down by itself is not reliable except in isolated cases.

Take, for example, that found in the nests of the Wigeon and the Shoveler ; it would be a bold ornithologist who would guarantee to separate the two were they mixed together. The clue is given by the feathers, those of the Wigeon being white sometimes with grey centres, and quite unmistakable. Then, again, the down in different nests of the same species is often so dissimilar that it appears to belong to different species. I think I shall be able to show that if the eggs, down, and feathers are all considered in relation to each other, identification, if not absolutely certain, becomes little doubtful.

Occasionally cases arise which are distinctly difficult. Last season, for instance, a beautiful nest of snow-white down, with white eggs and white feathers tipped with grey, was sent me from Ireland. The nest had been taken in heather, not from a hole, as might be expected from the colour of the down. It certainly belonged to no British, or even European, breeding duck, neither was it that of any foreign duck usually kept in confinement.

The female was eventually shot, and proved to be a speckled mongrel mallard ! Another interesting case was that of an unmistakable Wigeon's egg found in a Wild Duck's nest, with equally unmistakable down of the latter. The Wigeon had laid in the same nest after the larger duck had hatched off. I have known more than one instance of this, *Anas boscas* being an early breeder and the Wigeon much later.

In this article I have been asked to include the Golden-eye, Velvet Scoter, and Long-tailed Duck. Although these three have not as yet been *proved* to have bred in these islands, there is some evidence that at least two of the species named may have done so, and there seems no reason why they should not.

Most of the following remarks are from personal observation, and, with the exception of the three species mentioned above, the nests described have been found by the writer.

COMMON SHELD-DUCK (*Tadorna cornuta*).—The eggs of this bird could hardly be confused with those of any other British duck except, perhaps, the Goosander. They are creamy white, rather lighter and more glossy than those of the latter. The down is light pearl-grey, and larger than that of the Goosander, while the feathers render a mistake impossible, for they are tipped with black, or occasionally red-brown. There is no doubt that this handsome duck is greatly on the increase. The nests are generally placed in a hole in a sandbank, not far from the sea, but at Wolverton, in Norfolk, I have found them more than two miles from the shore. On this estate the keeper informs me that the female Sheld-duck may often be seen conducting her brood through the village in the early morning, *en route* for the sea, and they are sometimes noticed marching down the railway line to the same destination. So numerous are they in this carefully protected area, that whilst I was examining a nest twenty-three adult birds were counted in the air at the same time. On June 3rd this nest contained fifteen eggs about a quarter incubated, and underneath were many old egg-shells, showing that the hole had been occupied the previous season. In confinement I have known the " Burrow Duck " to nest in a hole in a tree. (Pl. I., Fig. 1.)

MALLARD (*Anas boscas*).—Though this species is usually an early breeder, nests may be found throughout the spring and summer months, from March, or even February, until well on into June. The eggs of this bird vary more than those of any other British duck, from greenish-blue they range through yellowish-cream colour to white. The down is large, brown in colour, with light centres, the points hardly lighter than the rest. The down might be confused with that of the Pintail, although the latter is smaller. The flank feathers found amongst the down are, however, larger, more pointed, and very different in pattern, as will be seen by reference to figures 2 (Mallard) and figures 7 (Pintail) in Plate I.

GADWALL (*A. strepera*).—This bird is now well established in certain parts of Norfolk and Suffolk. In the Thetford district it is one of the commonest ducks in winter, and a certain number remain to breed. The nests I have seen in the Eastern Counties and in Spain have never been far from water ; one was in a wood close to a river, and another in a reed-bed at the edge of a large lagoon. The eggs are buffish-white, with no *tinge of* green ; the down is very dark, with small light centres, and with distinct grey points. The feathers are small, light in colour, with irregular darker markings in the centre, but lighter towards the tips. It would be difficult to confuse them with those of any other duck. (Pl. I., Fig. 3.)

SHOVELER (*Spatula clypeata*).—This duck is probably far more common than is generally supposed, owing to the fact that numbers leave their breeding haunts after the young are able to fly. In some counties where they breed regularly I have never seen one in the shooting season. The nest is often placed on dry ground, some little distance from water, and they seem to show partiality for rough, rushy meadows. The eggs are distinctly greenish in colour, which at once separates them from those of the Wigeon, though the down closely resembles that of the latter, and except that it is slightly darker (probably varying in different individuals) I can find little difference. The feathers, however, are totally different, and render confusion between these two species impossible. The only British duck's eggs which at all approach the Shoveler's are certain varieties of the Pintail's, but the Pintail's down is larger and lighter, and the Shoveler's feathers (especially the large ones) are quite distinct, as will be seen by reference to Plate I., Figs. 4, 4. This species breeds fairly early, half incubated eggs were found by me on May 8th. Wigeon seldom nest until the latter part of that month.

PINTAIL (*Dafila acuta*).—The fact of this bird breeding on Loch Leven is now unfortunately common knowledge : there is, or was, a considerable colony on one of the islands. I once counted thirteen nests, and there were probably more. A few also nest in a certain spot in Orkney, and it is said to have bred in Ireland. Its breeding range extends as far south as Andalusia, and I found a nest there (from which the female was procured) in May, 1902. The eggs are generally described as greenish in colour, and this is, no doubt, usually the case, but there is at least one other variety which is almost as creamy-coloured as the egg of the Wigeon, and without any trace of green. As far as I know, the shape, which is oval, remains constant. The eggs might be confused with those of the Long-tailed Duck, but that bird's down is much darker, while its eggs are smaller, and the feathers are distinct, as will be seen by reference to the figures. The Pintail's nests observed in Scotland were placed on dry ground, and one was in an exposed situation on burnt grass. In Spain and Hungary I have found them in damp places, one on a marsh quite surrounded by water. It is an early breeder, and full nests may be seen by May 5th. (Pl. I., Figs. 7, 7.)

TEAL (*Nettion crecca*).—There is no difficulty about the identification of the eggs of this bird, as the only others that approach their small size are those of the Garganey. Teal's eggs are, however, slightly smaller than those of the Garganey, and they have a greenish tinge entirely absent from the eggs of the latter species. The down of the Teal is also darker than that of its congener, and has no white tips. The feathers are of a light stone colour, with broad dark patches extending almost to the tip of the feather (Pl. I., Figs. 5, 5).

GARGANEY (*Querquedula circia*).—In this case the down alone is quite sufficient for identification ; it is smaller than that of the Teal, and very distinctly white tipped. The feathers are light grey, with dark central patches which do not extend either to the tip or edges of the feather (Pl. I., Figs. 6, 6). The eggs are creamy without the green tinge. The Garganey is probably the rarest of the ducks which breed regularly in this country. It nests in the Broad district, and according to Howard Saunders* has been found breeding in Yorkshire, and its eggs have recently been discovered in Kent. The only nest that has come under my personal observation was found on May 18th, in a field of rank grass not far from one of the Broads. It contained seven fresh eggs.

WIGEON (*Mareca penelope*).—Probably owing to protection, and also to the numbers of these birds that are bred in semi-confinement and subsequently allowed their liberty, this bird has largely increased its breeding range. At one time confined to the North of Scotland, it has recently been known to nest in Perthshire, Dumfriesshire, and Yorkshire, whilst I have some evidence that eggs have been laid in Norfolk, and very young birds have been seen at Beaulieu, in Hants, during early August. It has not yet been known to breed in Ireland, and the statement to the contrary has been proved incorrect. Eggs are seldom laid before the latter part of May. The nest is often placed in heather some distance from water

* "Ill. Man. Brit. Birds," 2nd Ed., p. 435.

but frequently on an island. The eggs are cream-coloured, the down is dark, with no particular characteristic, but the feathers are unmistakable, being white sometimes with grey centres, which spread to the top of the web (Pl. I., Figs. 8, 8). The down of the American Wigeon is much darker, and the centres not so distinct.

COMMON POCHARD (*Fuligula ferina*).—The down of this bird is large and exceedingly soft to the touch ; the eggs are of a dirty greenish colour, and might easily be mistaken for those of the Scaup, or the Tufted Duck, although they are usually larger than the former, and *considerably* larger than the latter. Nests I have seen in Scotland were placed in thick dead rushes on or close to the edge of a loch, and they might almost have been mistaken for Coots' nests, for in two instances there was not a particle of down present, although the eggs were on the point of hatching.* In Spain we noticed a nest in the middle of a swamp, thickly lined with down, which was damp at the bottom. In the North, incubation commences about the middle of May. The feathers found in the nests are rather large and brownish in colour, slightly streaked from the centre upwards, and often tipped for a quarter of an inch with grey (Pl. I., Figs. 9, 9).

EXPLANATION OF PLATE I.

Figs.	Feathers from Nest of.	Where taken.	When taken.	By whom taken.
1	Sheld-duck	Norfolk	3.6.1903	H. Noble.
2, 2	Mallard	Berkshire	31.3.1901	,,
3	Gadwall	Norfolk	23.5.1901	,,
4, 4	Shoveler	Norfolk	8.5.1897	.,
5, 5	Teal	Norfolk	8.5.1897	,,
6, 6	Garganey	Norfolk	17.5.1899	.,
7, 7	Pintail	Scotland	18.5.1899	:.
8, 8	Wigeon	Scotland	26.5.1896	,,
9, 9	Pochard	Scotland	22.5.1899	,,

* Since the above was written, I have had particulars of seven more nests placed in thick rushes, in which no down was present. I have several times noticed that Mallards' nests have no down when placed in such positions. I should be very glad to know the experience of other readers of BRITISH BIRDS on this point.

(To be continued.)

ON THE MORE IMPORTANT ADDITIONS TO OUR KNOWLEDGE OF BRITISH BIRDS SINCE 1899.*

BY

H. F. WITHERBY AND N. F. TICEHURST.

PART X.

(Continued from Vol. I., page 350.)

FLAMINGO *Phœnicopterus roseus* Pall. S. page 395.

[On November 22nd, 1902, a Flamingo was shot on the Wash ; on November 5th, 1904, another was seen in Norfolk ; and in August, 1906, three were shot in the same county. In December, 1904, one was killed in Kent; but so many have been turned out at Woburn with only cut wings (*cf.* Vol. I., p. -91), and probably at other places, that we cannot regard these as genuine migrants.

We must here record our emphatic opinion that it is contrary to the interests of scientific ornithology to turn out birds of species which visit us or may be likely to visit us as genuine migrants.]

GREY LAG-GOOSE *Anser cinereus* Meyer. S. page 397.

SCOTLAND.—A young bird still unable to fly was obtained in the Tay area in the autumn of 1906, and the bird was considered to have been bred in the district (T. G. Laidlaw, *Ann. Scot. Nat. Hist.*, 1906, p. 237). Mr. Harvie-Brown records a decided increase in the numbers of this species in many parts of Scotland, and a distinct expansion of range to certain new haunts (*Fauna N. W. Highlands and Skye*, p. 221).

A bird received from Limerick November 23rd, 1901, has been assigned by Mr. F. Coburn (*cf. Bull. B.O.C.*, XII., p. 80, and *Zool.*, 1903, p. 46) to the supposed distinct eastern form which was separated by Hodgson under the name of *Anser rubrirostris*. Mons. S. Alpheraky, who examined a very

---

* As was explained in the first instalment of these articles (*vide* BRITISH BIRDS, Vol. I., p. 52), we refer here only to those records and observations which are additions to the Second Edition (1899) of Saunders' "Illustrated Manual of British Birds." It must also be pointed out that nothing which has already appeared in any part of this magazine is included in these articles, so that they must be read in conjunction with the magazine so far as published, as well as with Saunders' "Manual."

large series of this goose, does not, however, admit the
validity of this bird even as a geographical form (*cf. Geese
of Europe and Asia*, p. 29), and Mr. Coburn's arguments seem
to be set aside by the proofs of great variability in size and
colouring brought forward by M. Alpheraky.

WHITE-FRONTED GOOSE *Anser albifrons* (Scop.).
S. page 399.

There has been a great deal of discussion during the last
few years as to the validity of *Anser gambeli* of Hartlaub,
the American representative of the White-fronted Goose
(*cf.* J. H. Gurney, *Ibis*, 1902, p. 269 *et seq. ;* F. Coburn,
Zool., 1902, p. 337 ; H. W. Robinson, *t.c.*, 1903, p. 268 ; J.
A. Harvie-Brown, *t.c.*, p. 315, and S. Alpheraky, *Geese of
Europe*, etc., p. 45, etc.). Mr. Coburn thinks, with some
former authors, that the bird is distinct, and that specimens
which he says he received from Ireland belong to it ; M.
Alpheraky, on the other hand, unites the bird with *A. albifrons*.
It must be pointed out that specific characters founded on
specimens obtained outside the breeding area of the bird
are really of little value. The White-fronted Goose is
without question a variable species, and whether it can be
separated into geographical races or not, can only be
determined by a careful comparison of a large series of
specimens obtained within one breeding area, with a corresponding series obtained in another breeding area.

LESSER WHITE-FRONTED GOOSE *Anser erythropus* (L.).
S. page 400 (also *cf. B.B.*, Vol. I., p. 14).

NORFOLK.—An adult female was shot near King's Lynn
on January 24th, 1900, and sent to Mr. F. Coburn (*cf.* F.
Coburn, *Zool.*, 1901, p. 317 ; *Bull. B.O.C.*, XII., p. 15;
J. H. Gurney, *Ibis*, 1902, p. 269, etc.).

YORKS.—A male in the collection of the late Sir H.
Boynton was said to have been taken near York (T. H.
Nelson, *B. of Yorks.*, p. 413).

BEAN-GOOSE *Anser segetum* (J. F. Gm.). S. page 401.

OUTER HEBRIDES.—One was shot and two others were seen
in South Uist in March, 1903 (J. A. Harvie-Brown, *Ann.
Scot. Nat. Hist.*, 1903, p. 119). The " Manual " says that its
reported occurrence in the Outer Hebrides requires confirmation.

M. Alpheraky's separation (*op. cit.*) of the *Anser arvensis*
of Brehm from the *A. segetum* of Gmelin is supported by a
considerable amount of evidence. The chief characters
lie in the bill, that of *A. arvensis* being " longer and com-

paratively broader at the point, and far more depressed
behind the nail of the upper mandible (than that of *A.
segetum*). At the same time the lower mandible in *M.
arvensis* is less curved and comparatively less depressed in
the thickest part (looking at the shut bill from the side) than
in *M. segetum*. The nail is considerably shorter, but at the
same time also broader and more rounded, both longitudinally
and transversely." The colours of the bills of the two birds
are also different, but they do not seem to form a safe guide
owing to their variability. We have quoted the above passage
at length because Mr. F. W. Frohawk affirms that this goose
is the usual form of the Bean-Goose to be found in this country,
and that the true *A. segetum* is rare (*cf. Field*, 1902, p. 605;
Zool., 1903, p. 41). Mr. Einar Lönnberg in discussing the
question is inclined to think that variability accounts for the
differences, and that there are not two distinct species (*Zool.*,
1903, p. 164).

Mr. Frohawk considers (*Zool.*, 1903, p. 42, etc.) that the
bird shot at St. Abb's Head on February 25th, 1896, and
described by Mr. F. Coburn at length in the "Zoologist"
(1902, pp. 441–448), as *Anser paludosus* of Strickland, is
referable to *A. arvensis*. Mr. Coburn laid stress on the great
length of the neck of the bird he described, but the specimen
being a stuffed one no reliance can be placed on this feature.

The distribution of *A. arvensis* and *A. segetum* is incompletely known, but according to M. Alpheraky *A. arvensis*
is far more numerous than *A. segetum*, and the "region of its
nidification is larger both in longitude and latitude."

Another species of Bean-Goose, viz., the *Anser neglectus*
of Sushkin (*cf. Ibis*, 1897, p. 5) from Novaya Zemlia and
Kolguev, is suspected by M. Alperaky (*op. cit.*, p. 81), and
by Mr. Frohawk (*Field*, 1902, p. 1045) to occur in Great Britain.

Many diverse opinions have been expressed as to the specific
differences of these Geese and their occurrence in this country,
and it appears to us that before a definite decision can be
reached more observations and examination of larger material
must be made.

These birds undoubtedly vary greatly individually both in
size and coloration; moreover, they are usually shot by
sportsmen rather than naturalists, and consequently it is
difficult to get together a good series with careful notes as to the
colouring of the soft parts, which has, perhaps unfortunately,
been used as a character for the separation of the species.
Thus a bird, which the editor declared to be a Pink-footed
Goose (*Anser brachyrhyncus*), was sent to the "Field" from
Breconshire this year, and this specimen had *yellow* legs and

feet (*cf. Field*, 1908, p. 182, 410). On this point M. Alpheraky (*op. cit.*, p. 89) remarks that he can find but one record of such an occurrence in the wild bird (Payne-Gallwey, *Letters to Young Shooters*, 3rd Series, p. 69), although it has been recorded that Pink-footed Geese bred in captivity sometimes have both the bill and the feet yellow.

SNOW-GOOSE *Chen hyperboreus* (Pall.). S. page 405.

IRELAND.—A female in excellent plumage was shot in *co. Longford* on October 28th, 1903. It was in company with another bird, also shot, but not preserved, which was described as "dark in the plumage," and may have been a young bird of the same species (Williams and Son, *Zool.*, 1903, p. 459). Four were observed flying overhead within forty yards at Foxford, *co. Mayo*, on December 1st, 1903 (G. F. Knox, *Irish Nat.*, 1904, p. 76, and R. Warren, *Zool.*, 1904, p. 32). On December 30th, 1906, Captain Kirkwood saw a flock of fourteen (four white adults and ten greyish-coloured birds) at Bartragh, *co. Mayo* (R. Warren, *t.c.*, 1907, p. 72).

GREATER SNOW-GOOSE *Chen nivalis* Forster.
(*cf.* S. page 406.)

This form, which is only to be distinguished from the foregoing species by its larger size, inhabits Arctic America, whereas the smaller bird is apparently confined, as a breeding species, to eastern Siberia and the western shores of Arctic America (*cf.* Alpheraky, *op. cit.*, p. 15). A specimen of this bird was shot near Belmullet, co. Mayo (? date), and was exhibited by Dr. R. B. Sharpe on behalf of Mr. R. J. Ussher at the November, 1899 meeting of the Brit. Orn. Club (*Bull. B.O.C.*, X., xv.).

BRENT GOOSE *Bernicla brenta* (Pall.). S. page 411.

An adult female of the American Black Brent (*B. nigricans*) is said by Mr. F. Coburn to have been shot by a wildfowler, named Richardson, in the Wash "deeps" (Norfolk), on January 15th, 1907, and sent to him (*cf.* J. H. Gurney, *Zool.*, 1908, pp. 121 and 123 and Plate). Mr. Coburn informs Mr. Gurney that a male of the same species was shot by the same wildfowler near Lynn and sent to him on February 14th, 1902. If the occurrence of a bird new to the British list is to be accepted as authentic, it is far more satisfactory wherever possible that it should be examined in the flesh by two or more ornithologists, and recorded at the time, than that it should be recorded for the first time months and even years after it was obtained.

(*To be continued.*)

AQUATIC WARBLER IN CORNWALL.

AN Aquatic Warbler (*Acrocephalus aquaticus*) was killed at the Eddystone Lighthouse, off Cornwall, on October 11th, 1907, and a wing sent for identification.

C. B. TICEHURST.

WHITE WAGTAIL IN CORNWALL.

IT may be worth while to record that a specimen of the White Wagtail (*Motacilla alba*) was killed at the Eddystone Light, off Cornwall, on October 11th, 1906, and a wing sent for identification.

C. B. TICEHURST.

BLUE-HEADED WAGTAIL IN NOTTINGHAMSHIRE.

IN August, or early in September, 1907, a specimen of the Blue-headed Wagtail (*Motacilla flava*) was shot near Nottingham. The bird was seen at Rose's, the taxidermist, of Nottingham, by Mr. J. Musters, who had it sent to me, as he felt sure it was a Blue-headed Wagtail. Mr. G. Millais and Mr. H. E. Dresser have also examined it and pronounced it to be a specimen of *M. flava*.

J. WHITAKER.

[This bird was shown to Dr. C. B. Ticehurst and myself by Mr. Dresser, and as it has been suggested that the bird must have been bred near the place of its capture it is as well to point out that the plumage of the bird affords no proof that this was so. Had it been in the pipit-like juvenile plumage of the species, it might have been well said that the bird had been bred near by, for this plumage is retained but a very short time after the young has left the nest. But the specimen in question was already in its first winter plumage, and was therefore perfectly capable of flying from the Continent, or elsewhere.—H. F. W.]

AN ESCAPED NUTCRACKER.

I NOTICE in last month's BRITISH BIRDS (Vol. I., p. 388) a note to the effect that a Nutcracker was shot " in Kent " on December 29th, 1907. It would be interesting to know in *what part* of Kent this bird met its death, inasmuch as a Nutcracker escaped from my aviaries on December 26th, 1907, three days previous to the time when the bird recorded was shot. This house is about five miles from the Kentish border, so that it is not unlikely that the example was my bird. It was in perfect plumage when it escaped. I believe it was a male. It was very tame, and would feed from hand.

G. M. BERESFORD-WEBB.

THE BLACK WOODPECKER IN ENGLAND.

MANY are the records of the occurrence of this bird in England, but it has been clearly shown that even the apparently best authenticated instances are untenable. This is only what might be expected of a bird which, although of strong flight, is strictly an inhabitant of the pine forests "from the Arctic Circle to Spain," and is a most unlikely species to wander far from its natural habitat, while the localities in this country which are suitable to its habits are very restricted. But, on the other hand, some of the more recent records are so precise, and the bird itself is so remarkable in appearance, that they cannot be dismissed offhand. This particularly applies to the numerous reports of its appearance on the borders of Norfolk and Suffolk, in the neighbourhood of Thetford, Brandon, and Euston, perhaps the most suitable locality that could be found for it in East Anglia.

The Rev. E. T. Daubeny, in recording several instances of the supposed occurrence of this bird in Euston Park, Ixworth, and Brandon, in 1897, remarked that it was "well-known that Lord Lilford liberated some of these birds towards the end of the last century," but I could obtain no confirmation of this, and so the matter stood till the year 1903, when, much to my surprise, my friend, Mr. W. H. Tuck, then living at Tostock, now at Bury St. Edmunds, informed me that a friend of his, whose name he was not at liberty to mention, brought seven or eight young Great Black Woodpeckers from Sweden in the year 1897. These were seen by Mr. Tuck, who further stated that they were placed in an aviary near Brandon for about two months, after which they were allowed to regain their liberty. This fact will, doubtless, account for the presence of the birds reported to have been seen in that neighbourhood, and perhaps for others which may have wandered further afield.

Mr. Tuck was requested not to mention this fact for a specified period, which accounts for his silence till 1903, but he quite agrees with the writer that it is most reprehensible that birds or insects should be thus secretly introduced to the disturbance of the British fauna.

THOS. SOUTHWELL.

SHORT-EARED OWL BREEDING IN NOTTINGHAMSHIRE.

A GOODLY number of these very interesting birds came to some young coverts at Rainworth in November last. There are two newly-planted pieces, one forty-six acres, the other twenty-seven. There were about eight birds in each, and we

Nest and eggs of Short-eared Owl found at
Rainworth, Notts., May 1st, 1908.
(*Photographed by* H. E. FORREST.)

saw them when shooting, and in February I saw four or five. About the middle of March I again saw one in each wood, and on April 21st one in the smaller covert. I now thought they would nest, so on Saturday, May 2nd, I tied a hand-

kerchief on the end of a 20-ft. salmon rod and went to the plantation. I started where the trees are thinly planted, and where
there is much white grass, and waving my flag over the plants I
walked down the side of the piece, and soon flushed a Short-
eared Owl, which flew into a big tree and watched me. When
I got near him he flew round, and settled in a big ash
near where I started to beat. I find the male is slimmer and
lighter in plumage than the female. I turned and took a piece
back, and when I turned again for a third beat he left the tree
and came circling round over my head, calling " Keii, keii."
When I had gone about twenty yards further another Owl rose
about three yards to my left, and on looking I found a nest
at the foot of a small Scotch fir with eight eggs upon a thickish
bed of dry grasses. The eggs were not in a cluster, but rather
scattered. The nest was nine inches each way and two-and-a-
half inches thick. I need hardly say that I was full of delight,
for I had never seen the nest and eggs of this bird before, and
they are the first ever found in this county.

J. WHITAKER.

EIDERS OFF SOUTH DEVON IN APRIL.

ON April 22nd, off Bolt Head, and at the entrance of the
creek running past Salcombe up to Kingsbridge, South Devon,
I noticed a small flock of Eider Ducks (*Somateria mollissima*).
Their presence so far south at this time of year seems rather
remarkable. The explanation may be the extraordinary
weather we have just been experiencing, of which the birds
were the forerunners, as the snow followed next day. The
wind had been in the east and north-east, but changed that
day to north-west.

K. S. SMITH.

STOCK-DOVE NESTING ON BUILDINGS.

WITH reference to the Rev. F. L. Blathwayt's note on the
nesting of the Stock-Dove (*C. œnas*) on Lincoln Minster, it
may be of interest to record that during the latter part of
March, 1907, I frequently heard a Stock-Dove cooing in the
Close at Winchester ; and on April 1st I watched a pair of
these birds flying about the Cathedral, and twice saw one of
them enter a hole in the masonry, high up on the Cathedral wall.

In my experience the Stock-Dove is a bird which has of
late years become commoner in many localities, and perhaps
it is developing that taste for " town life " which is now so
noticeable in the Wood-Pigeon.

B. B. RIVIERE.

[We think it will be found that Stock-Doves frequently
nest on buildings.—EDS.]

SPOTTED CRAKE IN SUSSEX.

A SPOTTED CRAKE (*Porzana maruetta*) was captured at the Royal Sovereign Lightship on May 8th, 1906, and a wing sent for identification.

C. B. TICEHURST.

KENTISH PLOVER IN CHESHIRE.

A SANDY stretch of the shore of Marbury Mere, near Northwich, Cheshire, is a favourite halting place of passing migratory waders and of wanderers from the neighbouring Mersey Estuary, and on many occasions I have seen there small mixed parties of Dunlins and Ringed Plovers. On April 29th, 1908, I put up eight birds from the edge of the mere; six of these were Dunlins in summer dress, and the other two, at first sight, looked like small pale Ringed Plovers. There was, however, something in the flight or appearance of the birds which specially attracted my attention; I was sure they were strangers. After a short flight over the water the birds returned to the bank, where they settled and allowed me to approach to within a dozen yards. I then saw that the Plovers lacked the complete pectoral band of *Ægialitis hiaticola* or *Æ. curonica*, that they were lighter in colour, and were distinctly smaller than the former, for I was able to compare their size with that of the Dunlins. The birds were, I concluded, a male and female. The male had a short black band above his white forehead, black lores and ear-coverts, and a black patch in front of the wing below the white collar. In the female these black regions, with the exception of the lores, were a dark brown. The rest of the plumage was sandy-grey on the upper part, the female being noticeably paler than the male. The bill and legs of both were black, or so dark that they appeared black even at close quarters and in an excellent light. The black patches on the aural regions and sides of the neck did not cover quite so large an area as is represented in the figure in Yarrell's "British Birds" (4th Edition, Vol. III., p. 267). In both birds the collar, forehead, stripe above the eye, and underparts were white. In spite of the fact that the Plovers were slightly smaller than the Dunlins, they stood a little higher on their legs. I watched them for nearly three hours, making rough sketches of them and noting down the details of plumage, and was perfectly satisfied that they were Kentish Plovers (*Ægialitis cantiana*), an addition to the Cheshire avifauna.

In the West of England this species has only hitherto been recorded from Devon and Cornwall, but it has been met with further north than Cheshire on the East Coast. The birds were not noticed by the gamekeeper when he made his rounds on the 28th, but he saw them on the 30th. On May 1st and 2nd, when I visited the mere again, I could not find them, and a large number of passing migrants of other species, which I saw on April 29th—including many Common Sandpipers, five Common Terns, two White Wagtails, many Yellow Wagtails, and the six Dunlins—had also disappeared.

T. A. COWARD.

PURPLE SANDPIPER IN THE CHANNEL ISLANDS.

An example of the Purple Sandpiper (*Tringa striata*) struck Hanois Light, Channel Islands, and a wing was sent for identification on November 15th, 1906. This species, no doubt, has been overlooked in these islands (whose ornithology is very incompletely known), and is not mentioned in Smith's " Birds of the Channel Islands."

C. B. TICEHURST.

REDSHANK BREEDING IN WARWICKSHIRE.

In the spring of last year a pair of Redshanks (*Totanus calidris*)—locally called "Whistling Plovers"—nested in a boggy field in the district of Hampton-in-Arden. Four chicks were hatched, one of which was by some means killed ; the other three left with their parents in the autumn.

This spring, about the beginning of March, three birds, a cock and two hens, returned, and two nests were made near the previous one. Judging by the pieces of eggshell lying near the nest it would appear that in one case the four chicks have been successfully hatched ; if this is the case they emerged about April 27th. It is doubtful whether the young birds have survived the floods, not being yet of an age to fly. One egg, which proved to be addled, was taken from the second nest by the gamekeeper, and I think it very probable that the others are infertile, since there appears to be only one cock bird.

Last year Redshanks were recorded for the first time as breeding in Warwickshire (*vide B.B.*, Vol. I., pp. 158 and 191), Oxford, eighty miles away, being, so far as I know, the nearest place where they had previously been seen.

A. G. LEIGH.

BLACK TERN IN CHESHIRE.

On April 29th, 1908, I watched a small party of Terns on Marbury Mere, near Northwich, which consisted of five

Common and one adult male Black Tern (*Hydrochelidon nigra*). They were, together with half a dozen or more Black-headed Gulls, feeding on insects which were flying above the surface of the mere. The Common Terns repeatedly dived downwards towards the water but did not strike the surface; they swooped upwards before they reached the water, evidently having captured their prey. The Black Tern flew with more graceful sweeps and curves, never half closing its wings and shooting downwards, and occasionally just touched the water with its bill as it passed, apparently picking something from the surface. It repeatedly flew to and settled upon a stump which projects above water; when it was standing on the stump I could easily see the black head, and almost black throat, breast and belly, strongly contrasted with the white vent and slate back and wings. On May 1st, when I next visited the mere, I could see neither the Common nor the Black Tern, but on the 2nd I found that there were two mature Black Terns, one, probably a female, being noticeably lighter on the underparts.

Black Terns are occasional visitors to Cheshire on both spring and autumn migration. In June, 1900, three birds were seen by Messrs. F. S. Graves and P. Ralfe, and in September, 1903, Mr. C. Oldham and I saw one, and in August, 1905, two birds on this mere.

T. A. Coward.

BIRDS IN NORFOLK IN 1907.

MR. J. H. GURNEY contributes to the " Zoologist " for April his usual interesting annual report on the ornithology of Norfolk. The most notable events to which reference has not previously been made in our pages were as follow :—

DESERT WHEATEAR (*Saxicola deserti*).—A male was shot " near the sea " on October 31st. This is only the second recorded occurrence of this southern bird in England, although, curiously enough, three have been obtained in Scotland.

FIRE-CRESTED WREN (*Regulus ignicapillus*).—One was caught in the town of Yarmouth on October 31st.

YELLOW-BROWED WARBLER (*Phylloscopus superciliosus*).— One was shot at Cley on October 29th. This is its second occurrence in Norfolk.

RED-BREASTED FLYCATCHER (*Muscicapa parva*).—One was identified (not very satisfactorily) by Mr. E. C. Arnold on the coast on September 11th, and another was identified by another observer on October 29th.

ROSE-COLOURED STARLING (*Pastor roseus*).—One at Toftrees in April.

NUTCRACKER (*Nucifraga caryocatactes*).—One was seen at Gunton, near Lowestoft, on November 28th and 30th.

BARN-OWL (*Strix flammea*).—Mr. Gurney has a good deal to say about the luminous Barn-owls. There is, however, no information as to what causes the luminosity which is the point of real scientific interest, and curiously enough Mr. Gurney thinks it would be a reprehensible deed to shoot one of the birds, although this is obviously the best way of clearing up the mystery. It would certainly do no harm, and might advance scientific knowledge. The chief points of interest in Mr. Gurney's notes on the subject, are that the evidence that Barn-owls occasionally exhibit luminosity is incontrovertible, and that the " light " emitted is very much stronger than one would imagine possible.

PURPLE HERON (*Ardea purpurea*).—A young bird was captured in the streets of Kirkley, a suburb of Lowestoft, by a tram conductor on October 9th.

SPOONBILL (*Platalea leucorodia*).—The first seen on Breydon was on April 21st, and several were subsequently seen at intervals in May and June, and the last on August 6th. Two " very " young ones were noted on June 4th. None appear to have been shot, we are glad to say.

A PLAN FOR MARKING BIRDS.

IN volume I. of this Magazine several communications were published on this subject. The advantage to students of migration of knowing exactly where birds travel by observations on marked birds is obvious; but the difficulty of the plan is that so few birds which are marked are ever found again. If, however, great numbers were marked, no doubt a large enough percentage would turn up to make the results of value. Mr. C. Hawkins, of " Lyndhurst," Woodside Road, South Norwood, informs us that he has had made a number of suitable aluminium rings of various sizes, stamped with a registered address (" Avis, Wye, Kent "), and each bearing a separate number for identification purposes. He is willing to supply these rings to anyone who will undertake to place them on birds, at the price of 5s. per gross, or 6d. per dozen. Mr. Hawkins also undertakes to keep a register of the particulars supplied by his correspondents concerning the birds marked, and to publish the results from time to time. H.F.W.

MARKED BIRDS.

ON the same lines as Herr Chr. Mortensen, of Viborg (*vide British Birds*, Vol. I., page 298), I have this year been marking and liberating a number of birds of various species. The mark employed is an aluminium ring on which is stamped

" Ticehurst, Tenterden," and a register number, and the ring is
put round one of the legs. Should any of my birds be met
with by any readers of BRITISH BIRDS will they kindly return
the ring and the leg to me, stating the locality and the date
of capture ?

C. B. TICEHURST,
Hurstbourne, Tenterden, Kent.

* * *

A TIMETABLE OF BIRD SONG.—Mr. W. Gyngell, in a short
article (*Nat.*, 1908, pp. 181-4) gives the result of his observations
on the duration of the song of thirty-six species of birds in
the Scarborough district. The results are shown by means of
a table of curves, which, however convenient, does not give
sufficient detail to make the observations as valuable as they
might have been. Comparing this table with the Messrs.
Alexander's observations lately published in this magazine
(*cf.* Vol. I., pp. 367-372) it is interesting to note that Mr.
Gyngell generally records a shorter song-period for resident
birds in Scarborough than Messrs. Alexander in Kent and
Sussex. Summer migrants appear to sing later in the north
than they do in the south. According to Mr. Gyngell, also
most of the resident birds make a considerable break in their
song in the autumn, whereas Messrs. Alexander record
occasional singing at this period.

SHORT-EARED OWLS IN THE ISLE OF MAN.—In connection
with Mr. W. J. Williams' note (*ante* Vol. I., p. 358) with regard
to the influx of Short-eared Owls (*Asio accipitrinus*) into
Ireland in the autumn of 1907, it is interesting to note that the
bird was common in the Isle of Man in the same season, nearly
a dozen being put up in one turnip field (*cf.* P. G. Ralfe,
Nat., 1908, p. 169).

HONEY-BUZZARD IN NORTH WALES.—Mr. C. D. Head
records that he has a male *Pernis apivorus* shot at Abergele
on October 15th, 1907 (*Zool.*, 1908, p. 156).

BITTERNS IN HAMPSHIRE.—Mr. C. B. Corbin notes that
two or three examples of *Botaurus stellaris* frequented the
reed-beds of the Avon in the last winter (*Zool.*, 1908, p. 157).

SUPPOSED SPOONBILL IN CO. LIMERICK.—Mr. H. G. O.
Bridgeman writes that a local farmer near Foynes described
a bird which he had shot on the estuary of the Shannon in
the frost of 1905 as being " all white, and had legs long like
the crane (*i.e.*, Heron), and had a bill what got bigger as it
got out, and flat like " (*Irish Nat.*, 1908, p. 101). We can
but agree with Mr. Bridgeman that this sounds like a very
honest, if quaint, description of *Platalea leucorodia*.

BRITISH BIRDS

AN·ILLUSTRATED·MAGAZINE
DEVOTED·TO·THE·BIRDS·ON
THE·BRITISH·LIST

JULY 1,
1908.

Vol. II.
No. 2.

MONTHLY·ONE·SHILLING·NET
326·HIGH·HOLBORN·LONDON
WITHERBY & C°

TYPICAL FEATHERS FROM DUCKS' NESTS.

Nat. size.

(Reproduced direct from the feathers.)   *(For explanation see p. 41.)*

RED-THROATED DIVER : "She waddled to the Nest."
(From one of Mr. Bahr's Photographs.)

To Messrs. WITHERBY & CO., 326, High Holborn, London,
Or
To Mr..
                    BOOKSELLER.

....

Please send me.. cop............ T

HOME - LIFE OF SOME MARSH - BIRDS," for w

I enclose...

....

....

TYPICAL FEATHERS FROM DUCKS' NESTS.

Nat. size.

(Reproduced direct from the feathers.)

(For explanation see p. 41.)

RED-THROATED DIVER : "She waddled to the Nest."
(From one of Mr. Bahr's Photographs.)

To Messrs. WITHERBY & CO., 326, High Holborn, London,
Or
To Mr..
             BOOKSELLER.

....

Please send me   . .. .   cop .   of " THE

HOME - LIFE OF SOME MARSH - BIRDS," for which

I enclose..

....

....

BRITISH·BIRDS

EDITED BY H. F. WITHERBY, F.Z.S., M.B.O.U.

ASSISTED BY W. P. PYCRAFT, A.L.S., M.B.O.U.

CONTENTS OF NUMBER 2, VOL. II. JULY 1, 1908.

ON THE IDENTIFICATION OF DUCKS' EGGS.

BY

HEATLEY NOBLE, M.B.O.U.

(PLATE II.)

(Continued from page 23.)

TUFTED DUCK (F. cristata).—This species breeds in June, and is very numerous in certain localities. On one

ERRATUM.—In the first instalment of this article (ante p. 19, line 10 from the bottom) Wolferton was printed Wolverton.

island in a northern loch I once counted nearly one
hundred nests, containing from seven to fourteen eggs
apiece, the average number being nine (June 12th).
Incubation lasts twenty-three days, as proved by eggs
placed under a hen. The eggs are smaller than those of
the Pochard, or Scaup, and slightly lighter in colour ;
the down is dark and compact, without conspicuously
light centres ; the feathers are greyish-white, and very
small (Plate II., Figs. 11, 11).

SCAUP-DUCK (*F. marila*).—On June 14th, 1899, Captain
Sandeman and I were fortunate enough to find the
first authentic nest of this species in Scotland.* I believe
another nest was found last year in the Hebrides by a
competent ornithologist, who, I fancy, was satisfied with
the identification. We have often been blamed for not
procuring the female from our nest, but this course seemed
to me quite unnecessary, for we had watched the birds
(two females and a male) for days, and saw the bird both
going to and coming from the nest as we lay in the reeds
within a few yards. I had kept these birds in confinement, and they were so well-known to me that mistaken
identification was impossible. The eggs are much the
same colour as those of the Pochard, but slightly smaller.
Those with which they are most likely to be confused are
Tufted Ducks, but they are much larger ; the down is
lighter, with more conspicuous light centres, while the
feathers are quite distinct, being larger, sometimes
slightly speckled, especially towards the tip, and of
different markings (Plate II., Figs. 10, 10).

GOLDEN-EYE (*Clangula glaucion*).—This bird has not
yet been proved to breed in the British Isles, but there is
some evidence of its having done so, and on August 1st,
1887, I shot a young bird in Scotland some considerable
distance inland. The eggs when first laid are of a beautiful
green, which unfortunately soon fades. For the loan of
down and feathers I am indebted to Mr. P. C. Musters,
who took the nest from a hole in a pine tree in Norway,

* See "Ann. Scot. Nat. Hist.," 1899, p. 215.

on June 19th, 1897. The down is white, as are also the feathers, and should the nest be discovered in Great Britain it could not be mistaken for that of any European duck, with the exception of Barrow's Golden-eye *C. islandica*, which breeds in Iceland and Greenland, and has not yet been recorded as visiting this country (Plate II., Fig. 14).   ·

LONG-TAILED DUCK (*Harelda glacialis*).—This is another species which has never been *known* to nest in this country, though it may have done so. The eggs are green, rather smaller than those of the Pintail, and more pointed. The down is dense, small, and "Eider-like" in texture, quite unlike that of *D. acuta*, which is the only duck with which it could be confused. It will also be seen that the feathers are very unlike (Plate II., Figs. 12, 12).

EIDER DUCK (*Somateria mollissima*).—These eggs could not be mistaken for those of any other British duck. Always green, but varying in shade, there is an "Eiderish" look about them which would prevent the possibility of error. Howard Saunders gives the number of eggs as from five to eight, but the latter number must be very rare, at least in our islands. I have examined numerous nests in' Scotland, the Farne Islands, and Orkney, and only once noticed six eggs, far more often the female was sitting on four than five. The well-known down needs no description, but it may be mentioned here that the downs of the three "British" Eiders are very distinct. That of the Common Eider is light; of the King-Eider darker; and of Steller's Eider darker still. The eggs also graduate in size, those of the Common Eider being the largest, the King-Eider smaller, and Steller's Eider smallest (Plate II., Figs. 13, 13).

COMMON SCOTER (*Œdemia nigra*).—This bird breeds quite commonly in Caithness and Sutherland, more rarely in Ross and Cromarty, and probably Inverness. It is, however, very local in distribution. It has nested in Ireland, and Mr. Ussher kindly sent me an egg, down, and young-in-down, for identification. The nests are difficult to find, being often well concealed in rank heather, and at some considerable distance from the loch side, while they are not infrequently on islands. A clutch taken in Sutherland on June 17th consisted of seven eggs, advanced in incubation. They were buffish-white in colour, the down almost black, and the feathers with no distinguishing marks, but the nest and eggs were unmistakable (Plate II., Figs. 15, 15).

VELVET SCOTER (*Œ. fusca*).—-I spent six weeks in Norway in an unsuccessful endeavour to discover the nest of this species. Although plenty of birds were seen, it is doubtful if they were breeding in that particular year, and I was driven away by snow on July 11th. There appears to be some evidence that a pair or two have bred in the Highlands of Scotland, but at present proof is wanting. The eggs are larger than those of the Common Scoter, and rather lighter in colour. The down is also lighter and larger. I am indebted to Mr. Witherby for the loan of a beautiful nest taken in Lapland, and to Mr. Ogilvie-Grant for the feathers depicted (Plate II., Figs. 16, 16).

GOOSANDER (*Mergus merganser*).—In certain parts of Scotland, especially the Garve district, this bird is not uncommon in the breeding season. Nesting much earlier than the Red-breasted Merganser, the Goosander has its clutch of from nine to twelve, or even fifteen, eggs, generally complete by the first week in May, and on the 28th of that month I have found birds hatched a few days. Any kind of hole seems to suit the nesting requirements of this duck. In the experience of the writer, nests have been found in holes in trees, clefts in rocks, and under peat hags. The eggs are creamy-white ; the down is pearl-grey, and the feathers are white, with a tinge of yellow. The only other duck's eggs like those of this species are the Sheld-duck's, but the feathers in two nests are so different as to at once preclude the possibility of mistake (Plate II., Figs. 17, 17).

RED-BREASTED MERGANSER (*M. serrator*).—Much more

widely distributed than the last species, this bird breeds commonly on many rivers and lochs in Scotland, also on the sea coast in Ireland it is common, and in Orkney and the Hebrides numerous. The nest is well concealed, often in high heather, sometimes in dense reeds on an island, often in a rabbit-hole or cleft in a peat bank, but seldom far from water. The eggs, which number up to fifteen, and are not laid before the end of May, are stone-coloured, with just a greenish tinge, the down is dark grey (much darker than that of the Goosander), and the feathers (very much smaller than those of *M. merganser*) are quite white (Plate II., Fig. 18).

EXPLANATION OF PLATE II.

Figs.	Feathers from Nest of.	Where taken.	When taken.	By whom taken.
10, 10	Scaup-Duck	Scotland	14.6.1899	H. Noble.
11, 11	Tufted Duck	Scotland	12.6.1898	,,
12, 12	Long-tailed Duck	Norway	4.6.1899	Ramperg.
13, 13	Eider Duck	Scotland	2.6.1896	H. Noble.
14	Golden-Eye	Norway	19.6.1897	P. C. Musters.
15, 15	Common Scoter	Scotland	17.6.1899	H. Noble.
16, 16	Velvet Scoter	Petchora, Siberia.	6.7.1875	H. Seebohm and J. A. Harvie-Brown.
17, 17	Goosander	Scotland	25.4.1903	H. Noble.
18	Merganser	Scotland	6.6.1896	,,

SOME EARLY BRITISH ORNITHOLOGISTS AND THEIR WORKS.

BY

W. H. MULLENS, m.a., ll.m., m.b.o.u.

II.—RICHARD CAREW

(1555—1620).

WILLIAM CAMDEN (1551-1623), the celebrated author of " Britannia " (London, 1586, 1 vol., 8vo), at the conclusion of the " Account of Cornwall," contained in that work, wrote as follows :—

" But these Matters will be laid open more distinctly and fully, by *Richard Carew* of *Antonie*, a Person no less eminent for his honourable Ancestors, than his own Virtue and Learning, who is writing a Description of this Country,* not in little but at large."

Carew's work duly appeared in 1602, and was entitled " The Survey of Cornwall."

It was dedicated by its author to Sir Walter Raleigh, and in the dedication Carew describes his book as " This mine ill-husbanded Survey, long since begun, a great while discontinued, lately reviewed, and now hastily finished . . ." And in his preface to the reader he informs us that " When I first composed this Treatise, not minding that it should be published in Print, I caused onely certaine written copies to bee given to some of my friends, and put *Prosopopeia* into the bookes mouth. But since that time, master Camden's often mentioning this work, and my friends perswasions, have caused my determination to alter . . ."

---

* The original Latin word is " regionis," the translation is from James Woodman's edition of Carew's " Survey of Cornwall."

Through this fortunate alteration of his original purpose we are indebted to Richard Carew for a book of peculiar and lasting interest. Not only is the "Survey of Cornwall" one of the earliest works to deal with the birds of a particular county, but it contains a far fuller and more important description of them than do many of the so-called County Natural Histories, afterwards published in this country.

Carew's observations on Cornish birds have frequently been quoted by later writers, and as the first edition of the "Survey of Cornwall" is an uncommon book, we propose to give them at length.*

Our author begins his account of the birds on fol. 24 of the "Survey" as follows :—

"Among living things on the land, after beastes follow Birds, who seeke harbour on the earth at night, though the ayre bee the greatest place of their haunt by day.

"Of tame Birds, Cornwall hath Doves, Geese, Ducks, Peacockes, Ginney Duckes, China geese, Barbarie hennes, and such like.

"Of wild, Quaile, Raile, Partridge, Fesant, Plover, Snyte, Wood-dove, Heath cocke,† Powte, etc.

"But amongst all the rest, the Inhabitants are most beholden to the Woodcockes, who (when the season of the yeare affordeth) flocke to them in great aboundance. They arrive first on the North-coast, where almost everie hedge serveth for a Roade, and everie plashoote for Springles to take them. From whence as the moyst places which supplie them food beginne to freeze up, they draw towards those in the South Coast, which are kept more open by the Summers neerer neighbourhood : and when the Summers heate (with the same effect from a contrairie cause) drieth up those plashes, nature and necessitie guide their returne to the Northern wetter soyle again.

* *Cf.* Harting's edition of Rodd's "Birds of Cornwall," Introduction, pp. xiv.-xviii.

† *i.e.,* the Black Grouse and its "powte" or young.

"Of Hawkes there are Marlions, Sparhawkes, Hobbies,
and somewhere Lannards.* As for the Sparhawke,
though shee serve to flie little above sixe weekes in the
yeere, and that only at the Partridge, where the Faulkner
and Spanels must also now and then spare her extra-
ordinarie assistance ; yet both Cornish and Devonshire
men employ so much travaile in seeking, watching,
taking, manning, nusling, dieting, curing, bathing, carrying and mewing them, as it must needs proceede from a
greater folly, that they cannot discerne their folly therein.
To which you may add, their busie, dangerous, discourteous
yea, and sometimes despiteful stealing one from another
of the Egges and young ones, who if they were allowed
to aire naturally and quietly, there would bee store
sufficient to kill not onely the Partridges but even all the
good-huswives Chickens in a Countrie.

"Of singing Birds they have Lynnets, Goldfinches,
Ruddockes,† Canarie birds, Black-birds, Thrushes, and
divers other ; but of Nightingals, few, or none at all,
whether through some natural antipathie betweene them
and the soyle (as Plinie writeth that *Crete* fostereth not
any Owles, nor *Rhodes* Eagles, nor *Larius Lacus* in *Italy*
Storkes) or rather for that the Country is generally bare
of Covert and woods, which they affect, I leave to be
discussed by others.

"Not long sithence, there came a flock of Birds into
Cornwall, about Harvest Season, in bignesse not much
exceeding a Sparrow, which made a foule spoyle of the
Apples. Their bils were thwarted crossewise at the end,
and with these they would cut an Apple in two, at one

---

* It seems doubtful whether the Lanner, *Falco lanarius* (*cf.* Newton,
Dict. of Birds, p. 503) ever bred in this country. Turner makes no
mention of it doing so, and though Merrett (*Pinax Rerum* London, 1666,
1 vol., 12mo), gives it in his list of British birds as "Lanarius, the
Lanar" and states that it bred in various places in England, he was
most probably referring to some other species of Falcon. Willughby
also does not include it among the birds found in this country, on the
other hand Symon Latham in his "Falconry," 1618, distinctly informs
us that it did breed in England. (Book II., p. 112).

† *i.e.*, Robins.

snap, eating onely the kernels. It was taken at first for a forboden token, and much admired, but, soone after, notice grew, that Glocestershire, and other apple Countries, have them an over-familiar harme.

" In the West parts of *Cornwall*, during the Winter season, Swallowes are founde sitting in old deepe Tynne-workes, and holes of the sea cliffes : but touching their lurking places, Olaus Magnus* maketh a farre stranger report. For he saith, that in the North parts of the world, as Summer weareth out, they clap mouth to mouth, wing to wing and legge in legge, and so after a sweete singing, fall downe into certaine great lakes or pooles among the Canes, from whence at the next Spring they receive a new resurrection : and hee addeth for proof hereof, that the Fishermen, who makes holes in the Ice, to dip up such fish with their nets, as resort thither for breathing, doe sometimes light on these Swallowes, congealed in clods, of a slymie substance, and that carrying them home to their Stoves, the warmth restoreth them to life and flight : this I have seen confirmed also by the relation of a Venetian Ambassadour, employed in Poland, and heard avowed by travaylers in those parts : where-through I am induced to give it a place of probabilitie in my mind, and of report in this treatise."

Dealing next in order with fresh and salt water and the fish thereof, Carew comes in due course to the " sea-foule," of which he writes as follows :—

" Besides these flooting [*i.e.* floating] burgesses of the Ocean, there are also certaine flying Citizens of the ayre, which prescribe for a corrodie† therein ; of whom some serve for food to us, and some but to feed themselves. Amongst the first sort, we reckon the Dip-chicke (so named of his diving and littlenesse), Coots, Sanderlings, Sea-Larkes,

---

* Olaus Magnus, Archbishop of Upsala, whose " Historia de Gentibus Septentrionalibus " (Romae, 1555, 1 Vol., folio) Carew here quotes.

† Corrodie—an allowance, or right of sustenance. M. L.—Corrodium

Oxen and Kine,* Seapiès,† Puffins, Pewets,‡ Meawes, Murres,§ Creysers, Curlewes, Teal, Widgeon, Burranets,|| Shags, Ducke, and Mallard, Gull, Wild-goose, Heron, Crane, and Barnacle. These content not the stomacke, all with a like savorinesse, but some carry a rancke taste, and require a former mortification : and some are good to be eaten while they are young, but nothing tooth-some, as they grow elder. The Guls, Pewets, and most of the residue, breed in little desert Islands, bordering on both Coastes, laying their Egges on the grasse, without making any nests, from whence the owner of the land causeth the young ones to be fetched about Whitsontide, for the first broode, and some weekes after for the second. Some one, but not everie such Rock may yeeld yeere-ly towards thirtie dozen of Guls. They are kept tame and fed fat, but none of the sea kind will breed out of their naturall place : yet at Caryhayes, Master Trevanion's house, which bordereth on the Cliffe, an old gull did (with an extraordinarie Charitie) accustome, for divers yeares together, to come and feede the young ones (though perhaps none of his alliance) in the Court where they were kept. It is held that the Barnacle breedeth under water on such ships sides, as have been verie long at Sea, hanging there by the Bill, untill his full growth dismisse him to be a perfect fowle : and for proofs hereof, many little things like birds, are ordinarily found in such

* Oxen-and-kine was, according to Swainson (*Provincial Names of British Birds*, p. 195), the name given to the Ruff and Reeve at the end of the sixteenth century. In the present case, as Carew is here dealing with Sea-fowl, it probably means the " Oxbird " or Dunlin (*cf.* Harting's edition of Rodd's *Birds of Cornwall*, p. 17).

† Oyster-catcher.

‡ The Pewit Gull, *Larus ridibundus* (*cf.*, Plot's *Natural History of Staffordshire*, Oxford, 1686, p. 231). Willughby calls it the Pewit or Black-cap.

§ Murre, the Cornish name of the Common Guillemot, also the Razorbill (*cf.*, Swainson, p. 218).

|| Probably the Shelldrake. The Shelldrake is called Burgander or Bergander (*i.e.*, Burrow Gander) by Turner. *Cf.* also Charleton, *Onomasticon Zoicon*, London, 1668, 1 vol., folio, p. 98. " The Bergander or Burrow Duck." Willughby says the " Shelldrake or Borough Duck. . . . it is called Burrow-duck from building in Coney Burrows " (*The Ornithology*, p. 28).

places, but I cannot heare any man speake of having seen them ripe. The Puffyn hatcheth in holes of the Cliffe, whose young ones are thence ferretted out, being exceeding fat, kept salted, and reputed for fish, as comming neerest thereto in their taste. The Burranet hath like breeding, and, after her young ones are hatched, shee leadeth them sometimes over-land, the space of a mile or better, into the haven, where such as have leasure to take their pastime, chace them one by one with a boate, and stones, to often diving, untill, through wearinesse, they are taken up at the boates side by hand, carried home, and kept tame with the Ducks : the Egges of divers of these Foules are good to be eaten.

" Sea-fowle not eatable are Ganets, Ospray (Plynyes Haliæetos).*

"Amongst which Jacke-Daw (the second slaunder of our Countrie) shall passe for companie, as frequenting their haunt, though not their diet: I meane not the common Daw, but one peculiar to Cornwall, and there-through termed a Cornish Chough : his bill is sharpe, long, and red, his legs of the same colour, his feathers blacke, his conditions, when he is kept tame, ungratious, in filching, and hiding of money, and such short ends, and somewhat dangerous in carrying stickes of fire."

The full title of the book is as follows :—

" The / Survey of / Cornwall / written by Richard Carew / of Antonie, Esquire. / London / Printed by S. S. for John Jaggard, and are to bee sold / neere Temple-barre, at the signe of the Hand / and Starre. 1602."

1 vol. f. c. 4to.

Collation pp. 10 unnumbered + fol. 160 + pp. 6.

---

* Cf. "The Ornithology" of Francis Willughby (London, 1678, 1 Vol., folio). In the account of the Bald-Buzzard, p. 70, occurs the following : "At Pensans in Cornwal we saw one that was shot, having a Mullet in its claw : for it preys upon fish, which seems very strange and wonderful, sith it is neither whole-footed nor provided with long legs or neck."

Joshua Childrey, in his "Britannia Baconia" (London, 1661, 1 vol., 12mo) in his article on Cornwall observes (p. 20) "There are also Sprayes here, the same fowle that Pliny calls Haliaetos, but it is not eatable."

THE
SVRVEY OF
CORNWALL.

Written by Richard Carew
of Antonie, Esquire.

LONDON
Printed by S. S. for Iohn Iaggard, and are to bee sold
neere Temple-barre, at the signe of the Hand
and Starre. 1 6 0 2.

This, the first edition, which is rare, is described by
John Nicholson in his "Bibliotheca Topographica
Britannica" as "an exact and excellent survey." A
facsimile of the title page is given opposite.

This edition was followed by another in 1723, entitled:—

> "The / Survey / of Cornwall / and / an Epistle
> concerning the / Excellencies of the / English
> Tongue. / Now first published from the
> Manuscript. / By Richard Carew, of Antonie,
> Esq. ; / with / The Life of the Author, / By
> H*** C***** Esq. ; / London, / Printed for
> Samuel Chapman, at the Angel in Pallmall ; /
> Daniel Browne jun. at the Black Swan without
> Temple- / Bar ; and James Woodman, at
> Cambden's-Head in Bowstreet / Covent-Garden
> MDCCXXIII."
> 1 Vol. f. c. 4to.

Collation pp. xx. + pp. 8 unnumbered + fol. 159 + pp.
8, 'table of contents,' + pp. 14. The dedication is signed
by James Woodman.

This edition was reprinted in 1769.

And in 1811 appeared that of Thomas Tonkin. 1 vol. 4to.

Richard Carew was born at East Antonie "In the
Eastern Parts of Cornwall, within some Miles of
Plymouth," in the year 1555 (cf. Wood *Athen. Oxon.*,
Vol. I.). He was the son of Thomas Carew, and Elizabeth
Edgecomb, daughter of Sir Richard Edgecomb, of Mount-
Edgecomb in Devon.

In 1566 at the very early age of eleven, Carew "became
a Gentleman Commoner of Christ Church" Oxford, but
"had his chamber in Broadgate's Hall." While at Oxford,
Carew (according to Dr. Fuller in his *History of the Worthies
of England*, p. 203) "being but fourteen years old, and
yet three years standing, he was call'd out to dispute
extempore, before the Earls of Leicester and Warwick,
with the matchless Sir Philip Sidney." *

* Sir Philip Sidney was born in 1554 and was then, therefore, fifteen
years old.

After leaving Oxford, Carew seems to have proceeded to the Middle Temple, and according to Wood, was three years later " sent with his Uncle (Sir George Carew, as it seems) in his embassage unto the King of Poland ; whom when he came to Dantzick, he found that he had been newly gone from thence into Sweden, whither also he went after him." Richard Carew mentions his uncle, " Master George Carew," in his Survey (fol. 61), and refers to the embassy to Poland, but says nothing about accompanying his relative.

Carew, in due course, appears to have settled down at his ancestral seat of Antonie, and to have studied agriculture and husbandry to such purpose that " he was accounted among his Neighbours the greatest Husband and most excellent Manager of Bees in Cornwall." He became High Sheriff of his County in 1586, and in 1599 was " Colonel of a Regiment consisting of five companies, or 500 Men, armed with 170 Pikes, 300 Musquets and 30 Calivers,* appointed for Causam Bay."

In 1589, Carew was elected a member of the College of Antiquities, a Society which at that time was about to apply to Queen Elizabeth for a Royal Charter—" But as fair as the Hopes of this famous College appeared in its Bloom, they were soon blighted by the Death of that ever-memorable Princess " and all " their applications to his successor, proved vain and unsuccessful. But what else could be expected from a Man . . . whose Genius and taste were as low and mean as his Soul and Inclinations ! "

Richard Carew died on the 6th day of November, 1620, in the 63rd year of his age, and lies buried in the Church of East Antonie among his ancestors.

---

* *i.e.,* a light hand-gun fired without a rest.

ON THE MORE IMPORTANT ADDITIONS TO OUR KNOWLEDGE OF BRITISH BIRDS SINCE 1899.

BY

H. F. WITHERBY and N. F. TICEHURST.

PART XI.

(Continued from page 27.)

COMMON SHELD-DUCK *Tadorna cornuta* (S. G. Gm.). S. page 419.

NORFOLK.—A satisfactory increase is recorded in the Lynn and Hunstanton districts (J. H. Gurney, *Zool.*, 1903, p. 130).

KENT.—Breeds numerously in the marshes adjoining the tidal waters in the north of the county. For interesting details *vide* T. Hepburn, " Zool.," 1907, pp. 54 *et seq.*

RUDDY SHELD-DUCK *Tadorna casarca* (L.). S. page 421.

NORFOLK.—An adult female, " believed to have been shot in Norfolk," was sent to Mr. Cole for preservation August 18th, 1898 (A. Patterson, *Zool.*, 1900, p. 530).

Two (possibly turned out) seen on Foulmere by Mr. W. Clarke, April 13th, 1906 (J. H. Gurney, *t.c.*, 1907, p. 126).

GADWALL *Anas strepera* L. S. page 425.

HANTS.—In 1904 a number of pinioned birds were turned out on Beaulieu Manor (Heatley Noble, *Zool.*, 1904, p. 193).

Is supposed to have nested at Beaulieu (J. E. Kelsall and P. W. Munn, *B. of Hants.*, p. 226).

CORNWALL.—Has been procured at least six times, the two latest were a male near Bodmin, in January, 1905, and a female near Land's End, January 10th, 1907 (J. Clark, *Zool.*, 1907, p. 285).

SCILLY ISLES.—One was shot at Tresco on January 1st, 1900, the first recorded (J. Clark and F. R. Rodd, *t.c.*, 1906, p. 304).

SHROPSHIRE.—A drake was seen in Hawkstone Park on December 9th, 1906 (C. Oldham, *t.c.*, 1907, p. 32).

MERIONETH.—A male was shot at Ynysfor, on December

30th, 1890, and a female at the same place on December 14th, 1901 (G. H. Caton Haigh, *t.c.*, 1902, p. 112) ; while another was shot at the same place on December 20th, 1904 (H. E. Forrest, *Vert. Fauna N. Wales*, p. 277).

NOTTS.—One was shot at Besthórpe in November, 1906, and a pair at Clumber in December, 1906, and a few have been seen in recent years at Annesley (J. Whitaker, *B. of Notts.*, p. 196).

YORKSHIRE.—Three were obtained at the Teesmouth in October, 1896 (T. H. Nelson, *B. of Yorks.*, p. 451).

PEEBLESSHIRE.—A pair were reported to have nested near Broughton, and to have reared their brood in 1906 (H. B. Marshall, *Field*, 28, VII., 06).

In spite of its increase in Norfolk, the Gadwall seems, according to all recent accounts, to be still a rather rare visitor to the rest of Great Britain, and especially so in the west. It does not appear to have established itself as a breeding species in any county but Norfolk and Suffolk.

IRELAND.—Several were ·hot on Lough Key, co. Roscommon, in the winters of 190·ʋ-7 and 1907-8 (H. G. O. Bridgeman, *Irish Nat.*, 1908, p. 101).

The Gadwall is a scarce and irregular winter visitor to Ireland, and has not apparently been recorded from Roscommon previously, although it has occurred from time to time in most counties.

SHOVELER *Spatula clypeata* (L.). S. page 427.

LINCOLNSHIRE.—In August, 1902, Mr. Caton Haigh saw a few Shovelers at Tetney, and was told that at least one pair had bred there ; on August 14th, 1903, he saw two broods at the same place (G. H. Caton Haigh, *Zool.*, 1903, p. 368 ; 1904, p. 297).

NORFOLK.—Nearly thirty pairs were breeding at Hoveton in 1906 (J. H. Gurney, *t.c.*, 1907, p. 127).

SUFFOLK.—Breeds regularly in the north-èast of the county (F. C. R. Jourdain, *in litt.*).

ESSEX.—Mr. H. M. Wallis has found the nest on the coast, and Mr. Miller Christy has recorded it as breeding (*Vict. Hist. Essex*).

HERTS.—Nests regularly at Tring (O. V. Aplin, *t.c.*, 1902, p. 68). Near Tring two or three pairs have bred regularly for at least ten or twelve years (Rothschild and Hartert, *Vict. Hist. Bucks.*, I., p. 145 (1905)).

STAFFORDSHIRE.—Now known to breed regularly in several places in the Cannock district (F. C. R. Jourdain *in litt.*).

BEDFORDSHIRE.—Now known to breed regularly in several places (*id.*).

KENT.—A brood of nine young, with the parents, seen in Romney Marsh May 19th, 1900 (N.F.T., *Zool.*, 1900, p. 279). During the last seven years the birds in this locality have increased, and nests have been found every year (N.F.T.). In the north of the county it breeds numerously in the marshes of the Thames Estuary, *cf.* Mr. Hepburn's article (*Zool.*, 1907, pp. 52 *et seq.*).

HAMPSHIRE.—Increasing as a breeding species, especially in the valley of the Avon (J. E. Kelsall and P. W. Munn, *B. of Hants.*, p. 232).

DEVON.—A pair reared their young at Braunton, in 1904 (J. Cummings, *Zool.*, 1905, p. 112). A pair said to have bred in North Devon for the past three years (B. F. Cummings, *t.c.*, 1907, p. 22).

SHROPSHIRE.—At least one pair nested and reared a brood on the marshes at Minsterley in 1907 (H. E. Forrest, *Caradoc F. Club Rep.*, 1908, p. 30).

The Rev. F. C. R. Jourdain informs us that he knows that it has bred also in the following counties not mentioned in the "Manual," DORSET, SUSSEX, CAMBRIDGESHIRE (locally common) and LANCASHIRE.

NORTH WALES.—It breeds regularly in some numbers in *Anglesey*, and a pair nested in 1896 at Llyn Mynyddlod, in *Merioneth* (H. E. Forrest, *Vert. Fauna N. Wales*, p. 278).

SCOTLAND.—It nests not infrequently on the *borders* of Northumberland, especially on the Scottish side (A. Chapman, *Bird-Life Borders*, p. 97). *East Lothian.*—Although nesting, is by no means a common bird (H. D. Simpson, *Zool.*, 1904, p. 459). *Tay Basin.*—Still increasing. Becoming very generally distributed in suitable situations on the shallower and reedier lochs of the east (J. A. Harvie-Brown, *Fauna Tay Basin*, etc., p. 233). *Sutherland.*—Colonel Duthie found a nest, and saw three or four birds on Loch Canna, Assynt, in the west of the county (*id., Fauna N. W. Highlands*, etc., p. 231). *Outer Hebrides.*—Bred on South Uist (first time recorded) in 1903 (*id. op. cit.*, p. 237). Nest found, and several pairs seen in 1906 (N. B. Kinnear, *Ann. S.N.H.*, 1907, p. 82), and in 1907 "still on the increase," and two nests found (P. H. Bahr, *t.c.*, 1907, p. 213).

IRELAND.—*Co. Antrim.*—Two nests taken in May, 1901, near Belfast, were first recorded as being those of the Wigeon, but were afterwards proved by Mr. Heatley Noble to be Shovelers' (R. Patterson, *Irish Nat.*, 1901, p. 147, and 1903, p. 275). *Co. Mayo.*—Although scarce a few years ago they now breed on many of the lakes in North Mayo (R. Warren, *t.c.*, 1902, p. 247). *Co. Donegal.*—Has increased very much as a breeding species on Lough Swilly of late years (D. C. Campbell, *t.c.*, 1905, p. 263).

The Shoveler is evidently increasing, and extending its range, and ornithologists would do well to take most careful notes from year to year of the numbers of these birds wherever they are nesting, as well as of Pochards, Tufted Ducks, and other increasing species.

Sequence of Plumages.—Mr. J. L. Bonhote states that the drakes have an intermediate plumage between that of the "eclipse" and the full breeding plumage. This plumage succeeds the "eclipse" in September, and the full plumage is attained gradually during the course of the winter (J. L. Bonhote, *Bull. B.O.C.*, XVI., p. 64).

PINTAIL *Dafila acuta* (L.). S. page 429.

SCOTLAND.—*Berwick.*—A nest with seven eggs (five hatched out) was found near Hawick on May 17th, 1901 (*Ann. S.N.H.*, 1902, p. 133). *Selkirk.*—A female was flushed from her eggs and watched in the southern part of the county on May 15th, 1901 (W. Renton, *t.c.*, 1902, p. 120). *Argyll.*—Four or five were seen (? breeding) on June 4th, 1907, on Loch Tulla (C. H. Alston, *t.c.*, 1908, p. 119). *Inverness.*—In the British Museum there is a clutch of seven eggs from "Cromlit, Knockie," from the late Edw. Hargitt (F. C. R. Jourdain, *in litt.*). *Outer Hebrides.*—Broods were seen in S. Uist in 1902, and the species appears to be increasing as a winter bird in Benbecula (J. A. Harvie-Brown, *t.c.*, 1902, pp. 209-210). *Shetland.*—A pair with young birds identified June 4th, 1905, at Dunrossness (T. Henderson, Jun., *t.c.*, 1906, p. 53; *cf.* Harvie-Brown, *t.c.*, 1907, 115).

TEAL *Nettion crecca* (L.). S. page 431.

OUTER HEBRIDES.—Now breeds plentifully in the *Uists* and *Benbecula* (N. B. Kinnear and P. H. Bahr, *Ann. S.N.H.*, 1907, pp. 213 and 820; and J. A. Harvie-Brown, *t.c.*, 1902, p. 209). The first actual record of its nesting on *Lewis* was made in 1903 (*id.*, *t.c.*, 1903, p. 245).

GARGANEY *Querquedula circia* (L.). S. page 435.

DURHAM.—Bred at the Teesmouth between 1880-7 (T. H. Nelson, *B. of Yorks*, p. 457).

NORFOLK.—Nests estimated at two in 1898 in the Broad District (J. H. Gurney, *Zool.*, 1899, p. 115).

KENT.—Two nests found in Romney Marsh in May, 1900

(N.F.T., *t.c.*, 1900, p. 279). Seems to be on the increase in Romney Marsh, five pairs seen in 1907 at one locality (N.F.T.). In the North Kent marshes Mr. T. Hepburn believes that it nests, and has seen birds in April and May, but has not as yet been able to confirm the fact (*t.c.*, 1907, p. 48).

HANTS.—It appears to have nested near Fareham in 1897 (J. E. Kelsall and P. W. Munn, *B. of Hants.*, p. 228).

SCILLY ISLES.—Has been obtained seven times (J. Clark and F. R. Rodd, *Zool.*, 1906, p. 304).

ANGLESEY.—An adult male seen April 15th, 1905 (T. A. Coward, *t.c.*, 1905, p. 386).

SHETLAND.—A male was shot on April 14th, 1907 (T. E. Saxby, *Ann. S.N.H.*, 1907, p. 182).

WIGEON *Mareca penelope* (L.). S. page 437.

CUMBERLAND.—A nest with ten eggs reported to have been found in 1903 (*Field*, 25, VII., and 1, VIII., 03).

YORKS.—In addition to the nest found near Scarborough in 1897, a pair bred at Malham Tarn in 1901, and in a semi-domesticated state it breeds regularly at Thirkleby Park, and at Scampston (T. H. Nelson, *B. of Yorks.*, p. 460).

[NORFOLK.—A deserted nest, said to have been a Wigeon's from the appearance of the eggs and down, was found in Norfolk in 1904 (J. Whitaker, *Field*, 18, VI., 04). We believe that this certainly was not the nest of a Widgeon but that of a Gadwall.]

MERIONETH.—A pair nested on Llyn Mynyddlod in 1898. Two pairs were seen at the same place April 19-30th, 1902, and a young bird was shot there September 30th, 1904 (H. E. Forrest, *Vert. Fauna N. Wales*, p. 283).

SCOTLAND.—*Sutherland.*—There is evidence of Wigeon breeding on Loch Assynt in 1901 and 1902, and several pairs were seen around Loch Urigil in 1903. A nest was found at the latter in May, 1903, by Mr. Blathwayt, and is the first authentic record of the Wigeon breeding to the west of the Divide (J. A. Harvie-Brown, *Fauna N.W. Highlands*, etc., pp. 234, 235.) *Roxburgh.*—There is a certain amount of evidence that Wigeon have bred near Yetholm from time to time, but absolute proof is still wanting (A. Chapman, *Bird-Life Borders*, p. 90). *Outer Hebrides.*—A pair seen several times in June, 1906, but no nest found (N. B. Kinnear, *Ann. S.N.H.*, 1907, p. 82).

For some interesting notes as to the first records of the breeding of the Wigeon in Scotland *vide* "Ann. Scott. Nat. Hist.," 1902, p. 200, footnotes.

AMERICAN WIGEON *Mareca americana* (J. F. Gm.). S. page 439.

An adult male was shot on Benbecula, Outer Hebrides, on January 3rd, 1907, by Mr. E. M. Corbett (R. Bowdler Sharpe, *Bull. B.O.C.*, XIX., p. 57, *cf.* also *Ann. S.N.H.*, 1907, p. 116). This is the first authentic record of the occurrence of this rare wanderer in Scotland. The bird is now in the British Museum (Natural History).

RED-CRESTED POCHARD *Netta rufina* (Pall.). S. page 441.

YORKSHIRE.—One shot near Redcar, January 20th, 1900 (T. H. Nelson, *Zool.*, 1900, p. 483). Another was shot about February 10th, 1900, at Coatham, near Redcar (J. W. Fawcett, *Nat.*, 1900, p. 304).

NORFOLK.—Thirteen appeared on Breydon on September 4th, 1906, and nine of them were killed by a punt gunner named Youngs. A tenth was picked up dead soon after in the neighbouring marshes (A. H. Patterson, *t.c.*, 1906, p. 394). Another pair was shot at Hickling by Alfred Nudd, on September 8th (N. H. Smith, *Field*, 15, IX., 06) ; while two others were seen there on the 12th and escaped. They all appear to have been adult birds, the drakes being still in eclipse (J. H. Gurney, *Zool.*, 1907, p. 134).

SUFFOLK.—An adult pair shot at Thorpe Mere by the sea, January 16th, 1904, by Mr. F. G. Garrett (*id., t.c.*, 1905, p. 90 ; *Bull. B.O.C.*, XIV., p. 62).

COMMON POCHARD *Fuligula ferina* (L.). S. page 443.

DURHAM.—A pair nested successfully in 1903 in the south-east of the county, and attempted to nest again in 1904 (C. E. Milburn, *Nat.*, 1904, p. 216).

NORFOLK.—Mr. J. H. Gurney describes a female bird in his possession, which was caught in Saham Mere in 1904, and which he believes to be a hybrid with a Tufted Duck (*Zool.*, 1905, p. 268).

ESSEX.—Has bred since 1886 (M. Christy, *Vict. Hist. Essex*).

KENT.—Mr. T. Hepburn found a nest containing seven eggs, which he believes to have been Pochard's, in the marshes of north Kent, on April 19th, 1904 (*t.c.*, 1907, p. 48). In the same locality it has since been found nesting with certainty by Mr. Walpole Bond (*in litt.*).

HERTS.—Breeds in increasing numbers at Tring.

BERKSHIRE.—At least six pairs nested in Windsor Park in 1907 (Graham W. Kerr, *Zool.*, 1908, p. 139).

BEDFORDSHIRE.—Recorded as breeding (*Vict. Hist. Beds.*, Vol. I., p. 125).

STAFFORDSHIRE.—Recorded definitely as breeding at Gailey Pools in 1890 (*Rep. N. Staffs. F. Club*, 1905-6, p. 49).

HANTS.—Is said by Hart to have nested in the New Forest district since 1880, but has not been found nesting elsewhere in the county (J. E. Kelsall and P. W. Munn, *B. of Hants.*, p. 238).

SCOTTISH BORDERS.—Twenty to thirty pairs nesting on Hoselaw Loch in 1906. Whitrigg Bog, near St. Boswells, Roxburgh and Hule Moss, on Greenlaw Moor, Berwickshire, are now the only other localities on the borders where the Pochard nests (A. Chapman, *Bird-Life Borders*, p. 90).

OUTER HEBRIDES.—Now far from uncommon (J. A. Harvie-Brown, *Ann. S.N.H.*, 1902, p. 211).

FERRUGINOUS DUCK *Fuligula nyroca* (Güld.).
S. page 445.

YORKSHIRE.—In the spring of 1903 four birds frequented a sheet of water near Ackworth, and two, an adult male and a female, were shot. The others, which were a pair, remained there till the end of the year (W. B. Arundel, *Zool.*, 1904, p. 33).

NORFOLK.—In April, 1903, twenty birds in two flocks, frequented Rollesby and Hickling Broads. One flock was composed entirely of adult males (J. H. Gurney, *t.c.*, 1904, p. 207). An immature bird was shot on January 1st, 1906, on the Broads (M. C. H. Bird, *t.c.*, 1906, p. 75). Four were seen on the Broads on April 10th, 1906, the day following a N.E. gale (J. H. Gurney, *t.c.*, 1907, p. 126). A flock of five was "reported" on Hickling on December 27th, 1907 (*id., t.c.*, 1908, p. 135).

SUFFOLK.—One was shot at Culford on January 23rd, 1906, after a N.E. gale, and a second a few days later (*id., t.c.*, 1907, p. 123).

SURREY.—One in the Charterhouse collection is stated to have been shot at Bramley (J. A. Bucknill, *B. of Surrey*, p. 239).

CORNWALL.—An immature male was killed on the beach near Mylor, on March 11th, 1905, during very stormy weather. The first record for the county. (J. Clark, *Zool.*, 1907, p. 285.)

MONTGOMERY.—One was shot out of a party of seven at Machynlleth by Mr. Percy Lewis, on April 2nd, 1906 (H. E. Forrest, *Vert. Fauna N. Wales*, p. 286).

(To be continued.)

NOTES

THE NEST AND NESTLINGS OF THE BEARDED TIT.

WHILE spending the Whitsun holidays on Hickling Broad I had the good fortune to inspect two nests of the Bearded Tit (*Panurus biarmicus*) and the following notes thereon may prove of interest.

Each nest contained six nestlings just showing the first traces of feathers, but no trace whatever of nestling down —a point worth mentioning—further, they had passed the " blind " stage, the eyes being fully opened.

But what I was specially interested in was the coloration and form of the markings of the inside of the mouth, which differed from any description hitherto given, including my own (*cf. ante*, Vol. I., p. 130).

Briefly, these markings take the form of four rows of pearly-white, conical, peg-like projections, suggesting the palatal teeth of reptiles, two on either side of the middle line. These tooth-like bodies, which are well shown in the aecompanying photograph, were not of uniform size, and were set in a background of black surrounded by a rich carnelian red, the whole being framed in by the lemon-yellow gape-wattles, which are not very strongly developed. The tongue is black with a white tip, and a pair of white spurs at its base.

The first nest was quite normal in position, but the second had, unfortunately, been placed actually on the ground, and some five or six yards from the water. I first visited this nest on June 8th and photographed the young *in situ*. The photograph proving unsatisfactory, I returned on the 11th for the purpose of making another attempt but found, to my dismay, that a tragedy had happened. In the nest lay two dead and bleeding young, while around the nest lay the remaining four, all more or less mangled. The burying beetles had commenced their work of interment, and at first I wondered whether they had gathered in force and worked the mischief. Realizing how highly improbable this was, I removed the nest and, tearing away the grass on which it had rested, discovered, beneath, the runs of a mole! About these there could be no mistake, and we must assume, therefore, that a mole had worked the mischief—a not unprecedented event.

The nest I pulled to pieces on the spot—it was already greatly damaged—and found that while it was typical in its general conformation—leaves of the reed forming its outside, the flower-heads thereof its lining—it differed from all .the published descriptions I have so far met with in having a number of feathers interwoven with the lining. I detected feathers of Swan, Mallard, Water-Hen and Snipe.

W. P. PYCRAFT.

NUTHATCHES BREEDING AT LLANDUDNO.

In view of the fact that up to the time of publishing my " Fauna of North Wales " no authentic occurrence of the Nuthatch (*Sitta cæsia*) on the north coast of Wales was known, it is interesting to note that a pair took up quarters in Gloddaeth Woods, Llandudno, early in the present year, and bred there later on. They were first observed by Mr. R. W. Jones, who showed me the nest-hole on May 10th. I heard the bird calling close by at the time.

H. E. FORREST.

GOLDEN ORIOLE IN SHROPSHIRE.

Mr. J. A. JUCKES (Acock's Green, Birmingham) reports seeing a male Golden Oriole (*Oriolus galbula*) at Cleobury-Mortimer, Shropshire, on April 26th (*Birmingham Daily Mail*, May 2nd, 1908). The Golden Oriole has occurred previously on two or three occasions in Shropshire.

H. E. FORREST

WOODCHAT IN CHESHIRE.

ON May 2nd, 1908, I saw two Woodchats (*Lanius pomeranus*), I think male and female, on some furze bushes by the side of the river Dane, about two miles above Congleton. The reddish-brown head and conspicuous black and white plumage of the male, coupled with the unmistakable Shrike beak, struck me at once. The female was not so bright in colour. I watched them for about fifteen minutes. The birds were remarkably tame, and allowed me to approach within about three yards of them. They seemed to be hunting for something among the spines of the furze. Eventually they flew away. I have been to the spot on several occasions since, but have not seen them again. The Woodchat has not been observed in Cheshire on any previous occasion, but it has twice been recorded from Lancashire.

J. M. ST. JOHN YATES.

[Mr. T. A. Coward kindly substantiates the above record, which is rather wanting in detail. Mr. Yates described the birds fully to Mr. Coward, and we are quite satisfied that the identification was correct. Mr. Yates is, Mr. Coward writes, an enthusiastic bird observer, and knows the Red-backed Shrike well. We have only to add that it is a pity that those who observe rare wanderers and do not obtain them, do not always write down on the spot as full a description as possible of what they see.—EDS.]

HOOPOE IN SHROPSHIRE.

MR. G. H. PADDOCK saw a Hoopoe (*Upupa epops*) in his garden in Wellington, Salop, on the morning of May 29th, and watched it for some time. It was set upon by a number of Sparrows, which compelled it to fly away. Over a dozen previous occurrences in Shropshire have been recorded.

H. E. FORREST.

SHORT-EARED OWL BREEDING IN PEMBROKESHIRE.

I AM indebted to Lieut. W. M. Congreve for news of the finding of two nests of the Short-eared Owl (*Asio accipitrinus*) in the neighbourhood of Pembroke Dock. The first was discovered early in May, and was remarkable for the elongated shape of the eggs. The second—a fortnight later— contained eggs of the ordinary rounded type. The Rev.

Murray A. Mathew, in his book on the " Birds of Pembroke-shire," refers to this bird as having bred on Skomer Island, but appears not to have seen it there himself.

H. E. FORREST.

SUPPOSED WILD SWANS ON COLL.

THE following appeared in one of the Scottish papers, and was forwarded to me.

" Some of your readers will be interested in learning that a pair of Wild Swans are this year nesting on a small loch near the Parish Church in the Island of Coll. . . . etc.—W. A. G."

I visited the Island of Coll on June 14th and saw the birds in question; they are not " Wild Swans," but Mute Swans (*C. olor*). Five eggs were laid; two hatched, and the cygnets are now with the parent birds. These particulars are sent in case the mistake may be quoted at some future date.

HEATLEY NOBLE.

INLAND NESTING OF THE SHELD-DUCK.

MR. HEATLEY NOBLE in his interesting and useful paper " On the Identification of Ducks' Eggs " (*ante*, p. 19), calls attention to two facts regarding the nesting of the Sheld-Duck (*Tadorna cornuta*) in Norfolk which are of especial interest. First, as to the greatly increased number of Sheld-Ducks nesting with us, and, secondly, as to this bird fre-quenting localities distant from the sea-shore, where alone we have been accustomed to look for it at that time.

As to the increased numbers to be found nesting. We have the evidence of Sir Thomas Browne that in his time (1668) they were "not so rare as Turn[er]* makes them comon in Norfolk so abounding in vast and spatious warrens," but, like all other breeding birds, constant persecution reduced its numbers in pre-protection times to a very sad remnant, so that Stevenson writing about the year 1890 could only record that, at that time, only a few pairs nested in the sandhills on the north-west coast of the county. I can well remember how in the summer of 1853 I was surprised to find fragments of the egg-shells of this bird outside a burrow on the Wells " meals " from which a brood had evidently been hatched. Since protection has been extended to them their numbers (as well as those of other species of Ducks) nesting in Norfolk have increased amazingly and they are to be found nesting in most suitable localities on the north and west coasts.

* See " Turner on Birds," Evans' Edit., p. 25.

The fact of their occasionally nesting at a considerable distance from the sea-shore in former times did not escape Stevenson's notice, and he enlarges upon the subject in the "Birds of Norfolk" (Vol. III., p. 124), quoting Sir Thomas Browne (as above), who states that they bred "in cunny burrows about Norrold [Northwold] and other places" some eighteen or twenty miles from the sea. Stevenson also mentions that these birds had been known to nest in the heaths at Dersingham and Sandringham, and it is interesting to have Mr. Noble's statement that they still frequent the same neighbourhood for that purpose.

I have also been told, but I forget my authority, that the Sheld-Duck nests on the Twig Moor, the Lincolnshire breeding place of the Black-headed Gull. Can any of your correspondents confirm this? It is worthy of remark that seven of the species of Duck mentioned by Mr. Noble nest regularly in Norfolk and one other (the Wigeon) is suspected of having done so.

T. SOUTHWELL.

WANT OF DOWN IN MALLARDS' NESTS.

MR. HEATLEY NOBLE, in his interesting paper on the "Identification of Ducks' Eggs," asks the experience of readers as to the amount of down in Mallards' nests when placed among thick rushes. I have observed two such nests this spring, in one of which very little down was present, while in the other only a few bits could be found by lifting the eggs. One or two pairs nest every season in the same place, a large disused gravel-pit, overgrown with bulrushes, etc., and I always notice the same deficiency of down in these nests.

With reference to Sheld-Ducks, on June 9th, this year, I had, in this corner of Yorkshire (Hull district), a very similar experience to Mr. Noble's in Norfolk, viz., the sight of twenty-two adult birds on the wing together.

M. WINZAR COMPTON.

COMMON CRANE IN ANGLESEY.

A FEMALE Common Crane (*Grus communis*) was shot at midnight on the 16th May, 1908, at Rhosneigr, Anglesey, by the gamekeeper, on the estate of Colonel Thomas J. Long. The plumage and feet of the bird are in most perfect condition, and there is no indication that it had been in captivity. Judging from the colour of the plumage it had not quite reached maturity; but the ovaries were well developed, the largest being about the size of a pea, and the red wattles on the head were well marked. The measurements were as follows :—Length, 44 inches; wing, 21 inches; tarsus, 8¾ inches; weight, 11 pounds. The stomach was completely filled with equal parts of pebbles and grit, and the remains of the large tipulid larva (*Tipula oleracea*), of which fifty-four examples were almost perfect, the largest measuring 1½ inches in length. Besides these, there were also the remains of four Elaterid beetles (*Agriotes* sp.) and a freshly caught larva belonging to the same group, but not, apparently, of the same genus; there were also fragments of the dung beetles *Aphodius fimetarius* and *Geotrupes* sp., and two oat glumes. Colonel T. J. Long has very generously presented the specimen to the Grosvenor Museum, Chester, where it is highly valued, and forms an extremely interesting addition to the local collections preserved in this institution.

ALFRED NEWSTEAD (Curator).

Cranes of various kinds are often kept in semi-captivity with cut, and not pinioned, wings (*cf. ante*, Vol. I., p. 91), and frequently escape when they grow new quill feathers, and then show no signs of captivity. We have, therefore, asked Mr. Newstead to make a more critical examination of the contents of the stomach of the bird above recorded, in the hopes that this might prove its origin. Unfortunately, the contents of the stomach do not greatly help us.

Mr. Robert Newstead kindly writes :—" As to the insects taken from the stomach, I can only confirm what my brother has stated in his letter to you. I have given these a most criticial examination, and find that they are all indigenous species ; and the majority had been captured by the bird within a few hours of its death. With the exception of the *larva* of the Agrotid beetle, they are all common and widely distributed species ; and are as abundant in Anglesey as in any other part of the British Isles."

Mr. J. Lomas (lecturer in Geology at the University, Liverpool) has very kindly examined the stones, and finds them to represent the following :—

				SPECIMENS.
Quartzite and vein quartz white ..	..	..	..	485
Flints	..	..	..	40
Quartzite and sandstone ..	..	..	..	39
Chalcedony ..	..	..	..	19
Pottery and porcelain	..	..	..	10
Slates	..	..	..	5
Granites, hornblendic	..	..	..	6
Mica schist	..	..	..	1
Micaceous sandstone	..	..	..	1

606

Mr. Lomas adds : " It would seem that the Crane selects the stones on account of their brightness. In a general assemblage of stones they are all such as would strike the eye at once, 516 of the stones being white in colour. It is difficult, if not impossible, to state where the stones come from. The quartzite are universally distributed. The only distinctive ones are the granites, and they certainly do not come from Anglesey. The flints are brown, and resemble southern types.

They are not Irish. The slates and mica schist might come from Anglesey, the Isle of Man, or Scotland. We must not forget that the island is covered with glacial deposits which contain erratics from many localities, and that makes it increasingly difficult to trace their origin. I am sorry the examination of the stones does not lead to definite results from your point of view."—EDS.

COMMON TERNS ON THE HOLYHEAD SKERRIES.

IT is generally supposed that these birds do not breed on the Skerries, and that the rocks are occupied during the breeding season exclusively by Arctic Terns and a few Roseate Terns (*cf.* H. E. Forrest, *Vert. Fauna N. Wales*, p. 375). That this is not the case has recently been proved by her Grace the Duchess of Bedford, who has been good enough to forward me a Common Tern (*Sterna fluviatilis*), which killed itself against the telephone wire whilst she was visiting the colony. Her Grace added, " several were seen."

HEATLEY NOBLE.

ABNORMAL EGGS OF BLACK-HEADED GULL.

ON May 15th, 1908, I took a nest, with its clutch of three eggs, of the Black-headed Gull (*Larus ridibundus*) from a lake island in Ireland where this Gull was nesting in numbers. It is well known that the ground colour and markings of their eggs vary to a considerable extent, these are of a pale greenish blue—the two largest almost without markings, the smallest with a few brownish blotches. The contents of the eggs were fresh and, as far as I could judge, normal in appearance, as are also the shells, with the exception of the smallest, which appears to be rough and of a somewhat friable nature at its pointed end. They measure 1·4 × 1.1 in., 1.6 × 1.2 in., 1.7 × 1.2 in. Patten in his " Aquatic Birds" gives the average measurements of the eggs of the Black-headed Gull as 2.2 × 1.5 in. I believe complete clutches of abnormally small eggs of this species are not common.

HERBERT TREVELYAN.

INCUBATION-PERIODS IN SEA-BIRDS.

DURING the spring and summer of 1907 Mr. F. G. Paynter made some experiments as to the duration of the incubation-period of certain sea-birds at the Farne Islands by placing the eggs in a hot-air incubator. Mr. Paynter describes his experiments in " Country Life" (March 21st, 1908, p. 409), and we give below the results arrived at, and add for the sake of comparison the observations recorded by Mr. William Evans (*Ibis*, 1891, pp. 52–93, 1892, pp. 55–58) :—

EIDER DUCK (*Somateria mollissima*).—Thirty-one days. In Mr. W. Evans' experiments an egg hatched in an incubator on the twenty-seventh day, and under a hen on the twenty-eighth day.

RINGED PLOVER (*Ægialitis hiaticula*).—Twenty-five days. Mr. Evans' hatched in an incubator on the twenty-second, twenty-third, and twenty-fifth days.

OYSTER-CATCHER (*Hæmatopus ostralegus*).—Twenty-one days. Mr. Evans gives twenty-three to twenty-four days from observations on two nests watched by Colonel Duthie.

SANDWICH TERN (*Sterna cantiaca*).—Twenty days. Not given by Mr. Evans.

ARCTIC TERN (*S. macrura*).—Twenty days. Not given by Mr. Evans, but the Common Tern is given as hatching two on the twenty-first, and one on the twenty-second day, in an incubator.

HERRING-GULL (*Larus argentatus*).—Twenty-one days. Mr. Evans gives the twenty-sixth day under a hen.

LESSER BLACK-BACKED GULL (*L. fuscus*).—Twenty-one days. Not given by Mr. Evans.

RAZORBILL (*Alca torda*).—Twenty-five days, Mr. Evans gives the thirtieth day in an incubator. Both Mr. Evans and Mr. Paynter remark that they cannot be absolutely certain that the eggs were fresh.

COMMON GUILLEMOT (*Uria troile*).—Thirty-two days. Mr. Evans gives the thirtieth and thirty-third days for two eggs in an incubator, and thirty-first day for one under a hen.

PUFFIN (*Fratercula arctica*).—Thirty-six days. In this case Mr. Paynter took eight eggs from different nests, and those which hatched out last he took as giving the correct period. Mr. Evans also gives the thirty-sixth day in an incubator.

It will be seen that the results arrived at by Mr. Paynter and Mr. Evans differ considerably in many cases. Both authors took great care that the eggs used should be perfectly fresh, although in cases where the bird lays only a single egg this is somewhat difficult to ensure. There is no doubt considerable individual variation, and we shall hope that other ornithologists will take up the subject, so that sufficient observations may be made on this interesting subject to enable us to strike a reliable average of the incubation period in various species.

In the same article Mr. Paynter makes some interesting remarks on the way in which Gulls, hatched in an incubator, practised flying by allowing the wind to lift them a few feet, and then dropping down again. This was practised for a week and even ten days before the birds were able to balance themselves well enough to fly any distance.

H.F.W.

* * *

THE NIGHTINGALE IN DERBYSHIRE.—Mr. W. H. Walton writes that since he recorded the occurrence of several nightingales in 1901 at Mickleover and Ockbrook, near Derby, they have not been seen or heard until this year. On May 8th one was noticed at Chellaston, about four miles south of Derby, and was, up to May 13th, attracting large numbers of listeners (*Field*, 1908, p. 831). Nightingales, the Rev. F. C. R. Jourdain tells us, breed sporadically in south Derbyshire, north of the Trent, almost every year, but the above is a good deal further north than usual. Any records of the distribution of this bird towards the borders of its range are interesting.

SUPPOSED WOODCHAT IN CORNWALL.—Mr. G. H. Coles records (*Field*, 1908, p. 831) that he watched within forty feet with strong binoculars a Woodchat Shrike (*L. pomeranus*) on the downs near Sennen (Land's End) on May 13th last. The only description he gives of the bird is : " It was a male bird in brilliant plumage, and the chestnut colour of the back of the head and neck was particularly bright." It is very possible that the bird was a Woodchat, but it is really impossible to accept such records as authentic unless better descriptions are given. The Woodchat has so many distinguishing characteristics in the field that there is really no excuse in this case.—H.F.W.

HABITS OF THE CUCKOO.—Mr. F. Banister writes in the " Field " (6, VI., 1908, p. 932) to the effect that he watched a Cuckoo visit a Hedge-Sparrow's nest, containing three eggs, in a hedge. The Cuckoo went to the nest and emerged in about a minute with one of the Hedge-Sparrow's eggs in its bill. This it proceeded to break up and, apparently, eat. On going to the nest, Mr. Banister found one Cuckoo's egg and two Hedge-Sparrow's. The author thinks that the Cuckoo laid the egg in the nest, but this does not seem to be proved in this case. The Cuckoo was at the nest for a very short time, and the egg might have been carried inside the mouth without attracting the attention of the observer. Observations on the actual depositing of eggs by Cuckoos, being few, are always welcome (*cf. B.B.*, Vol. I., p. 283).—H.F.W.

REVIEWS

The British Warblers—A History, with Problems of their Lives. By H. Eliot Howard, F.Z.S., M.B.O.U. Parts I. and II. Coloured and photogravure plates. (R. H. Porter.) 21s. net per part.

THIS work promises to be of quite unusual interest and importance on account of the original observations on the habits of many of the birds of which it treats. On this account, and also for the plates depicting various seldom-seen attitudes, it is to be highly commended. The plates—some in colour and some in photogravure—represent the best work we· have yet seen from Mr. Grönvold. Those showing various attitudes assumed during courtship are especially lifelike, and these have been drawn from Mr. Howard's originals.

Part I. is concerned with the Sedge-Warbler and the Grasshopper-Warbler, and Part II. with the Chiffchaff and the Yellow-browed Warbler. The observations on the habits of the first three mentioned species should be read by everyone interested in bionomical questions. To enable him to make such detailed studies as are here set forth on the daily life of these secretive little birds Mr. Howard must be endowed with a patience beyond most men, and it is evident that he must also be a persistently early riser. There are, too, several thoughtful passages on evolutionary subjects—such as sexual selection, and the plasticity of instinct—which deserve careful perusal.

We may here draw attention to a few of the points brought out by Mr. Howard's observations. In the three species mentioned, the males appear to arrive at the breeding place a week or ten days before the females. The area in which the nesting operations are to take place is apparently chosen by the male, and he spends much of his time in guarding this area from all other males of his species. It has often been noted that the same nesting site has been used for many years by a pair of the same species. We believe that this is much more generally the case than is supposed, and in such instances it may be concluded that if either of the pair dies during the winter the survivor brings a new mate to the nesting site the next spring. If only the males choose the nesting territory then it is puzzling how it occurs that the same place is occupied for many years in succession, unless, indeed, the heir returning to the locality of his birth finds his

father no more and steps into his shoes. Birds are creatures
of habit, as Mr. Howard demonstrates, and if the males come
back to the same territory then the females do also, and if so,
surely the same two birds are mated as long as both live.
Would not this account for the apparent absence of choice
by the female of any particular male (see Grasshopper-Warbler,
p. 14) ? Is there always an absence of choice, or has it so
happened that Mr. Howard has watched previously mated
birds, and not those which have never before been mated ?
We hope that Mr. Howard, with his great powers of observation, will give us in a future instalment the result of his
observations on this point, for it seems to us most unlikely
that birds choose a new mate each spring.

We have space only to allude to some of the many other
interesting facts so pleasantly recorded in these pages. A
curious feature in the courting display of these three species
is that the male frequently picks up a leaf or stick and holds
it in its beak while following the female. The females do
most of the building of the nest and the feeding of the young ;
the fæces of the young are sometimes swallowed by the parent
bird, as they are almost invariably by the Thrushes ; the
song of the Grasshopper-Warbler almost ceases after pairing
is over ; the nestlings of the same species leave the nest when
only a few days old, and some time before they are able to
fly.

There is so much that is good in this book that we are
somewhat unwilling to criticize. We must, however, express
the opinion that the general plan of the work appears to us
to be unwise. The descriptions of the plumages are most unsatisfactory in that they add little or nothing to our knowledge,
which is a great pity, for we know little of the sequences
of the plumages of these birds, and the moults they go through.
Then in species such as the Yellow-browed Warbler, with
which presumably the author has no acquaintance, no account
of habits is given. Thus the work is incomplete, and in no
sense a monograph. It seems to us a pity that the author
did not confine himself to a description of the habits only
(with the plates illustrating them) of those species which he
had observed. The work as at present planned is expensive.
The valuable original observations ought to have been made
accessible to all ornithologists, and might have been so
without any loss to science by the omission of what is not
valuable. If we may make a further criticism, it is that the
parts should appear at shorter intervals. Part I. was published in February, 1907, Part II. in March, 1908.

H.F.W.

BRITISH BIRDS

AN·ILLUSTRATED·MAGAZINE
DEVOTED·TO·THE·BIRDS·ON
THE·BRITISH·LIST

AUGUST 1,
1908.

Vol. II.
No. 3.

MONTHLY·ONE·SHILLING·NET
326·HIGH·HOLBORN·LONDON
WITHERBY & Co.

BRITISH·BIRDS

EDITED BY H. F. WITHERBY, F.Z.S., M.B.O.U.
ASSISTED BY W. P. PYCRAFT, A.L.S., M.B.O.U.

THE WOOD-PIGEON DIPHTHERIA.

THE RESULTS OF THE "BRITISH BIRDS" ENQUIRY.

BY

C. B. TICEHURST, M.A., M.R.C.S., L.R.C.P., M.B.O.U.

DURING the past autumn and winter Wood-Pigeons in
this country were ravaged by the disease known as
"Wood-Pigeon diphtheria." This disease has been
known for some years by gamekeepers and sportsmen

as occurring during acorn or beech-mast years, though
the *causa morbi* has not been so generally understood.

In BRITISH BIRDS (Vol. I., p. 243) I gave an account
of the micro-organism which was responsible for the
disease, the naked-eye appearance of birds dead of the
disease, and other facts as far as were known. In order
to ascertain more facts concerning this matter, at the
suggestion of the Editors of BRITISH BIRDS, I drew up
a schedule of questions which was sent round to all
readers of the Magazine for the purpose of securing help
from those who were interested in the subject, and the
results of this enquiry I now set forth. I must here
express my great indebtedness to all those who filled
in schedules, and so kindly supplied the information
upon which the following conclusions and suggestions
are based.

I have thought it better to group the facts under the
following headings :—

    I.—Geographical distribution.
    II.—Migratory flocks.
    III.—Transmission of the disease.
    IV.—Duration and course of the disease.
    V.—Relation to food supply.
    VI.—Transmission to other animals.
    VII.—Post-mortem appearances.

I.—GEOGRAPHICAL DISTRIBUTION.—From the accompanying map it will be clear that the reports show
that the disease was almost entirely confined to those
counties which border the Thames Valley. The only
positive returns received from other more distant
counties were from Yorkshire, Cumberland, Norfolk,
Essex, and doubtfully from Devon. Now, from all
these counties the reports seem to show that the
disease was local, or confined to isolated birds. For
instance, in Norfolk no disease was noted until the first
fortnight in February, when only one or two birds with
disease are recorded, whilst there is definite evidence
that the only occurrences in Essex were those from a

migratory flock which arrived in the last week of January
stayed a week, and in which nearly every bird wa
diseased. From Cumberland it was reported that ther

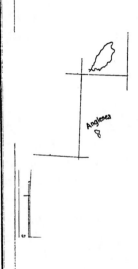

ENGLAND & WALES

■■■ *Positive, many.*
▨▨▨ *Positive, few.*
▨▨▨ *Sporadic cases.*
ⅢⅢⅢ *Negative.*

Anglesea

Sketch Map to show the distribution of Wood-Pigeon Diphtheria i
the Winter of 1907-8.

were a few diseased birds, but one would have liked t
have had more details on the distribution in this count

From Yorkshire the disease was only noted in two places, and those only in single examples, though presumably there must have been others. The counties in which the disease was most prevalent were : Wiltshire, Buckinghamshire (S.), Berkshire (N.), Oxfordshire (S.), and Hampshire ; to a lesser degree in West Sussex, West Kent, and Surrey ; and it seems undeniable that the centre of the disease lay in an area covered by beech wood in the Thames Valley in the counties of Berkshire, Oxfordshire, and Buckinghamshire, and in one place on the borders of these three counties the number of diseased Pigeons reached such extraordinary proportions that 2242 were burnt, and it is estimated that another 2000 were disposed of. From Scotland and Ireland only negative reports were received.

II.—MIGRATORY FLOCKS.—Nearly all observers agreed that the migratory Pigeons arrived at the end of October or beginning of November, that they increased in numbers during November, and decreased towards the end of January, and that by the end of February most of the migratory birds had gone.

These migratory flocks probably come from Central Europe, as large migratory flocks appear in Holland in the autumn ; it has been often suggested that these flocks come from Scandinavia, but as this bird is only found in southern Scandinavia, and is not particularly numerous there, this theory is untenable.

That the disease occurred mainly in these migratory flocks most observers agreed, some, indeed, asserting that the resident birds were never attacked, but this is very difficult to prove.

There is little doubt that the disease was found where there was a great increase of Pigeons in the autumn, and, as a rule, where there were few birds no disease was noted, though in one or two exceptional cases there were large numbers and no disease.

III.—TRANSMISSION OF THE DISEASE.—Very little light seems to have been thrown by schedule returns on the transmission of the disease. It can easily be understood that transmission of a disease from member to member in a vast horde, as in any crowded community, can take place in a number of different ways. For example, the exudation from a diseased bird's throat may easily be " coughed " out (and, as breathing gets more difficult, it is natural that it should be), and may fall on to the plumage of a neighbouring bird which, in turn, preens itself and takes up, of course, the bacilli.

Another way, which I suggested in BRITISH BIRDS, was that a diseased Pigeon after eating an acorn regurgitated it, and this, being picked up by another Pigeon, transmitted the bacilli of the disease. This suggestion seems to have been quite misunderstood by some people, for one writer in the " Field " shortly afterwards wrote to say, that " except at the time they are feeding their young, Pigeons do not regurgitate . . . and there is no need to assume regurgitation, or to attribute to them a habit which has not been observed "—which is absurd, since if there is any obstruction in the gullet, solid food will be regurgitated immediately that obstruction is reached.

Moreover, I have Mr. A. H. Patterson's evidence that he had a Pigeon sent him by a friend who had shot it, and that on the ground where the bird was sitting there lay an acorn which it had evidently tried to swallow and had regurgitated.

Both young and old birds are affected, though the occurrence of the disease in nestlings requires confirmation.

IV.—DURATION AND COURSE OF THE DISEASE.—No definite observations were made on this subject, though it was inferred that from the condition of the birds that the malady sometimes ran a quick course and sometimes a lingering one.

Through the kindness of Dr. Eyre and Mr. Leeming, bacteriologists of Guy's Hospital, I was enabled to have two Pigeons inoculated, and so obtain direct evidence of the length and course of the disease. The Pigeons inoculated were a wild Wood-Pigeon and a blue " racer."

In both, the inside of the throat was pricked, and some " membrane " from a Pigeon dead of the disease was rubbed on. The next day both had contracted the disease, as manifested by a white spot the size of a pin's head, the neighbouring parts being reddened. These spots remained apparently of the same size for about five days, the birds feeding and looking well, but at the end of a week, whereas the patch in the Wood-Pigeon had noticeably increased, that in the blue "racer" had noticeably decreased, and by the next day had entirely gone, and the temperature of this bird, which had risen from 104.8° F. (normal) to 107° F. during the period of infection, dropped again to its normal. The Wood-Pigeon's temperature, on the other hand (whose normal temperature is 108°), went up to and remained between 109° and 110° F. At the end of ten days the patch was greatly increased, and by twelve days it had extended across the middle of the throat. The Pigeon still ate peas and corn fairly well, and kept in good feather, but was much thinner. The patch continued to grow, and towards the end of the third week it had almost blocked the gullet, though it left the windpipe free. At this time it did not feed so well, and its temperature fell to several degrees below normal, and remained so until it died on the twenty-first day of the disease.

The recovery of the blue " racer,'" after taking the disease, is worthy of note, though from a single observation it could not be said that the disease cannot be fatal to it, though it suggests that it has a better resisting power than has the Wood-Pigeon.

V.—RELATION TO FOOD SUPPLY.—The idea amongst gamekeepers, as well as amongst other people, was that the disease was caused by the food supply. Of course, he of the beech woods said it was due to the excess of beech-mast, whilst he of the oak was equally confident that it was due to the plentiful supply of acorns, and they both agreed that the disease was prevalent during beech-mast or acorn years. This difference of opinion is strong evidence that the disease is not directly due to the kind of food supply, but due to massing.

In any crowded community the incidence of a contagious disease is always high, and where in a less crowded one a disease may be endemic, in a greatly crowded one it will become epidemic. This rule applies in no less degree to Wood-Pigeons, and it is the abundant food supply which accounts for the massing, and the massing which favours the spread of the disease.

From the returns it would appear that where there was any disease there was, in most cases, a plentiful supply of either acorns or beech-masts, and in a few cases a plentiful supply of either corn or green crops.

VI.—TRANSMISSION TO OTHER ANIMALS.—That this particular disease is transmissible to other animals seems certain, for Löffler, in his orginal researches on this micro-organism inoculated, with mild results, fowls and rabbits ; guinea-pigs and rats suffered more severely ; while Wood-Pigeons and Sparrows succumbed.

From observations sent in there is little to record. Two observers noted the disease in Stock-Doves, one in a Tawny Owl, and one or two affirm that they have seen it in Pheasants and Partridges. In no case, however, were any of these birds sent in for examination bacteriologically, and therefore there must always be some doubt as to whether the disease was the same as that under consideration. On the other hand, I have made several enquiries as regards Pheasants and Partridges being affected in quarters where it might be expected, but have always received a negative answer, and on one estate where about 4000 Wood-Pigeons were destroyed last winter, and where 3000 Pheasants are shot every year, no case of " diphtheria " in Pheasants had ever been known.

The only other evidence I have on this matter is of a negative character, namely, that on this same estate

Rooks, Crows, rats, and ferrets fed largely on the diseased Pigeons without apparently contracting the disease; also that numbers were eaten by labourers, etc., without any ill effect accruing; but, of course, in the latter case the Pigeons were cooked, and this would kill the micro-organisms.

The question has often been asked whether this Pigeon disease is the same as diphtheria in the human subject. This is an intricate bacteriological subject, and a discussion on the *pros* and *cons* would be quite out of place here; suffice it to say that the causative bacilli of the two diseases are different in character, and as yet there is no proof that the characteristics of *Bacillus diphtheriæ columbarum* change to the characteristics of *Bacillus Klebs-Löffler* (human diphtheria bacillus) on transplantation from the Pigeon to the human throat.

Dr. Sambon, in an interesting paper in the "Lancet" (April 18th, 1908, p. 1143) on the "Epidemiology of Diphtheria," in order to account for the increased amount of diphtheria on the eastern seaboard (where, as he says, Pigeons mass together in the autumn and winter) favours the suggestion of the transmissibility of Pigeon diphtheria to the human subject. Unfortunately, he takes only the deaths from the disease, and not the incidence of the disease, which will be found to be quite a different thing. Whatever it may have been in the years in which his statistics were made up (1855–80), this year, at any rate, as I have shown, there was practically no Pigeon disease in those counties, the disease being practically confined to inland counties bordering the Thames—the very counties which he shows to have the lowest diphtheria death-rate. The returns for the last nine months from these counties are not yet made up, so there are not yet any statistics to show whether there has been any corresponding rise in the incidence of human diphtheria.

VII.—POST-MORTEM APPEARANCES.—The most in-variable appearance after death is the presence of a cheesy yellow "false membrane" over the hard palate, fauces, and base of tongue, and the glands in the neighbourhood enlarged. The mass is sometimes so large as to block entirely the gullet, though it is much rarer to find the windpipe pressed on to any extent ; the parts around the "membrane" are inflamed. With the formation of this false membrane death of the underlying tissues takes place, even the bones being affected ; thus in advanced cases it was common to find the base of the skull reduced to a cheesy mass. In a few cases the membrane extended down the gullet into the crop, and in one instance it had perforated the proventriculus, and the bird had died from peritonitis which had resulted.

In most cases the crop was empty. The condition of the birds varied, some being very wasted, others being in good condition. Those which had contracted a virulent type of the disease, or had a low resisting power, succumbed quickly, and so had not time to waste, while those which had lingered long with the disease, or had the gullet occluded, partially or wholly, were correspondingly thin.

Acrocephalus phragmitis

VARIATION IN THE NESTS OF THE ARCTIC AND COMMON TERNS.

BY

F. B. KIRKMAN, B.A., OXON.

IN a recent number of BRITISH BIRDS* Mr. W. P. Pycraft contributed a highly stimulating paper on the subject of nests, with special reference to those of the Ringed Plover. It dealt not only with variation in site and material, but also with the origin of the nest-building

FIG. 1.—Nest of Common Tern.

instinct. The present paper supplements his observations on the first of these two subjects. It is based on the examination of some fifty nests of Arctic and Common Terns, and will, it is trusted, help to throw further light on what is a very obscure problem.

* Vol. I., pp. 373-380.

The Arctic Terns' nests were found, during the summer of 1905, in the protected area at the south end of Walney Island, off Barrow-in-Furness. I made careful notes of thirty, of which thirteen lay on the patches of bare sand in between the sandhills, eleven in the shingle patches that alternated with the sand, four on the beach, and two among the bent, a rough stringy grass growing abundantly at Walney, and serving to cord up the wind-made shifting sandhills, thus rendering them more or less permanent.

FIG. 2.—Pebble-paved Nest of Arctic Tern on Beach.

There were four distinct types of nest with intermediate forms. The majority, eighteen in number, representing the first type, were made of varying amounts of bent. Nearly all these nests proclaimed the individuality of their architects. Some consisted of an outer circle of bent, the inside being left bare. One, evidently the work of a bird with a geometrical turn of mind, showed a semi-circle of bent, and a semi-circle of sand, while another was adorned with an oyster-shell and a feather, if not put there, in any case left unremoved. To some

bits of wood had been added.   In half a dozen instances
the birds were seemingly content to preserve appearances by merely placing or leaving two or three quite
useless stalks of bent around or across the nest.   None
were as complete as the Common Tern's shown in the
first of my photographs (Fig. 1).   It marks the highest
form of this type of nest.

The second type was represented by three nests paved
with pebbles (Fig. 2).   If the use of these pavements

FIG. 3.—Arctic Tern's Nest in Sand.

is to keep the eggs dry by raising them above the level
of soil liable to become damp, then we must deny the Terns
in question any sense of the meaning of their acts, for
their pavements were placed either on loose grit (Fig. 2)
or on loose sand, through both of which water would
rapidly sink.   How unnecessary the pavement was is
shown by the third type, represented by two scraped
depressions in the bare sand.   I kept the one shown in
Fig. 3 under close observation in foul weather and fair,
and had the pleasure of watching the owner (Fig. 4) hatch
out both her young successfully.

The fourth type was of a somewhat transitory nature. It was represented by three nests placed on the beach in the high-water mark seaweed, the one photographed (Fig. 5) being enlivened, accidentally, perhaps, with a crab's claw and a cork. These builders showed more originality than discretion, two of the nests being destroyed by the sea : one of the eggs, in an unbroken state, going to form part of the stranded drift.

FIG. 4.—Arctic Tern sitting on Nest in Sand.

The remaining four nests were highly instructive, being a combination of the first two types. They were all paved with pebbles, to which bits of bent (Fig. 6*), and in one case a complete outer circle of bent, were added.

It is worth noting that there was no definite relation between site and material, except in respect to the two nests in the bent, which were made of bent, and those in the seaweed. The pebble-paved, the bent nests, and

* This figure will appear in the second instalment of this article.

the bare scraped depressions were to be found both among the shingle and on the sand patches.

On referring to the " Manual " of the late Howard Saunders, I find that the Arctic Tern, besides laying " in· a depression of the sand, or on scanty herbage," will place its eggs " on the bare rock, just out of reach of the waves." Here we have, then, a fifth type, which might have been represented at Walney if there had been any rocks.

FIG. 5.—Arctic Tern's Nest in high-water mark drift.

The half-dozen nests of the Common Tern that I examined at Walney, were, as already noted, of the type illustrated by Fig. 1. But the late H. A. Macpherson, visiting the place in 1891, has left on record, in his " Fauna of Lakeland " that some of these birds, building on the upper beach, started a new fashion : their nests being lined with rabbit bones. This innovation appears to have died out, the rabbits presumably not seeing their way to provide the necessary material in adequate quantities.

(To be continued.)

ON THE MORE IMPORTANT ADDITIONS TO OUR KNOWLEDGE OF BRITISH BIRDS SINCE 1899.

BY

H. F. WITHERBY AND N. F. TICEHURST.

PART XII.

(Continued from page 57.)

TUFTED DUCK *Fuligula cristata* (Leach). S. page 447.

SUSSEX.—Recorded as breeding in the county (J. G. Millais, *Vict. Hist. Sussex*, Vol. I.).

HAMPSHIRE.—First known to nest in the county in 1890, since then its breeding range has rapidly increased, and six or seven localities are enumerated where nests have been found (J. E. Kelsall and P. W. Munn, *B. of Hants*, p. 233 *et seq.*).

SOMERSET.—Nested at Blagdon Reservoir in 1906 (F. L. Blathwayt, *Zool.*, 1908, p. 114).

BUCKS.—Breeds at Weston Turville (Rothschild and Hartert, *Vict. Hist. Bucks.*, p. 146).

DERBY.—They first began to resort regularly for breeding to Osmaston Manor lake in S.W. Derbyshire, in 1886. Since then a brood or two have been reared almost every year, and at least two hatched off in 1899. [Now there are seldom fewer than seven or eight pairs to be found there in the breeding season (F. C. R. J., *in litt.*, 1908).] From Osmaston they have spread to neighbouring ponds, where they have bred regularly since about 1888. They were observed on the Ashbourne Hall pond in 1892, and a pair bred at Sturston Mill in 1895 (F. C. R. Jourdain, *Zool.*, 1899, p. 476). They have also established themselves at Bradley, further to the east (*id.*, *t.c.*, 1900, p. 429), and, still more recently, at Norbury in 1907 (*id.*, *Derby. N.H.S. Journ.*, 1908).

STAFFORDSHIRE.—Besides the Weston Park colony referred to below, this species first bred at Calwich Abbey in 1906 (F. C. R. Jourdain, *in litt.*).

SHROPSHIRE.—Has bred regularly at Weston Park, on the borders of Shropshire and Staffordshire since 1880. Its numbers have steadily increased, and in 1890 there were

about twenty pairs nesting on five or six ponds. At Sandford
Pool, near Whitchurch, on the Cheshire border, four pairs
nested in 1891, and have continued to do so since, but have
not increased. One or two pairs have bred since 1855, and
still do so, at Hatton, near Shifnal (H. E. Forrest, *Zool.*,
1900, pp. 506 *et seq.*).

NORTH WALES.—Breeds in Anglesey, and possibly in
Merioneth (H. E. Forrest, *Vert. Fauna N. Wales*, p. 286).

YORKSHIRE.—Increasing as a nesting species (T. H. Nelson,
B. of Yorks., p. 467).

CUMBERLAND.—First bred in 1888 (*Zool.*, 1888, p. 330).

SCOTTISH BORDER.—Within the last twenty years they
have begun to nest at nearly every suitable loch, or large sheet
of water, on either side of the Border, *i.e.*, in Northumberland,
Berwick, and Roxburgh (A. Chapman, *Bird-Life Borders*,
p. 92).

SCOTLAND.—The increase and extension of range of the
Tufted Duck in Scotland is one of the most interesting events
in British ornithology. Mr. J. A. Harvie-Brown has written
two admirable papers on the subject (*Ann. S.N.H.*, 1896,
pp. 3–22; *Proc. Royal Phys. Soc. Edin.*, Vol. XIII., pp.
144–160), and to these all who are interested in the subject
must make reference. Not many records have been published
since the date of these papers, but the following may be
noted :—

Solway.—Has spread through the area since 1887 until
now every suitable loch has at least one pair (R. Service,
Ann. S.N.H., 1897, p. 222). *Ayrshire.*—A parent bird with
young was seen on Loch Kilbirnie in 1905 (*t.c.*, 1906, 198).
Although very common in East Renfrewshire it appears
slow in spreading to Ayrshire. *West Lothian.*—Bred in 1906
and 1907 (S. E. Brock, *t.c.*, 1907, 185) ; has bred regularly
in the eastern part of the district for the last ten years (B.
Campbell, *t.c.*, 1907, 249). *Tay Basin.*—Has increased
enormously since the first record of its nesting was reported.
Now it is " one of the commonest ducks on all suitable lochs
throughout the central and east portions, and just outside the
S.W. boundary of the area in Forth " (J. A. Harvie-Brown,
Vert. Fauna Tay Basin, etc., pp. 240 *et seq.*).

North of latitude 56° it is rare at all seasons on the west
coast, but all over the lowlands of Caithness and the extreme
east of Sutherland it is exceedingly abundant (*idem*).

Outer Hebrides.—On the increase ; actual nesting took place
in South Uist in 1903 (*id., Ann. S.N.H.*, 1903, p. 245). Macgillivray states that it was formerly a common bird in the
Outer Hebrides, but it is quite certain that they almost en-

tirely disappeared for a long time, indeed since he wrote (*id.*, *Fauna N.W. Highlands*, etc., p. 237). Four pairs seen in 1906, and one in 1907 by N. B. Kinnear and P. H. Bahr (*Ann. S.N.H.*, 1907, pp. 83 and 213).

IRELAND.—Several pairs were seen, and a nest found on Lough Conn, co. Mayo, in 1905 (R. Warren, *Irish Nat.*, 1905, p. 165). Ten or twelve broods were observed, a young bird was shot and an egg taken on Lough Mask, co. Galway, in 1906 (A. R. Nichols, *t.c.*, 1907, p. 184). Both these records are extensions of its previously-known breeding range in Ireland. Major A. Trevelyan informs us (*in litt.*) that on May 13th last he saw a pair on Lough Derg, co. Donegal, in which county we believe it has not yet been recorded as nesting.

SCAUP-DUCK *Fuligula marila* (L.). S. page 449.

SUTHERLAND.—A pair of Scaups was watched, and the nest found in rushes five feet from the water, on a small island in a loch in 1899, by Mr. Heatley Noble. The nest contained three eggs. It was left for a week, and the female bird was then seen to leave the nest and was clearly identified. The nest now contained nine eggs (J. A. Harvie-Brown, *Ann. S.N.H.*, 1899, p. 215).

OUTER HEBRIDES.—It nested in one of the islands south of the Sound of Harris in 1897, 8, and 9, and three pairs in 1900. Probably also again in 1901, and certainly in 1902 (J. A. Harvie-Brown, *t.c.*, 1902, p. 211). A nest with nine eggs was found on an island in a loch in one of the Uists in 1906, and the bird was seen to leave its nest (N. B. Kinnear, *t.c.*, 1907, p. 82; *cf.* P. H. Bahr, *t.c.*, p. 213).

The Rev. F. C. R. Jourdain points out that the first authentic record of the breeding of the Scaup in Scotland was that of the late A. C. Stark, who found a nest with eleven eggs at Loch Leven, Fifeshire, in 1880 (*cf. Proc. Roy. Phys. Soc. Edinb.*, VII., p. 203). The eggs were sold at Stevens' for £2 7s. 6d. on June 19th, 1902. This record appears to have been overlooked by Howard Saunders.

EIDER DUCK *Somateria mollissima* (L.). S. page 459.

CHESHIRE.—An immature bird was seen at Leasowe, on December 31st, 1905 (C. Oldham, *Zool.*, 1906, p. 75). It is rare on the north-west coast of England, and has only once before been recorded from Cheshire.

SCILLY ISLES.—Six examples are recorded (J. Clark and F. R. Rodd, *t.c.*, 1906, p. 304).

SCOTLAND.—Previously unknown on the west coast of

Sutherland, they were present in large numbers in 1901 and 1902, and perhaps for a *few* years before in every suitable place from Cape Wrath to Hansa and the Badcall Islands. A large increase is also noted along the eastern side of South Uist. Eider are only now (1904) beginning to push their distribution to any points between Loch Nevis and Badcall (*i.e.*, W. Ross, and parts of Sutherland and Inverness) (J. A. Harvie-Brown, *Fauna N.W. Highlands and Skye*, pp. 244–248).

IRELAND.—A young male was shot in Malahide Estuary, on the Dublin coast, in November, 1902 (E. Williams, *Irish Nat.*, 1903, p. 112). To Ireland the Eider is a rare and irregular winter visitor.

KING-EIDER *Somateria spectabilis* (L.). S. page 461.

FIFESHIRE.—A male was shot on a moor in Fifeshire on June 15th, 1899 (B. B. Riviere, *Zool.*, 1902, p. 27).

ORKNEY.—An adult female was shot by Mr. S. Sutherland off Graemsay, on February 21st, 1906 (F. Smalley, *t.c.*, 1906, p. 113).

ISLAY.—One was observed by Mr. A. Ross near Kintra on July 25th, 1906 (*Ann. S.N.H.*, 1907, p. 198).

IRELAND.—A mature male was shot on November 10th, 1897, in the Foreland Bay, off Donaghadee, co. Down, by Mr. W. H. Shaw (R. Patterson, *Irish Nat.*, 1901, p. 50).

JERSEY.—Mr. H. Mackay states that he examined a female bird, and identified it as a King-Eider, which had been shot at La Roque. He gives no date (H. Mackay, *Zool.*, 1904, p. 380).

COMMON SCOTER *Œdemia nigra* (L.). S. page 465.

IRELAND.—Major H. Trevelyan communicated to the " Field " (15, VII., 05) an account of the nesting of this bird on one of the larger loughs in Ireland, the exact locality being suppressed. Between June 11th and August 18th, 1904, a pair of birds were constantly observed. On May 24th, 1905, the pair were again observed in the same locality, and on June 13th the female was found on the nest under a small bush on an island. There were eight eggs, five of which hatched between June 28th–30th, and the old bird was seen on the lough with the five young ones on July 1st. One of the young was obtained on July 3rd, and afterwards submitted, with an egg and some of the down from the nest, to Dr. Bowdler Sharpe, who confirmed the identification. The egg, down, and feathers from the nest, and the young bird, were also submitted to Mr. Heatley Noble, who likewise confirmed the identification (*Irish Nat.*, 1905, p. 200). It is much to Major Trevelyan's credit that he was thus able to authenticate this most interesting discovery of the first nesting of this species in Ireland without the destruction of the parent birds. The birds bred again in 1906 (R. J. Ussher *in litt.*).

In 1907 one male and two females were observed on the lough, but no nest was found. On June 4th, 1908. a nest with eight eggs was found well concealed in a furze bush on an island in the same lough, and this year also there appeared to be a second and solitary female (H. Trevelyan, *Field*, 4, vii., 08, p. 3).

SURF-SCOTER *Œdemia perspicillata* (L.). S. page 469.

SCILLY ISLES.—Has been obtained twice (J. Clark and F. R. Rodd, *Zool.*, 1906, p. 304).

CORNWALL.—An adult male was killed with two Velvet Scoters on the Helford River on December 16th, 1906 (J. Clark, *t.c.*, 1907, p. 285).

ORKNEY.—Young birds are of commoner occurrence than most people suppose, hardly a winter passes without one or more being seen among the Velvet-Scoters when they first arrive. The adult birds are much rarer. An adult male was seen inside Stromness Harbour between December 14th and 21st, 1905 (H. W. Robinson, *Field*, 17, II., 06).

GOOSANDER *Mergus merganser* L. S. page 471.

SCOTLAND.—A pair was identified off the north end of *North Uist* on October 31st, 1905 (A. Elfrish, *Ann. S.N.H.*, 1906, p. 53). A male was seen off Barra on May 22nd, 1906 (N. B. Kinnear, *t.c.*, 1907, p. 83). The bird is of rare occurrence in the Outer Hebrides.

SMEW *Mergus albellus* L. S. page 475.

NORFOLK.—An adult male was shot on Breydon on January 30th, 1907 (B. Dye, *Zool.*, 1907, p. 111). Adult males are rare.

YORKSHIRE.—An adult was shot at Skelton in the winter ot 1900 (T. H. Nelson, *Birds of Yorks.*, p. 486).

SHETLAND.—One was shot on February 14th, 1901, at Sconsburgh (*Ann. S.N.H.*, 1902, p. 134). It is a rare visitor to the Shetlands.

(*To be continued.*)

LARGE-BILLED REED-BUNTING (*EMBERIZA PYRRHULOIDES PALUSTRIS*) IN KENT.

A NEW BRITISH BIRD.

BY

M. J. NICOLL, M.B.O.U.

AT the meeting of the British Ornithologists' Club, held on June 17th last, I exhibited a male specimen of the South European Large-billed Reed-Bunting (*Emberiza pyrrhulo·des palustris*).

This bird, which is new to the British fauna, was obtained near Lydd, in Kent, on May 26th last. I was away from home at the time it was shot, and was thus unable to see it in the flesh, but I examined it shortly after it was mounted.

The occurrence of this Reed-Bunting in the British Islands is of interest, not only on account of its being a new British bird, but also because the commoner and typical *E. p. pyrrhuloides* is the form one would expect to occur in England, as it has done so on Heligoland.

The example obtained in Kent agrees exactly with specimens in the British Museum from South Italy, in which country, as well as in Southern France and Spain, the bird is resident. It may be distinguished at once from the common Reed-Bunting by its large thick bill.

The typical form of the Thick-billed Reed-Bunting inhabits the coasts of the Caspian Sea from the foot of the North Caucasus to the Volga, Transcaspia and Turkestan, and has occurred once on Heligoland. In coloration the former is very much paler, the broad white edges to the feathers of the upperparts, and the pale grey rump, give the bird an almost silver-grey appearance: a great contrast to the more sober coloration of the bird obtained in Kent.

It is somewhat difficult to account for the appearance of some South European birds in the British Islands. The present species and all other stragglers which have occurred, may have joined parties of other species, and thus found their way to our shores.

In the autumn southerly gales may be the cause of the visitation of rare Chats and other birds. But some have occurred in our islands during the height of the

Male Large-billed Reed-Bunting obtained near Lydd, Kent, on May 26th, 1908.

summer, and these visitations can, I think, only be accounted for by the supposition that these birds had lost their mates, or that their nesting had been interfered with in some way, and that, following the migratory "impulse," they had pushed northwards and thus reached the British Islands, far to the north of their usual limit.

CURIOUS SITE FOR A ROBIN'S NEST.

ROBINS have frequently been recorded as nesting in curious places, but I do not think they have been known to choose such a remarkable site as the following. A pair relined an old Blackbird's nest, situated in a thorn bush, four feet from the ground, with moss, and were successful in hatching a brood. A. G. LEIGH.

[The Rev. F. C. R. Jourdain writes :—" The habit of breeding in old nests of other species is common in the case of the Pied Wagtail and occasional in the Tits, Spotted Flycatcher, and other birds. The habit is, however, rarely recorded of the Robin. I have a note of one found in an old Swallow's nest, and one in an old Hedge-Sparrow's nest is recorded by Mr. J. E. Harting (*Birds of Middlesex*, p. 38)." In a list of such occurrences published in the " Zoologist " (1905, p. 33), Mr. R. H. Read records a Robin's nest in a Thrush's nest, and three nests of the Robin one over the other, the top one containing eggs and the middle one stale eggs of the previous season. Mr. T. T. Mackeith also records (*t.c.*, p. 69) a Robin's nest built upon a Swallow's nest of the previous year. Many instances of other birds utilizing the old nests of other species are on record.—EDS.]

GREY-HEADED WAGTAIL IN SUSSEX.

IN the spring of 1869 or 1870 a Grey-headed Wagtail was shot at Lancing, in Sussex, not far from the sea, which has, I am sorry to say, remained until now unidentified. Having, at the request of Mr. Witherby, submitted it to Mr. N. F. Ticehurst, that gentleman writes : " It is in my opinion undoubtedly *M. f. borealis.* . . . It is not nearly white enough on the throat, and is too dark on the head for *M. f. cinereicapilla.* It is even darker on the head than most of my *M. f. borealis*, but I take it that is due to wear."

J. H. GURNEY.

NESTING OF THE GREY WAGTAIL IN BERKSHIRE.

ON June 13th I noticed a pair of Grey Wagtails (*Motacilla melanope*) close to one of the locks on the Kennet and Avon Canal. After a very short search I found their nest (empty, but apparently ready for eggs) in the broken woodwork of the side of the lock. On June 18th the nest was visited by the keeper, and found to contain four eggs. On June 24th I again visited the place, but unfortunately a barge had been through the lock in the meantime, and the nest had been swamped and the eggs washed away. A careful search in the hole behind the nest showed one broken shell, while the birds were still near by. The nest was so placed, that when the lock was filled (in order to enable a barge to pass) it must have been quite three feet under water.

Mr. Heatley Noble records a nest in the " Victoria History of Berkshire," while I believe another one has been found more recently by Mr. F. C. Selous, near Newbury.

W. NORMAN MAY.

AN EARLY RECORDED WAXWING.

THE following extract from a letter which I have lately received from Mr. R. D. Roberts, of St. Asaph, North Wales, should prove of interest because it refers to one of the first Waxwings recorded as visiting this country, and it is remarkable that the specimen, although now 120 years old, should still be in good condition. Mr. Roberts writes : " The quotation in your ' Vertebrate Fauna of North Wales ' from Pennant's ' British Zoology,' under the heading ' Waxwing ' (page 130), is interesting to me inasmuch as the bird referred to is in my possession, and though shot in 1788 is in perfect condition. The account on the back of the case being nearly illegible through age I recently had copies printed, and enclose one." The label reads as follows :—

Bohemian Chatterer or Waxwing.
(Bombycilla Garrula.)

Kill'd during the cold Frost in December, 1788, at Garth-Meilio, in the County of Denbigh, by Mr. William Dod, of Edge, in Cheshire.

It was perching in one of the Fir Trees in the Avenue to the House.

H. E. FORREST.

LESSER REDPOLLS NESTING IN SURREY.

DURING a couple of hours spent among the birch trees on Wimbledon Common on July 12th, I saw a nest of a Lesser Redpoll (*Linota rufescens*), in a small birch, with four well-fledged young ones, another brood on the wing, and at least five pairs of old birds. Indeed, on this July morning, when but few birds were singing, the Redpolls uttering their characteristic call-notes, as they passed from place to place with undecided, wavering flight, high above the tree-tops, were the outstanding feature in the bird-life of the common.

CHARLES OLDHAM.

[For former records of the Lesser Redpoll nesting on Wimbledon Common and in other places in Surrey, see Vol. I., p. 184.—H. F. W.]

HAVE STARLINGS INCREASED BEYOND THE CAPACITY OF NESTING SITES ?

LATE last year two packs of Starlings of about 1000 each took possession of a young larch plantation near Ross, Herefordshire. Every evening the numbers increased until it was impossible to make any estimate of them, but to give some idea, I saw on one occasion a field of about four acres so covered with them that it was difficult to see any soil at all. Their movements, as night drew near, were a source of great interest to me. At times they would wheel for half-an-hour in the air, each battalion independent of the other ; at others they would settle in high trees and maintain a ceaseless chattering ; while, once or twice, being apparently still hungry, they would settle on ploughed ground and busily search for food.

The majority have distributed and paired, but there are still a number of small packs, ten to twenty in each, which pass my house every night on their way to roost, and I have noticed for the last three years that many remained unpaired. These are neither old birds which have done nesting, for as yet (June 7th) I have seen no young birds abroad ; nor are they for the same reason young birds. What, then, is the reason for these bachelor habits ?

It cannot be that there are greater numbers of either sex, because Nature's balance is very even ; nor can it be that they do not breed until the second season.

Is it possible that there are not enough suitable nesting places for so many ?

FRED. A. HERBERT.

NUTCRACKER IN KENT.

IN Vol. I., p. 388, we referred to a Nutcracker which had been reported by Mr. N. F. Richardson as having been shot in Kent on December 29th, 1907. On page 28 of the present volume Mr. G. M. Beresford-Webb suggested that this might have been a bird which escaped from his aviaries. Mr. Richardson has very kindly submitted the bird, with full particulars, to Mr. Beresford-Webb for examination. and that gentleman writes us that " as far as it is possible to see the bird appears to be similar to the one which escaped." It was shot only six miles from Mr. Beresford-Webb's house, and three days after his bird had escaped, and little doubt remains that it was in fact his bird and not a wild one.—EDS.

CLIMBING MOVEMENTS OF THE GREEN WOODPECKER.

THIS season a pair of Green Woodpeckers (*Gecinus viridis*) made a hole in a decaying cherry tree in the orchard next our garden, in the village of Burwash. I could, from a garden seat, watch them within fifty yards. The hen bird appeared to be sitting by the middle of May. The male bird was constantly bringing his mate food, and would fix himself on the trunk for ten minutes at a time, partly supporting itself by the stiff pointed tail-feathers, his head just level with the orifice opening into the nesting-hole, often uttering his plaintive cry. The female would now and again come up and greet him by putting out her bill through the opening hole. Unfortunately the orchard became the scene of carpet-beating operations, which disturbed the Woodpeckers, and in the intervals of absence a pair of Starlings commenced an attack, and by rapidly throwing in bits of foreign material, made the hole untenantable for the Woodpeckers, who consequently deserted it. The late Professor Newton, in the " History of British Birds " (Vol. II., p. 458), remarks in a footnote that " Selby says he had repeatedly seen it descend trees by moving backward. The editor has not been so fortunate, though he thinks he must have enjoyed more frequent opportunities of observing the bird." I can confirm this statement of Selby; the male bird of the pair I allude to searched the bark of the decaying cherry tree in which the hole was placed, with great assiduity. On reaching a sufficient elevation it would descend backward with as great rapidity as in its ascension. I was so close to the birds on many occasions, that with the aid of glasses, I saw that during the backward descent the points of the tail-feathers were kept about an inch off the bark of the tree, though the tail and back retained the curve, associated with the ascending bird.

H. W. FEILDEN.

MARSH-HARRIERS IN NORFOLK.

DURING a short stay in " Broad-land " recently I had the good fortune to have under observation for some considerable time first a single specimen, and a day or two later a pair,

of that handsome bird the Marsh-Harrier (*Circus œruginosus*).
I shall never forget the majestic wheeling flight of these birds
as they quartered the ground in search of prey. The exact
locality it will perhaps not be wise to name for the present,
but of the identity of the birds there can be no doubt.

W. P. Pycraft.

DUCKS' EGGS AND DOWN.

I have read with considerable interest Mr. Noble's article
on Ducks in the June issue of British Birds. In the main
I am in entire agreement with his remarks, especially as regards
the futility of attempting to identify by the down alone.
I take slight exception, however, to his remarks on the
Gadwall. I have observed, perhaps, a dozen nests of this
species in Norfolk, and in two cases at least there are numerous
white feathers which are indistinguishable from those of
the Wigeon, and which, in conjunction with precisely similar
down and precisely similar eggs, render identification
extremely difficult.

I have seen the cream-coloured variety in eggs of the Pintail,
but I wonder if Mr. Noble has come across the variety where
the eggs are as vivid a green as those of the Golden-eye. As
regards downless Pochards' and Mallards' nests, my experience
of the former is small, being confined to one locality, where
the nests are always floating structures, and down is not
abundant, but I have found, perhaps, half-a-dozen Mallards'
nests built *close to water*, and amongst thick sedge, which were
without a particle of down.

Norman Gilroy.

[The Editors have kindly allowed me a view of the above
note from Mr. Norman Gilroy. Doubtless the number of
proprietors over whose estates Mr. Gilroy has leave to prosecute
his investigations is larger than those who have extended a
like permission to myself ; in any case, I cannot lay claim
to having examined so many nests of the Gadwall as Mr.
Gilroy has. I have seen white feathers in Gadwalls' nests,
but in my opinion they are not typical, and as the object of
the article was *identification*, they were excluded. I have
not yet seen a nest containing only white feathers, and were I
to find such a nest, I should consider it necessary to procure
the female bird before identification could be established.
Some of the patterned feathers have always been present in
the nests I have examined, and it is these feathers which give
the key to the solution.—Heatley Noble.]

INLAND NESTING OF THE SHELD-DUCK, AND NESTING OF POCHARD, SHOVELER AND TEAL IN LINCOLNSHIRE.

In answer to Mr. Southwell's question about the Sheld-Ducks on Twig Moor, Lincolnshire (*supra*, p. 62), I may say with confidence that the birds breed there. On May 22nd, 1907, I saw five or six pairs there, evidently breeding ; and again on June 2nd, 1908, several birds were on the ponds, and a brood of ducklings had just been hatched off. The Sheld-Duck has also nested on the Gull-ponds on Scotton Common, six miles south-west of Twig Moor, and a pair or two perhaps do so every year. I have seen young in all stages there, and on June 22nd, 1903, there was a brood only a few days old, some of which I managed to catch in my hands, but quickly released. This species is very common on the Somerset coast, near Burnham and Weston-super-Mare ; in winter I have seen more than five hundred on the sea in one flock, and hundreds nest among the sand dunes in the neighbourhood. The water-bailiff of Blagdon Reservoir, about ten miles from the Somerset coast, states that a pair remained to nest there a few years ago. He is a most intelligent observer, and is not likely to have been mistaken. I may mention also that several pairs of Pochards breed on the Twig Moor Gull-ponds, and I saw quite young broods both in 1907 and 1908. A good number of Teal and Shovelers breed on Scotton Common, a fact which I can state from personal observation, having found their nests and seen their young broods there on several occasions during the last few years.

F. L. Blathwayt.

[Mr. Clifford Borrer has also written to us to the same effect as the above.—Eds.]

NESTING OF THE SHOVELER IN STAFFORDSHIRE.

Early in April, 1908, Mr. R. C. Thomas, of Bloxwich, told me that some Shovelers (*Spatula clypeata*) (at first two drakes and a duck) were on a " swag "—a piece of water formed by subsidence of land caused by mining operations—at one of their collieries. The one on which the Shovelers were seen is about an acre in extent, and is adjacent to a coal-pit, which is not now worked. On *May* 12th Mr. Thomas found the nest, which then contained six eggs, built in a depression, about fifty yards from the " swag." When on the nest the duck drew towards herself the tall grasses growing near, and thus formed a kind of canopy, a small opening being left at the side nearest the water. There is a footpath near, and, at this time, she took no notice of anyone passing by unless they stopped to look at her, when she hurriedly left the nest, ran a few yards, and took flight to the " swag."

On *May* 17*th* the clutch of ten eggs was completed, and on *May* 28*th* I accompanied Mr. Thomas to see the b rds and their nest. As we approached the place my friends told me that it was uncertain whether we should see the drake, as he some-times disappeared for a whole day ; however, to our great delight, we found the beautiful bird on the water. We had cautiously approached within some forty yards, when he rose and flew behind the " pit-bank " at the south side of the " swag." Our attention was next directed to the duck, which hastily took to flight when we were within a yard of the nest. The grass had grown very much and now completely hid it. There were ten eggs—of a greenish-cream colour, much soiled—laid on dried grasses and down, and not covered—perhaps owing to our arrival at the nest being rather sudden. We then walked on to another "swag," nearly three acres in extent, and about five hundred yards away. Here men were loading coal from " pit-tubs " into carts, on a wharf, close to which, in company with a number of domestic ducks, were the Shovelers. Although they had apparently taken no notice of the men who were at work, on our arrival they instantly took flight, fortunately only to the other side of the " swag." Very quietly we walked to the shelter of a tree, from which we watched the birds for a considerable time, during which they left the water and preened their feathers on the opposite bank.

On *June* 9*th* the ten eggs hatched out safely, after an in-cubation of twenty-four days, and the same day a second drake put in an appearance, but, after a fight, was driven off. The Rev. F. C. R. Jourdain, whom I told of this occur-rence, informs me that a pair of Shovelers had bred at Wyrley Grove, which is two miles from the locality of the nest I have described, in 1906 and 1907 ; from inquiries I have since made I find there has not been a nest there this year.

The best thanks of Staffordshire ornithologists are due to the Messrs. Thomas, who were draining the " swag " when the Shovelers were first noticed, and who instantly stopped operations when they found that they were nesting. They have since taken every precaution that the birds should not be disturbed.

W. WELLS BLADEN.

POCHARD NESTING IN SOUTH-WEST KENT.

ON May 16th I was punting with a friend in Romney Marsh

along a wide dyke, which ran along one side of a bed of high
reeds. On nearing an angle of the reed bed I noticed a female
Pochard (*Fuligula ferina*) swimming hurriedly away as if
she had just left her nest. I got out of the punt and searched
the corner of the reed bed, when I very soon discovered the
nest containing seven young Pochards, just hatched, one
duckling being not yet dry. While examining and trying
to photograph the nest the duck flew round quite close, so I
was certain of her identity. The nest was about eighteen inches
high, composed of pieces of dry reed, and had practically
no down or feathers in it ; in fact, except for the ducklings
and the broken eggshells, one might have supposed that the
nest was that of a Coot.

R. SPARROW.

UNUSUAL NESTING SITES AND INCUBATION PERIOD OF THE TUFTED DUCK.

ON June 17th, 1908, I found on a small island rock a nest of
a Tufted Duck (*Fuligula cristata*) in a water-worn crevice,
having cover from all sides except the south, with an overhanging rock giving partial cover from above. The nest was
made of dry rush, grass, moss, and a few green fern leaves ;
there were five eggs, on which the bird was sitting, but there
was little, if any, down.

On June 20th, 1908, on a wooded island, I found a Tufted
Duck's nest among bushes, with a dead branch overhanging
one side of it. South-west of it was a rock ; north of it a large
stone ; east of it a small stone ; to the south of it a sallagh,
probably *Salix caprea*. The floor of the nest was made of
dead leaves, the sides of it were almost entirely of down, with
a very few dead leaves and small dead twigs, and the occurrence of these two latter may have been accidental. As at
the bottom of the nest there was the skin of an egg, it seems
probable the site had been used before. There were ten eggs.
I have by no means infrequently found the nest of this species
under bushes, but I do not remember one placed as this one
was, right inside a covert.

There is no doubt birds adapt themselves to their surroundings, but it seems curious that they should select an unusual
site without immediately at hand the usual materials for a
nest, when plenty of such ground is to be found close by.

On May 29th last on a lake island in Ireland I found a
nest of the Tufted Duck with eight eggs. I am all but certain
I put the bird off it, but the one egg I took from the nest
when blown showed no trace of incubation. I replaced, within

a few minutes, this egg with one of a domestic Duck. At some time subsequent to this, and prior to June 22nd, I visited the nest and found her sitting on eight eggs, *i.e.*, seven of her own and the one of the domestic Duck. On July 1st my boatman visited the nest and found her sitting on five eggs only, *i.e.*, two of her own had gone as also the one of the tame Duck. On July 5th, on visiting the nest, he found the eggs had hatched out. Needless to say, I do not consider these observations by any means crucial, but the evidence, such as it is, points in the direction of the incubation period of the Tufted Duck being more than twenty-three days.

On June 16th, on a small lake island, I found a Tufted Duck's nest with sixteen eggs. One egg was on the top of the other fifteen, they were warm and evidently being incubated.

H. TREVELYAN.

TEAL AND PHEASANT LAYING IN THE SAME NEST.

I MET with rather a curious case of a mixed clutch of eggs a few days ago which may be worth recording, viz., a nest with both Teal's and Pheasant's eggs. Both birds laid in one nest, but when I saw it the Teal was in possession, sitting very close. CHARLES E. PEARSON.

[A number of records show that in the case of Game-birds and Ducks "joint" clutches of eggs are by no means rare.— H. F. W.]

PALLAS'S SAND-GROUSE IN ENGLAND.

AN invasion of Pallas's Sand-Grouse (*Syrrhaptes paradoxus*) into this country was not unexpected since the bird appeared numerously in European Russia in the latter half of April, and has been reported from several parts of Germany (*cf. Orn. Monats.*, 1908, pp. 100 and 132). The following have been reported in England :—*Yorkshire.*—Three flying high between Burley and Ilkley on May 20th (" Lichen Grey," *Country Life*, 13, VI., 08). *Hampshire.*—Five, said to be of this species, were seen near East Liss about the middle of April (" M. I.," *Field*, 20, VI., 08). Five were clearly identified by Mr. A. O. Lyon, near Burley, New Forest, early in August (*in litt.*). Two were seen flying N.E. over Havant on July 8th (B. Roper, *t.c.*, 18, VII. 08). *Berkshire.*—One was picked up near the River Kennet on June 6th (H. D. Astley, *t.c.*, 20, VII., 08). *Essex.*—A pair was seen several times near Southend-on-Sea in the last week of June (J. Seabrooke, *t.c.*, 4, VII., 08). *Surrey.*—Three were observed at Holmwood on June 28th (L. Mortimer, *l.c.*). *Norfolk.*—Two were seen at Brancaster on June 28th (F. H. Partridge, *l.c.*).—H. F. W.

BLACK-TAILED GODWIT AND SPOTTED REDSHANK IN KENT.

ON May 12th last I was walking alongside a "fleet" in Romney Marsh when a bird which was strange to me rose from the edge of some shallow water. I at once got my binoculars on to it, and by the long straight bill, white wing-bar and white rump and light brown back, I identified it as a Black-tailed Godwit (*Limosa belgica*), and from its size I should say it was a female. It flew some distance, and I thought I had marked it down, but on going to the spot I failed to flush it again. On the 16th I visited the same ground, but did not see the Godwit, so it had evidently continued its migration.

The same evening my attention was called to a strange bird flying overhead by hearing a whistle something like that of a Redshank. For a moment I thought it was a Golden Plover with black breast, but on looking at it through my glasses, I noticed it flew very like a Common Redshank, and had a beak as long as a Redshank, and was black all over, with white speckles. I at once decided it was a male Spotted Redshank (*Totanus fuscus*) on migration.

R. SPARROW.

CHANGE OF NESTING SITES THROUGH HUMAN INFLUENCE.

LAST January the Black-headed Gull was removed from the list of egg-protected birds. As a consequence its regular nesting-places were raided by collectors of eggs for local consumption, or for despatch to London as Plovers' eggs, and the result was that the birds, seeking fresh quarters, formed two new colonies on Wedholme Flow and Rockliffe Marsh, near Carlisle. The Redshank has been subjected to similar persecution. At one time the commonest of our shore-birds, its numbers suffered such depletion that it was put on the list of egg-protected birds by the Cumberland County Council. On the marshes in North Lancashire its eggs have been largely gathered for substitution as Plovers' eggs in the metropolitan market, and the result has been a notable exodus of the birds to the Yorkshire dales for security, In the neighbourhood of Bentham, as Mr. Murdoch, a capable naturalist, reports, Redshanks have been nesting freely, and in a Yorkshire dale several miles further inland I have noticed a remarkable development. With the sequestered and beautiful dale named Kingsdale, I have been familiar from boyhood, and have fished its trout stream for more than fifty years. I can vouch for it that such a bird as a Redshank

was never seen in that dale until very recent years.   Five
years ago there was, to my surprise, a pair of the birds ;
at the beginning of June this year Redshanks were so
numerous and noisy as to produce the illusion that I was
on a Cumberland marsh.   The addition to the avian life
of the dale was very pleasing.

T. HARRISON.

* * *

LESSER REDPOLL NESTING IN MIDDLESEX.—Lt.-Col. H.
Meyrick records that he has found two nests of the Lesser
Redpoll (*Linota rufescens*) on Hampstead Heath this year,
and that he suspected them of breeding there last year (*Zool.*,
1908, p. 227).

LITTLE OWL IN WILTSHIRE.—An example of *Athene noctua*
was shot near Avebury in November, 1907, and is now in the
Marlborough College Museum (*Rep. Marl. Coll. N.H. Soc.*,
1908, p. 76).   This may be a forerunner of a still greater
extension of this bird in a south-westerly direction from
Lilford than has yet been traced (*cf. ante*, Vol. I., p. 335
et seq.), or it may have been liberated locally.   The members
of the College Natural History Society would do well to make
a search for the Little Owl in the neighbourhood.

SCOPS OWL IN CUMBERLAND.—A specimen of a Scops Owl
(*Scops giu*) is reported by Mr. P. W. Parkin (in whose
possession the bird is) to have been shot on November 6th,
1907, at Broomrigg, near Armathwaite, by Captain W. H.
Parkin (*Field*, 13, VI., 08, p. 982).

BITTERN IN YORKSHIRE.—A Common Bittern (*Botaurus
stellaris*) was seen by the watcher at Kilnsea, Holderness,
Yorkshire, on May 6th last (R. Fortune, *Nat.*, 1908, p. 202).

GADWALL IN SOMERSET.—A male was shot near Bridgwater
on February 10th, 1908 (H. Whistler,*Field*, 20, VI., 08,p. 1030).

WOOD - PIGEON NESTING ON A HOUSE.—Two Wood-
Pigeons are said by Mr. F. Mansell to have nested and reared
their young on a window-sill in Highbury this year (*Field*,
20, VI., 08, p. 1030).

INCREASE OF TERNS NESTING IN IRELAND.—Mr. A.
Williams writes to the " Irish Naturalist " (1908, pp. 119–122)
that protection has greatly increased the Gulls and Terns in
co. Dublin.   At Malahide Island the numbers of Common
and Arctic Terns nesting is described as being incalculable.
This colony a few years ago numbered only a couple of pairs.
A rough idea of the number of birds at the present time is
given by the fact that Mr. Williams counted 211 nests, but
his search was by no means exhaustive.

RITISH

IRDS

STRATED·MAGAZINE
D·TO·THE·BIRDS·ON
HE·BRITISH·LIST

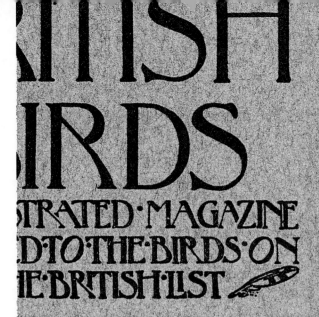

R 1,

Vol. II.
No. 4.

LY·ONE·SHILLING·NET
GH·HOLBORN·LONDON
THERBY & C°·

HOW TO
ATTRA
AND
PROTE
WILD BIF

A FULL DESCRIPTI
SUCCESSFUL MET

WITH MANY ILLUSTR

(For details see over.)

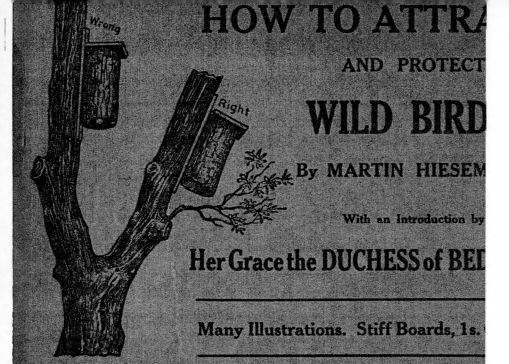

LITTLE TERN ON ITS NEST. ROMNEY MARSH.

(Photographed by F. B. Kirkman.)

BRITISH·BIRDS

EDITED BY H. F. WITHERBY, F.Z.S., M.B.O.U.
ASSISTED BY W. P. PYCRAFT, A.L.S., M.B.O.U.

CONTENTS OF NUMBER 4, VOL. II. SEPTEMBER 1, 1908.

VARIATION IN THE NESTS OF THE ARCTIC AND COMMON TERNS.

BY

F. B. KIRKMAN, B.A., OXON.

(PLATE III.)

(Continued from page 82.)

AT Romney Marsh I examined this summer (1908)
fifteen Common Terns' nests. Of these twelve were in
the shingle, and three on soil among herbage ; one made
of lichen and grass was shaded by a foxglove, and em-
bowered in white campion—a charming picture. Only

one, lying in the shingle, was without lining of any sort.
The material used in the case of the others consisted of
small twigs, chiefly broom, coarse stems, dry grass and
lichen.   By using one of these materials, or combining
two or three, the fourteen birds in question managed
to produce seven variations, of which one is shown in
Fig. 8.   This species, also, according to Mr. Kearton, lays
its eggs on bare rock.   As in the case of the Arctic

FIG. 6.—Arctic Tern's Nest, with Pebbles and Bent.

Terns, no definite relation between site and material could
be traced.

A word about the Little Tern.   The late Howard
Saunders stated that it uses no material for its nest.   Mr.
Fred. Austen, the watcher at Romney Marsh, endorses
this, adding that the hen alone constructs the nest, which
she does by the simple process of working her body round
and round in the pebbles, much like a dog preparing
its bed for a nap.   But the late H. A. Macpherson describes two nests (*t.c.*, p. 418) one lined with "dry stems of

grass," the other with "fine pebbles." While no doubt the species generally dispenses with a lining (Fig. 7), further observation may show that variations are not infrequent. One may note in passing that the preference of the Little Tern for unlined nests may possibly account for the comparative scarcity of this species.

Let us turn now to consider the bearing of the above facts, beginning with the variation in sites. It seems reasonable to assume that if there is any truth in the theory of protective coloration the normal (*i.e.*, the safest) nesting site of the Arctic and Common Tern, and, indeed, of all the grey and white Terns and Gulls, is the closely packed shingle such as one finds on the beach or the vast stretches of Romney Marsh. Sitting amid the vague outlines of black, white, and grey stones, a Tern is practically invisible. Something of this correspondence of colour is visible in Plate III. There is no reason to think that the invisibility is necessary to the safety of the Tern itself, for it is the last bird to be caught napping on its nest. But its advantage as a means of protecting eggs and young from discovery is obvious.

By placing its eggs among herbage, on the bare sand patches, on rocks, or even in the shingle beds among the sandhills, where the stones are seldom closely packed (Fig. 6), the Tern sacrifices all the advantages to be derived from its coloration. In such sites it is a conspicuous object. That it should be able to effect the change with comparative impunity seems to argue that the species has no longer many egg-stealing foes to fear. Under any circumstances it would require a bird of no mean courage or strength to pillage the nests of Terns, for they have an unpleasant habit of descending almost vertically, with the velocity of a bolt, upon unwelcome intruders, and striking with the beak. At Walney the young Black-headed Gulls, whose mottled brown plumage evidently caused their identity to be mistaken, suffered severely from this practice, often being struck down in mid-flight, and the more easily as they were

ignorant of the Corvine device of turning bodily in the
air and presenting beak and claws to an assailant from
above.   Howard   Saunders   records   that   a   flock   of
Arctic Terns " has been seen to mob and drown a
Hooded Crow.'"   On the other hand, it is stated that
in the Farne Islands, a Greater Black-backed Gull
forced to keep on the ground by a broken wing, relieved
the monotony of its existence by prolonged feasting upon

FIG. 7.—Lesser Tern's Nest in Shingle at Romney Marsh.

the eggs and young of the large colony of Arctic Terns.
A mile off one could see the whirling canopy of white
wings that marked the spots where the invalid paused
for refreshments.   No doubt a Raven could also exact
heavy toll.   But these and other large egg-eaters are
now no longer common.

The Arctic Terns of Walney were, however, far from
effecting with impunity the change from the normal
site.   Though they shifted their nesting grounds at least

once, and though they continued laying and re-laying up to the middle of July, it is doubtful whether they hatched out more than a dozen chicks. An almost clean sweep was made of the eggs, the marauder being no other than the humble rat. Traces of these creatures' feet and tails were to be seen leading from nearly every nest to the nearest tuft of bent, where the broken shells of the eggs told the story of the theft. The deep narrow furrow made in each case by the tail in the sand seemed to show that the rat had used this appendage as a support while it hopped along on its hind legs with the eggs clasped in loving fashion to its breast. That it escaped being murdered by the parents is remarkable.

If the Terns had nested in the packed shingle of the beach would their eggs have escaped the rats ? Arguing *a priori* it is at least clear that they would have been much more difficult to locate. Perhaps some other observer can throw light on this point.

So far I have shown, or attempted to show, that the Arctic Terns are quitting the safer shingle site for others more exposed. I suggest that this has been done with comparative impunity (assuming the rat ravages to be exceptional) owing to the greater scarcity of enemies powerful enough to take advantage of the new conditions. The same applies to a number of other sea-birds. Indeed, it is more than likely that the extermination of the larger birds of prey helps to explain much that is anomalous in the habits of British birds. But to show that the absence of enemies has rendered the change possible is not to explain why it took place. It may be that, owing to the spread of vegetation, or the invasion of sand, it is the nesting sites that are altering their character, and not the birds their choice. But this is mere conjecture.

When we quit the subject of variation in site for that of nesting material, we find ourselves face to face with the two questions already put by Mr. Pycraft in respect to the Ringed Plover, the first being : How comes it about that certain individuals of the species provide

their nests with a lining when other individuals dispense
with it altogether ?   The explanation may lie in the
undoubted capacity for imitation that birds possess.
This would account for the presence of a lining where it
was superfluous, or where, as has been shown to be
frequently the case, it is too rudimentary to be of the
least use.   Imitation, even among human beings, is often
quite unintelligent.   Or it may be due to a tendency

FIG. 8.—Common Tern's Nest of Broom at Romney Marsh.

inherited from some remote ancestor living under different
conditions.   Before, however, we can balance probabilities,
we must decide whether the lined nests are to be regarded
as the beginning of an advance, or as a survival, persisting
not by virtue of necessity, but simply because it is
harmless.   The first step towards a solution is to find
out whether the mortality among the chicks before or
immediately after quitting the egg-shells is due to the
absence or presence of a lining.

The second question may be stated thus : How comes

it that those individuals of the species which line their
nests differ in their choice of material ? Let us note that
this choice is limited by two conditions, the most obvious
being the accessibility of the material. But though
accessibility limits choice, it does not necessarily determine
its direction. The material of the Terns' nest above
described was equally accessible to all. A more striking
illustration is, however, provided by the following details
of the material used in six Thrushes' nests, all built in
gorse bushes within fifty yards of one another, one only
being old : (1) gorse ; (2) an old nest of grass, with a
small amount of gorse and twigs, and one bit of wool ;
(3) grass, moss, stalks ; (4) grass, moss, bracken ; (5) grass,
a little wool ; (6) grass, moss, hairs. Obviously, the most
accessible material for all was gorse, yet this was utilised
by not more than two.

The choice of materials in the Terns' nests described
appears to have been limited only by accessibility, but
in the vast majority of cases, if not in all, there is a
second quite distinct limiting condition which may briefly
be called the law of species. For instance, the Thrush,
though using a wide variety of materials in the normal
construction of the outside of its nest, appears to be
debarred from using twigs only. The material of the
inside lining is subjected to still greater restrictions,
being confined, " freak " nests apart, to dung or mud
studded with bits of rotten wood. This limitation is,
no doubt, to be explained, as Mr. Alfred Russel Wallace
has pointed out, by some peculiarity in the structure or
habits, past or present, of the species. It would be
interesting to know what this is in the case of the Thrush.
What led it to scorn the Blackbird's addition of grass,
or the wool, hair, feathers, used by other species ? Put
more generally, the question is one of the origin of
specific as distinct from *individual* variations.

Within the limits thus imposed upon it by accessibility
and the law of the species, why does one individual select,
say, twigs, another pebbles ? It may be that the young

bird constructing its first nest takes the first " lawful " material that accident places in its way, and so contracts the beginning of a habit that leads it to use normally the same material in the construction of all subsequent nests, even though the search for it demands much more time and labour than taking any other " lawful " material that happens to be near the adopted site.   This theory, which would apply equally to cases in which both sexes took part in building, is at first sight plausible enough. It depends, however, upon a question of fact.   It should not be difficult either by marking birds or watching their behaviour in captivity to find out whether they tend to continue the use of nesting material once adopted. Perhaps someone already has the facts.   If so, let him write and deliver.

Facts it is that are wanted, and as far as nests are concerned, it should not be difficult to collect a large number.   Those who are prepared to co-operate in this work will at least have the satisfaction of feeling that their time is being put to good use.   The question of variation, specific or individual, structural or functional, occupies to-day a place in the foremost rank of scientific problems, because it takes us to the roots of the evolution theory.   It has no mere academic importance.   Human progress depends upon human control of natural forces. There can be no control of these forces except by under-standing the laws that govern their operations.   And these laws can be reached only through a persevering accumulation of seemingly trivial facts.   If there is one thing that Darwin, Wallace, and their successors have made clear, it is the immeasurable importance of the unimportant.   They have shown us that from the spectacle of the humble Thrush collecting a beakful of rotten wood for its nest there is but one step to the brink of the un-plumbed depths that hide the answer to the riddle of the universe.

SOME EARLY BRITISH ORNITHOLOGISTS AND THEIR WORKS.

BY

W. H. MULLENS, M.A., LL.M., M.B.O.U.

III.—CHRISTOPHER MERRETT

(1614—1695).

THE first printed list of British Birds is that contained in the " Pinax Rerum . . . Britannicarum " of Christopher Merrett, or Merret.   This small 8vo work was published in London in 1666, and was, as its name denotes (*Pinax* = a list, or index) an attempt on Merrett's part to catalogue the vegetables, animals, and minerals, of Great Britain.   Of the 223 pages of which the book consists, 165 are devoted to botany, 42 to zoology, and the remainder to minerals.   In making his list Merrett was content, at any rate as regards the birds, to do little more than enumerate those which he considered he had identified from the descriptions of Ulyses Aldrovandus, whose twelve books on birds, largely founded on the work of Gesner, appeared between 1599 and 1603, and of Johannes Jonstonus, a Scotsman by descent, but by birth a Pole, the first edition of whose " History of Birds " appeared in 1650.* The English names are added in many cases, but the few short notes are rarely original, and Merrett does not seem up to this time to have devoted much personal attention to the observa- tion or study of birds ; indeed, the chief object of his book was to replace the " Phytologia " (London, 1 vol., 8vo) of William Howe (1620–1656) a " Flora " which had appeared

* Merrett's references to Gesner and Belon, both, as authors, far more accurate than the two above-mentioned, are, unfortunately, but few.

in 1650, and speedily passed out of print. Meagre and imperfect as Merrett's efforts must now appear, his work was at any rate the first of its kind, and was held in high estimation by his contemporaries. Sir Thomas Browne (1605–1682), the celebrated author of the " Religio Medici," with whom Merrett had entered into correspondence when he was contemplating a new and enlarged edition of the " Pinax," thus writes :—

" July 13th, 1668.

" To Dr. Meret.

" most honoured Sir,—I take ye boldnesse to salute you as a person of singular worth and learning and whom I very much respect and honour. . . . I should be very glad to serve you by any observations of mine against yr. second edition of your Pinax which I cannot sufficiently commende." (*cf.* Southwell's *Notes and Letters on the Natural History of Norfolk*, etc." London, 1902. 1 vol., 8vo, p. 57).

Dr. John Fleming, the author of " A History of British Animals " (Edinburgh, 1828, 1 vol., 8vo) also appreciates the " Pinax," and describes it as " This small work, which, though it claims little more than the merit of a catalogue, exhibits many proofs of great diligence, and rises in importance, when viewed as a first attempt at the construction of a British Fauna," a far juster criticism than Pulteney's (Richard Pulteney, 1730–1801, author of " Historical Sketches of the Progress of Botany," 1790) that it was " extremely superficial."

As regards the book itself, a *facsimile* of the title page of the first edition (1 Vol., small 8vo) is here given.

The Collation is :—pp. 2, Title & Imprimatur, + pp. 7, Epist. Dedicat. + pp. 21, Epist. ad Lect., + pp. 231 + p. 1.

This edition (of 1666) is very rare, many copies having presumably been destroyed in the Great Fire of London of that same year, either at the printer's or at Merrett's house.

In the next year, 1667, there appeared two editions, or re-issues of the " Pinax," similar in contents to the original, but with different title pages, as below, one entitled " Editio Secunda," as follows :—

PINAX

Rerum Naturalium

BRITANNICARUM,

CONTINENS

VEGETABILIA, ANIMALIA

ET

FOSSILIA,

In hac insulà repperta inchoatus.

AUTHORE

Christophoro Merrett

Medicinæ Doctore utriusque Societatis Regiæ
Socio primoque Musæi Harveani custode.

Μή τῶ λόγω μοῦννον ἀλλὰ
ἔργω δἔινομίζεαθαι τὸς ἰήτρͽς.
Hipp.

Londini Impensis *Cave Pulleyn* ad Insigne *Rosæ*
in *Cameterio Divi Pauli*, Typis *F*. &
T. Warren, Anno 1666.

Pinax / Rerum Naturalium / Britannicarum, / continens / Vegetabilia, Animalia, / et / Fossilia, / In hac Insula reperta inchoatus / Editio Secunda. / Auctore / Christophoro Merrett,/ Medicinæ Doctore utriusque Societatis / Regiæ Socio primoque Musæi Har- / veani Custode. / (quotation from Hippocrates) Londini, / Typis T. Roycroft, Impensis Cave Pulleyn, Prostat apud / Sam Thomson in vico vulgo dicto Duck lane, 1667. / 1 vol., small 8vo.

Collation : pp. 2, Title & Imprimatur. + pp. 10, Epist. Dedicat. + pp. 20, Epist. ad Lect. + pp. 223 + p. 1.

The other, a mere reprint of the original edition :—

Pinax / Rerum Naturalium / Britannicarum, / continens / Vegetabilia, Animalia, / et / Fossilia, / In hac Insula reperta inchoatus. / Auctore / Christoporo Merrett, / Medicinæ Doctore utriusque Societatis / Regiæ Socio promoque Musæi Har- / veani Custode. / (quotation from Hippocrates) Londini, / Typis T. Roycroft, Impensis Cave Pulleyn. / MDCLXVII. 1 vol., small 8vo. Collation as above.

It will be noticed that the date of this last edition, unlike that of the first, and the " editio secunda," is in Roman, not Arabic, figures.

(Engelmann gives an Edition of 1704. N.S.)

Although there appears to be no evidence that Merrett published any edition of the " Pinax " later than 1667, he certainly contemplated doing so, and in August, 1668, he writes Sir T. Browne that : " Besides those mentioned in ye pinax I have 100 to add & . . . I doe entreat this favour off yu to inform me fuller off those unknown things . . ." and in response to this request Sir Thomas Browne placed at his disposal the notes which he had prepared " of many animals in these parts whereof 3 years agoe a learned gentleman of this Country wished me to give him some account, which while I was doing ye gentleman my good friend died."

Christopher Merrett, who, like so many of the earlier ornithologists, was by profession a physician, was born at Winchcomb, in Gloucestershire, on Feb. 16th, 1614. In 1631 he became a member of Gloucester Hall, Oxford, and removed to Oriel College in 1633. He took his B.A. degree in 1635, and then, devoting himself to the study of medicine, graduated M.B. in 1636, and M.D. in 1643. He afterwards settled in London, became a Fellow of the Royal College of Physicians in 1651, and Gulstonian Lecturer in 1654. Through the influence of his friend, Dr. William Harvey (1578-1657) Merrett became the first librarian of the College. He resided at Amen Corner, and is stated by Wood (Athen. Oxon.) to have acquired a considerable practice. The bulk of the library belonging to the College, and Merrett's house were, however, destroyed in the Great Fire, and Merrett lost his appointment. He thereupon brought an action against the Royal College of Physicians, in which he claimed that he was entitled to his office for life. In this claim he failed, and was ultimately in 1681 expelled from his Fellowship, nominally for non-attendance. He died at his house in Hatton Garden, August 19th, 1695, and was buried " 12 feet deep in the church of St. Andrew's, Holborne." (Wood.) Merrett was the author of numerous other works, chiefly on medicine, and he also contributed several papers on " vegetable physiology " to the " Philosophical Transactions." His name is commemorated in Botany. S. F. Gray having in his " Natural Arrangement of British Plants " (1821), given the name of *Merrettia* to a group of unicellular *Algæ*.

We here print Merrett's list of birds verbatim, adding, with the assistance of Mr. W. Warde Fowler, a few short explanatory notes, which are placed within thick square brackets. The pages of the original text are enclosed within ordinary square brackets.

[Page 170] *Aves Britannicæ.*

Terrestres Carnivoræ.

Aquila, *the Eagle*, I. 10, tab. 1. 2. Ald. 110. G. 149. quandoq ; huc migrat ex Hibernia ubi abundat.

["Migrates out of Ireland where it abounds." Merrett has here, as elsewhere, availed himself freely of information contained in Giraldus Cambrensis' (1146–1223) " Topography of Ireland."* Giraldus in his ninth chapter, which deals with

* First printed in 1577 (Anderson).

"The Eagle and its Nature," informs us that "Eagles are as numerous here (*i.e.*, in Ireland) as kites are in other countries."]

Accipiter, *the Hawk*, I. 20. tab. 7. Ald. 225. 228. G. 3.

Haliætus, *the Sea Eagle*, vel Osprey Turn. Quandoq ; conspicitur in Cornubia, I. 12. tab. 2. Ald. 188. 190. G. 177. sine icone.

[The Osprey, *cf.* Turner (Evans' edition, pp. 35, 37, 193-195).* Giraldus seems to have been responsible for the idea, freely copied by later writers, that "By an extraordinary contrivance of sportive nature, one of their feet spreads open, armed with talons, and adapted for taking their prey, the other is close, harmless, and only fit for swimming." Merrett's statement that it is seen in Cornwall is, no doubt, taken from Carew's "Survey of Cornwall " (1602, Fol. 35).]

Lanarius, *the Lanar*, mas vocatur, *the Lanaret*, Ald. 381.382. I. 24. tab. 9.—*in Shirwood Forest*, in agro Notinghamensi, *and in Dean Forest*, in agro Glaucestrensi.

[The name " Lanar " has been applied to various species of Falcons (*cf.* Newton, *Dict. Birds*, p. 503). It is doubtful if Merrett here means *Falco lanarius*—probably this bird never bred in the British Isles—but *vide* Latham's " Falconry " (1618, Book 11, p. 112), and Hollingshead " Description of England " (1577, Ch. V., p. 227) to the contrary.]

Accip. Palumbarius, *the Goshawk*, mas dicitur *the Tassel*, Tertiolus, G. 43.

[The Goshawk, *cf.* Willughby (p. 85). " Tassel," or Tercel, the term applied by falconers to the male of the Goshawk and Peregrine.]

Accip. Fringillarius, & Nisus, *the Sparrow-Hawk*, I. 22. tab. 8. Ald. 346. 347. G. 44. mas appellatur, *the Muschel*, In plerisq ; locis sylvaticis.

[" The male is called Muschel." The male of the Sparrow-hawk was termed in falconry the Musket—*cf.* " Diary of Master William Silence " (p. 151), and " Merry Wives of Windsor " (3.3.21) : " How now, my eyas-musket."]

Tinnunculus mas & fæmina, *a Stannel, or Stonegall*, I. 22. tab. 8. Ald. 358. *a Keshrel, or Kastrel*, in tractibus Austral. G. 46.

[" Stannel " = Kestrel, *cf.* Swainson " Provincial Names of British Birds " (p. 140).]

* References to Turner are from Mr. A. H. Evans' edition. Cambridge. 1903. 1 vol. 8vo.

SOME EARLY BRITISH ORNITHOLOGISTS. 115

Falco, *the Faulcon*, I. 30. tab. 12. speciem nescio, *in Pembrook-shire*.

[The figure in Jonstonus is seemingly that of the Peregrine, *cf.* Willughby (pp. 76 and 79) for Peregrine.]

Coccyx, Cuculus, *the Cuckoe, or Guckoe*, I. 24. tab. 10. Ald. 414, 416. sub medium Aprilis nos advolat.

Lanius, *the Butcher, or murdering Bird*, I. 24. [Page 171.] tab. 10. Ald. 389. G. 520. vidi juxta *Kingsland*, æstivo tempore ter. quaterve. Lanius Cinereus Anglicè, *a Skreek*, G. 520.

Laniorum duas alias species observavit nobilis vir D. *Willoughby* totius naturæ diligentissimus Callentissimusq ; scrutator non solum per Britanniam sed maximam partem Europæ.

[Most probably the Red-backed Shrike, since he saw it in summer. The two other species of Shrike observed by Willughby were (1) " The greater Butcher-Bird, or Mattagesse, and in the Peak of Derbyshire after the German name Wierangel, or Werangel, Lanius Cinerus Major " ; and (2) " The Wood-chat, Lanius Minor Cinereo-ruffus," *cf.* " The Ornithology " (p. 21).

The word " Shreek " was applied to the Mistle-Thrush and also to the Barn-Owl in old English vocabularies of the eleventh and fifteenth century. Willughby seems to infer that Turner was responsible for the name Shrike, as applied to the Butcher-bird, and John Ray, in " A Collection of English Words " (London, 1674, 1 vol., 12mo, p. 83) confirms this.]

Milvus, *the forked tail'd Kite*, I. 24. tab. 11. Ald. 395. G. 549. Turn. *a Glede, a Puttock*.

[*Cf.* Turner (Evans' Ed., p. 117) " milvus, in English, a glede, a puttock, a kyte." The name " puttock " was also applied by Willughby to the Buzzard (p. 70).]

Subuteo, *the ring-tail'd Kite*, I. 24, tab. 9.

[The Ringtail was the old name for the female Hen-Harrier. *Cf.* Willughby (p. 21). Merrett seems to have added Kite by mistake.]

Buteo Triorchis, *the Buzzard*, Ald 367.

[Willughby makes the curious statement (p. 21) that this bird is a great destroyer of conies.]

Peronos, *the bald Buzzard, or Kite*.

[Turner applies the name " Bald-Buzzard " to the Marsh-Harrier, which he says the English call " Balbushard "(*cf.* Evans'

Edition, p. 33). Merrett here again wrongly introduces the word Kite.]

Noctua, *the Night, or little grey Owl*, I. 48. tab. 18. Ald. tom. 1. 544. Bubo Turn. *a like Fowl.*

[Turner says "bubo, in English alyke foule" (p. 47). It is difficult to determine what Owl Merrett here refers to. Can it be the Little Owl ? The Short-eared Owl is called to this day the "grey yogle" in the Shetlands (*cf.* Swainson, p. 129). It is not, however, a night Owl. Charleton in his "Onomasticon Zoicon" (1668, p. 70) calls "Noctua" "the Common Grey Owl."]

Ulula, *the white hooping Owl, or Owlet, or Howlet*, I. tab. 19. Ald. 538. G. 700.

[Ulula = the Barn-Owl.]

Strix, *the Screech, or Screeching Owl*, I. tab. 19. Ald. 563.

[The Screech Owl = the Tawny Owl (*cf.* Swainson, p. 129.)]

Corvus, *the Raven*, I. 38. tab. 16, Ald. 694. in ulmetis juxta-ædes nobilium, G. 294.

Corvus, I. 38. tab. 16. *our common or Carrion Crow*, G. 282. Cornix nigra, Ald. 736. & Cornix simpliciter Turn.

Cornix frugilega, spermologus, *a Rook*, I. 40. tab. 17. Ald. tom. 1. 753.

Cornix aquat. Hanc videt Turn. apud Morpetenses in ripis fluminum, G. 293. suspicor esse, *the* mur Cornubiensium.

[This is the Water-Ouzel, or Dipper. Merrett has been misled by Turner's use of the Northumbrian name "Watercraw" (Evans' Edition, p. 23), and has placed it among the *Corvidæ*. He has further confused the matter by suspecting it to be the Cornish "Mur." The word "murre" is used in Cornwall to designate the Razor-Bill, called also the Sea-Crow (*cf.* Swainson, p. 217).]

Cornix Cinerea, *the Royston Crow*, I. ubi supra, Ald. 755.

[Formerly also spelt Roiston Crow (*cf.* Ray, *op. cit.*, p. 83, and Cotgrave's *Dictionary*). Willughby (p. 22) says : "Common in Cambridgeshire about Newmarket and Royston."]

[Page 172.] Graculus vel Monedula, *a Jackdaw, a Chough*, Turn. *a* Caddo, *a* Ka, I. 38. tab. 16. Ald. 771. G. 467.

[Jackdaws were sometimes called Choughs (*cf.* Harting's *Ornithology of Shakespeare*, p. 119).]

Coracias Arist. *the Cornish Chough*, I. 38. tab. 16. Ald. 768. In omnibus oris maritimis a Cornubia ad Doroberniam.

[Charleton (*op. cit.*) says its Cornish name was " the Killegrew " (*cf.* Swainson, p. 74).]

Pica Glandaria, *a Jay,* I. 40. tab. 17. Ald. 789. Garrulus avis, G. 634.

Pica, *the Magpie, Pyot, Py-anet,* I. 40. tab. 17. Pica varia seu caudata, Ald. tom. I. 85. G. 628.

[Ray (p. 84) has Pianet (*cf.* Swainson, p. 75).]

Pica Marina, *the Sea Pye,* I. ut supra Ald. tab. 792. 794.

[The Oyster-Catcher (*cf.* Turner, Evans' Edition, p. 199 ; and also Swainson, p. 188). The *Pica Marina* of Aldrovandus is the Roller (*cf.* also Willughby, p. 132). Ray properly places the " Sea-pie " among the Waders (p. 80).]

Vespertilio, *a Bat, Flittermouse, Rearmouse,* I. 52. tab. 20. Ald. 574. G. 604. vesperi apparet æstate. Hyeme vero latet in cryptis, & rupibus.

[Merrett, following the example of Gesner,* Belon,† Aldrovandus,‡ Jonstonus,§ and Lovell,‖ has placed the Bat in his list of birds. Charleton follows Merrett, and Albin as late as 1738 includes the Bat in his " History of Birds " (*cf.* Linnæus, *Fauna Suecica,* p. 7). Turner, avoiding this error, makes no mention of the Bat in his " Avium . . . Historia." Rearmouse = Reremouse, *cf.* Bartholomew (*de Proprietatibus Rerum,* Berthelet's Edition, 1535, Book XII., Fo. 38), and Shakespeare (M.N.D., II., 2.4.) :—
" Some war with rere-mice for their leathern wings
To make my small elves coats."]

Loxias, *the Shell-Apple,* Ald. 2. 877. I. 46. sine Icone in agro Warwic. in Pomariis, Mr. *Willoughby.*

[The Shell-apple = the Crossbill (*cf.* Carew, Fol. 25, Willughby, p. 248, and Swainson, p. 67).]

Caprimulgus, *the Goat-sucker,* I. 52. tab. 20. Ald. 568. G. 215. Hunc cœpit Dominus *Cole,* in agro Hantoniensi, an. 1664. rara admodum avis.

[It is strange that Merrett should describe the Goatsucker as a rare bird. Turner also does not mention it as a British bird, but relates that he made enquiries concerning its habits in Switzerland.]

* Conrad Gesner (1516—1561), " De Avibus," 1555.
† Pierre Belon (ob. 1564), " L'Histoire des Oyseaux," 1555.
‡ Ulyses Aldrovandus (1522—1605), " Ornithologiæ," 1599.
§ Johannes Jonstonus (1603—1675), " De Avibus, 1650.
‖ Robert Lovell (1630—1690), " History of Animals," 1661.

Aves Granivoræ non canoræ.

Pavo, *the Peacock*, I. 52. tab. 22. Ald. 219. G. 594.

. . . Fæm *the Peahen*, I. 56. tab. 22. Ald. ib. plo.

Gallo pavo, *the Turkey-cock*, I. 58. tab. 24. Ald. tab. 2. 39. G. 426.

Phasianus, *the Pheasant*, I. ut supra Ald. tom. 2. 49. G. 619. Horum pulli vocantur Pouts, Est albus & alter fuscus.

[Page 173.] Urogallus, *major Cock of the Wood*, I. 60. tab. 25. Urogallus seu Tetrao major Ald. 2. 64. in Hibernia occurrit.

[The Capercalzie (also as Capricala, p. 179). Merrett's statement that it occurs in Ireland is derived from Giraldus Cambrensis' " Topography of Ireland " (Chapter X.), " wild peacocks here abound in the woods," *cf.* also Willughby (p. 23), " This is not found in England, but in Ireland there be of them," and Ussher and Warren, " Birds of Ireland " (p. 330).]

Gallina Coryllorum, *the Hasel Hen, Grous*, I. 60. tab. 25. Ald. 2. 82. Bonosa Albert, G. 203.

[Merrett here differs from Aldrovandus, who figures what is apparently the Francolin, under the title Attagen, and states that it was also called the Hazel Hen. *Gallina coryllorum*, Aldrovandus calls Räb-hun. There is some confusion here, as in many other of Merrett's statements, the Hazel Hen, as far as we know, never having inhabited Great Britain.]

Gall. Africana, *the Guiney Hen*, I. 58. tab. 24. Gallina Guinea, Ald. tom. 2. 337. meleagris vel Gallus Numidicus, G. 424.

Otis, Tarda, Bistarda, *the Bustard*, I. 62. tab. 26. Ald. 288. G. 430. *On Newmarket Heath*, & in Campestribus Sarisburiensibus.

[Turner (p. 167), " in English a Bustard or a Bistard " (*cf.* Willughby, p. 178).]

Attagen, *a Godwit*, I. 62, tab. 26. Ald. 275, in agro *Lincoln*.

[Turner's " Attagen " (p. 45) is the Godwit, *cf.* also Willughby (p. 292).]

Perdix Ruffa, *the Partridge*, I. 62, tab. 27. Ald. 2. 139. G. 606.

Coturnix, *the Quail*, I. 62. tab. 27. Ald. tom. 2. 153. G. 311.

(To be continued.)

BIRD ROOSTS AND ROUTES.

BY

BRUCE F. CUMMINGS.

THE following paper does not pretend to be an exhaustive one, but is the result of my own observations during the past winter in the district of Barnstaple, North Devon.

All birds show considerable care in the choice of a secure roosting site, and in order to spare labour in looking for a fresh one every night, they frequently return to the same place continuously.

A great many of the small species roost in company, " cuddling," or keeping close together in a bunch for warmth. I have found four Wrens roosting in this way in a hole in a tree, and have disturbed several sleeping in their " cock " nests, but as far as my notes go, these are generally vacant. On one occasion last summer I noticed several Long-tailed Tits (probably a brood) on the top of their nest, which had become quite flattened and was covered with droppings. I expect, therefore, that they returned to the nest every night, and when they got too large, roosted on the top of it. Wrens up to the number of thirty at a time, Long-tailed Tits, and Golden-crested Wrens are recorded as roosting together in this " bunching " fashion by Mr. G. A. B. Dewar (in the *Birds of Our Wood*). One night I saw two Blue Tits embracing each other in this way in an apple tree. They looked like one large bird, so close to each other were they. This is not, however, the usual habit of this Tit, for it generally roosts in holes.

The Sparrow, as is well known, will occupy an old

House-Martin's nest, or will line a hole in a thatch with feathers. Partridges roost on the ground, while Pheasants and fowls prefer to roost in trees.

A Hedge-Sparrow which I had under observation, returned every evening last winter with the utmost regularity to a cranny among dead ivy on an elm. When driven out it would return in a few moments, first-pitching on a branch of the tree, and then swiftly sneaking into the cranny, so that its return very frequently escaped my notice entirely.

Kestrels roost at the same spot, in a quarry for example, for many consecutive weeks.

The Pied Wagtail and the Grey Wagtail in the Barnstaple district collect in some numbers every evening, and roost in reed beds, like the Starlings. They drop in from all directions, but do not come from more than a mile distant. As a rule they collect on the ground, or telegraph wires, near the reed bed, before disappearing into the reeds, calling, and flying short distances in one flock. This flock increases as the birds come up one by one, and finally they drop into the reeds, where they are joined by Robins and Wrens.

A great many species of birds roost in company, notably Starlings. Others are : House-sparrows, Carrion Crows (especially in Devon and Somerset), Magpies (which I have observed near Barnstaple), Rooks, and Wood-Pigeons.

In North Devon, in the colder months of the year, the Rooks never roost in their rookery during, at all events, the months of November, December, January, February, and part of March, but they collect in large numbers and roost in a wood, perhaps two or three miles away from the rookery. In the morning the roost breaks up, and the members of each community make away, with the utmost regularity, to their respective rookeries. At the rookeries they stand about " talking," perhaps till nine o'clock, and then they disperse to feed and meet again in the evening at the roost. If the morning is a frosty one they stay on the rookery trees longer than usual.

At Tapely Park, Instow, Jackdaws collect in prodigious quantities, numbering many thousands (though it is extremely difficult to judge the number), and roost in the beech trees. A roost of Rooks occupies the same group of trees. The interesting feature connected with these Jackdaws is that the birds, in going to and from their roost, always take exactly the same route. A large flock which, during part of its course, is forced to fly over the town of Bideford, always flies across exactly the same part of the town every evening. It was by watching and following up for several days another big flock (numbering 200 or 300), which fed daily in the fields at Braunton (about three and a half miles from the roost) throughout the whole of last winter, that I finally discovered this large roost. Every morning and every evening this flock as regularly as a Royal Mail performs this journey. They follow very carefully the same line of flight, even to the barest detail, but occasionally they fly very high, and they then appear to follow a more direct course, for it is noteworthy that these birds do not, as a rule, make a bee-line by any means. The reason why they sometimes fly at a great height I cannot imagine. I do not think that it has anything whatever to do with wind or weather. Arrived at the roost, the birds " rocket " down perpendicularly, dropping like plummets through space, and commence to " chock " for an hour or more before darkness falls. Starlings and Wood-Pigeons when dropping in to roost, " rocket " down in this same eccentric way, and many birds behave similarly at times, when they may be said to be " at play." The habit with the roosting birds is, however, a constant one, and takes place every evening. I have found another big Jackdaw roost at Eggesford—in a very wooded district.

Far more striking evidence as to the use of flight-lines in these " miniature migrations " is to be seen in the

case of the Starling.   A large Starling roost is a very
imposing sight, and has attracted the attention of a great
many writers.   The very remarkable turns of flight
displayed by these birds at roosting time constitute,
perhaps, one of the most striking phenomena which
British bird-life has to show.

In the Barnstaple district there are four or five such
roosts.   I have not discovered the birds travelling more
than six miles to and from a roost.   I have repeatedly

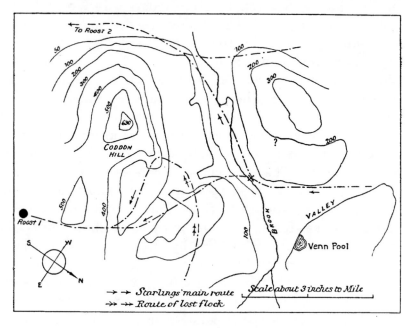

noticed how strictly the birds keep to their arbitrarily
prescribed line of flight.   The best instance I can give
is shown in the accompanying map.

The flocks sweep along this main course with astonish-
ing regularity every night, flock succeeding flock, and
each separate flock pursuing the same course, as a rule
dividing at x, one half going to one roost, and the other
half to another roost.   They fly high—well above the
neighbouring hills and valleys—although it will be noticed
that they follow a valley for some distance ; this route,
moreover, was not merely roughly followed, but the birds

came accurately along a mathematically straight line, as far as x.

On February 19th I was at this spot watching the Starlings. I was particularly interested in one flock which never arrived along the usual, main, flight-line, but cut into it at right angles (as indicated in the sketch map). This flock, on this particular evening, however, appeared to have lost its bearings, for it wandered about, as I show in the sketch, as if trying to cross Coddon Hill, which the birds never did at any time ; finally, it seemed to perceive its whereabouts, doubled back and went on, crossing the 400-foot ridge. On the 22nd, this same flock was making for the roost, flying against a heavy westerly gale. Hard weather and frost seems to make no diminution in numbers at the roosts. I may mention here that on every occasion that I have visited a Starling roost last winter (about seven times) there was always a Sparrow-Hawk flying close at hand, and I have repeatedly seen this Hawk harrying flocks as they came in to roost.

Individual flocks, when perhaps three miles away from their roost, and out of the main stream of " migration," followed, I found, in the few cases I had under observation, the same route every night. One small flock, for example, always crossed the River Taw at a certain point near a signal box, for several weeks last winter. Routes, however, like these, on the extreme periphery of the system, vary when the particular flock changes its feeding quarters.

Possibly some of the foregoing will have to be modified after more prolonged observation, but the main point will hold—the universal use of flight-lines by Starlings in going to and from their roost.

Whether birds, with their large semi-circular canals, have a sense of direction or whether their migrations are carried out by the aid of the sun or by the earth's magnetism or any other power is moot, yet one thing seems certain and that is that they possess a powerful memory. I feel sure that however the migrational movement as a whole is effected, the way in which the Swallow returns year after year to the same old beam in the same old barn is simply memory—topographical knowledge of the chief natural features and the general mould of the country in the neighbourhood of its nesting home.

ON THE MORE IMPORTANT ADDITIONS TO OUR KNOWLEDGE OF BRITISH BIRDS SINCE 1899.

BY

H. F. WITHERBY AND N. F. TICEHURST.

PART XIII.

(Continued from page 87.)

STOCK-DOVE *Columba œnas* L. S. page 481.

DURHAM.—Two nests in drains underground entered by gargoyles in walls (H. B. Tristram, *Vict. Hist. Durham*, vol. 1).

NORTHUMBERLAND.—First seen in 1878, now a regular resident (A. Chapman, *Bird-Life of the Borders*, p. 31.)

SCOTLAND.—*Caithness.*—A young bird was shot near Castletown, Thurso, on December 4th, 1901. Believed to be the first record for the county (T. E. Buckley, *Ann. S.N.H.*, 1902, p. 53). *Ayrshire.*—A nest was found near Darvel in May, 1902 (J. Paterson, *t.c.*, 1902, p. 184). *Bute.*—Nests were found in 1906, (*t.c.*, 1907, p. 199). *Shetlands.*—One at Halligarth, June 22nd–25th (T. E. Saxby, *t.c.*, 1905, p. 117).

" I cannot consider their appearance anywhere on the west side of the backbone of Scotland (*i.e.*, anywhere north of Clyde) as anything but phenomenal " (J. A. Harvie-Brown, *Fauna N.W. Highlands and Skye*, p. 260). A very full account of its arrival and spread in the east is given in " Fauna of Tay Basin and Strathmore " (pp. 259–266).

ISLE OF MAN.—Nests in small numbers (P. Ralfe, *B. of Isle of Man*, p. 178).

IRELAND.—Extending its range. Breeds in Leinster, parts of Ulster and Munster to the Shannon (R. J. Ussher, *in litt.*).

TURTLE-DOVE *Turtur communis* Selby.

SHROPSHIRE.—A marked increase of late years (H. E. Forrest, *in litt.*).

CHESHIRE.—Now steadily increasing in numbers ; it was practically unknown in the county about fifty years ago (Coward and Oldham, *B. of Cheshire*, p. 180).

NORTH WALES.—Is spreading westward, especially along

the north coast, where it breeds as far as Bangor. Has just begun to penetrate to the western side of Montgomery and Merioneth (H. E. Forrest, *Vert. F. N. Wales*, p. 304).

YORKSHIRE.—It is extending northwards. " At the present time its nesting area may be defined as being on the eastern side of a line passing through the centre of the county by Ripon, Harrogate, Leeds and Wakefield, to Sheffield." The most northerly point at which it is known to have nested with certainty is Scarborough, where a nest was found in June, 1900, and again at Wykeham by Mr. R. Fortune in 1905 (T. H. Nelson, *B. of Yorks.*, pp. 496 and 497).

SCOTLAND.—*Shetlands.*—One at Lerwick on December 4th, 1905 (*Ann. S.N.H.*, 1906, p. 199). One on May 28th, and a good many in the second and third weeks of June, 1902, were seen at Dunrossness (*t.c.*, 1903, p. 153). *Caithness.*— One on June 23rd at Barriedale (*t.c.*, 1900, p. 83). *Argyll.*— One seen on August 29th, 1900, at Dhuheartach (*t.c.*, 1901, p. 139). *Outer Hebrides.*—One was shot on the Flannan Isles on September 14th, 1900 (*t.c.*, 1901, p. 139). A young bird appeared at Eoligary on August 18th, 1901, and was caught on September 29th. Another older bird appeared on September 25th (J. A. Harvie-Brown, *t.c.*, 1902, p. 215). *N.W. Highlands and Skye.*—Only an occasional visitor (*id.*, *Fauna N.W. Highlands and Skye*, p. 263).

ISLE OF MAN.—A rare straggler (P. Ralfe, *B. of Isle of Man*, p. 183).

IRELAND.—A female was shot on May 24th, 1904, near Hillsborough, co. Down. It had eggs in the ovary, and showed no trace of having been in captivity (N. H. Foster, *Irish Nat.*, 1904, p. 155). Messrs. Ussher and Warren record thirty-three occurrences in May and twenty in June, on migration (*B. of Ireland*, p. 227), so that Mr. Foster's record cannot be taken as an addition to the breeding records in Ireland, which are only two of many years ago.

PALLAS'S SAND-GROUSE *Syrrhaptes paradoxus* (Pall.). S. page 488.

1899.—From the last week of January to March 23rd, a flock of thirty or so was seen on the north wolds of Lincolnshire in the same field in which they appeared in 1888. A single bird was seen in the same district on May 19th, and a small flight was observed in the Spurn district (Yorkshire) on May 13th (J. Cordeaux, *Ibis*, 1899, p. 472).

1904.—A flock of eighteen was observed in the second week of February flying northward over Millington, Yorkshire (T. H. Nelson, *B. of Yorks.*, p. 503).

1906.—A flock of six or seven was seen (in May) on some
" well-known links " in East Lothian (C. E. S. Chambers,
Field, 2, VI., 06, p. 901). A flock of about twenty was
seen by Mr. R. Vincent on June 11th, in Norfolk, and ten were
seen by Mr. D. Annison at Somerton on June 17th, while
some were reported in Yorkshire in July (J. H. Gurney, *Zool.*,
1907, p. 130).

CAPERCAILLIE *Tetrao urogallus* L. S. page 491.

SCOTLAND.—*Midlothian.*—Two were seen in the autumn
of 1906 at Bavelaw (H. N. Bonar, *Ann. S.N.H.*, 1907, p. 52).
Mr. W. Evans has six records, including one shot in Bavelaw
fir-wood nearly " twenty years ago " (*l.c.*). *Dumfriesshire.*—
Three were seen in November, 1905, in the N.N.W. of the
county (H. S. Gladstone, *l.c.*). *Wigtownshire.*—Two were
shot about 1874 (H. Maxwell, *t.c.*, 1907, p. 116). *Ayrshire.*—
A female was killed on December 14th, 1905, near Tarbolton
Moss (H. S. Gladstone, *t.c.*, 1906, p. 116). *Aberdeenshire.*—
A female in full male plumage was shot in January, 1906,
in the north of the county (E. T. Clarke, *t.c.*, 1907, p. 117).

HYBRID. — A hybrid between this species and the
Pheasant was obtained at Stronchullin, Blairmore, Argyllshire,
in September, 1897. This bird and the three previously
known specimens of such a hybrid are fully described (W. E.
Clarke, *t.c.*, 1898, pp. 17–21).

BLACK GROUSE *Tetrao tetrix* L. S. page 493.

CORNWALL.—Now almost extinct (J. Clark, *Vict. Hist.
Cornwall*, vol. 1).

IRELAND.—Bones discovered in the Ballynamintra Cave,
co. Waterford, prove the former existence of this species in
Ireland (G. E. H. Barrett-Hamilton, *Irish Nat.*, 1899, pp.
17 and 37).

HYBRIDS.—Willow Grouse ♂ × Greyhen (*P.Z.S.*, 1904,
Vol. I., p. 411, figure). Black Game × Pheasant—Fifty-
five specimens in Great Britain recorded (F. C. R. Jourdain,
Zool., 1906, pp. 321–330 and 433 ; *Ann. S.N.H.*, 1906, p.
238 ; *cf.* also *Bull. B.O.C.*, XVI., pp. 54 and 55). " Since
these papers were written I have received notes of several
other occurrences " (F. C. R. J. *in litt.*).

INTRODUCTION.—*Surrey.*—Those introduced in 1875 on
Witley Common, and which did good for some time in helping
to keep the old stock going, are believed now (1900) to be
practically extinct (J. A. Bucknill, *Zool.*, 1901, p. 253).
According to Mr. G. W. Swanton, two pairs bred in 1905 in

" a certain wild tract of country," and a single Greyhen was
seen in the spring of 1906 (L. B. Mouritz, *t.c.*, 1907, p. 93).
Norfolk.—The experiment of turning out Black Game at
Thetford by Mr. W. Dalziel Mackenzie has been continued,
and thirty were turned out in 1900–1901. Broods hatch off
regularly, but seem to disappear in some unaccountable
manner, and the numbers, in spite of fresh introductions,
steadily decrease (Heatley Noble, *t.c.*, 1903, p. 155). *Herts.*—
A Greyhen was shot on December 1st, 1906, near Watford—
the only record for the county (W. Bickerton, *in litt.*). *Hants.*
—In the New Forest district they are almost extinct (H. F. W.)
(A useful article on the distribution of this bird in English
counties, by Mr. J. E. Harting, appeared in the *Field* for
September 8th, 1900, p. 387.)

RED GROUSE *Lagopus scoticus* (Lath.). S. page 495.

[CORNWALL.—One reported to have been shot near Tintagel
on December 1st, 1906 (J. Clark, *Zool.*, 1907, p. 286).]

HYBRID.—Red Grouse ♂ × Bantam Fowl ♀ , exhibited
by J. G. Millais (*Bull. B.O.C.*, VIII., p. 36).

INTRODUCTION.—*Shetland.*—Some six hundred birds were
liberated on the mainland in September, 1901 (T. E. Saxby,
Zool., 1902, p. 113). Two were seen at Balta Sound,
November 16th, 1902 (*id.*, *t.c.*, 1903, p. 157). *Suffolk.*—A
few brace were turned out about 1903 at Elveden, and they
have increased to about 150 birds (" Head Keeper " *in litt.*
to J. Green, February 2nd, 1908). *Surrey.*—Details regarding
early introductions (J. A. Bucknill, *Zool.*, 1902, p. 68).

PTARMIGAN *Lagopus mutus* (Montin).

Bones of this species were found amongst others in the
Shandon and Ballynamintra Caves, co. Waterford (G. E. H.
Barrett-Hamilton, *Irish Nat.*, 1899, p. 17).

COMMON PARTRIDGE *Perdix cinerea.* Lath. S. page 501.

A brood of twelve a few days old was discovered at
Stratton Strawless, Norfolk, on January 31st, 1906. They
were reduced to two by February 22nd, and these apparently
did not long survive (J. H. Gurney, *Zool.*, 1907, p. 123).

RED-LEGGED PARTRIDGE *Caccabis rufa* (L.). S. page 503.

For some evidence of its migrating to the coast of Norfolk
and Yorkshire (*cf.* A. Patterson, *Zool.*, 1905, p. 186, and W.
J. Clarke, *t.c.*, p. 314).

SPOTTED CRAKE *Porzana maruetta.* S. page 509.

SCOTLAND.—*Argyll.*—One was taken in August, 1900, at Dhuheartach (*Ann. S.N.H.*, 1901, p. 140). *Perthshire.*—One was shot at Murthly on November 2nd, 1903 (a late date for so far north) (T. G. Laidlaw, *t.c.*, 1904, p. 55). *Dumfriesshire.*—One was killed on September 3rd, 1903, at Noblehill (R. Service, *t.c.*, 1904, p. 69). *Shetland.*—One was shot in Spiggie Marsh on September 25th, 1901 (*t.c.*, 1902, p. 135). This may be the same as the bird referred to in the "*Zoologist,*" 1901, p. 391. *Orkney.*—One was shot at Stornoway on November 24th, 1906 (*Field*, 1906, p. 908).

IRELAND.—One was heard calling several nights early in May, 1900, in a swamp at Cappagh, co. Waterford (R. J. Ussher, *Irish Nat.*, 1900, p. 160); it has twice been recorded as breeding in Ireland. One was shot on October 8th, 1904, near Templepatrick, co. Antrim (W. H. Workman, *t.c.*, 1904, p. 261).

CAROLINA CRAKE *Porzana carolina* (L.). S. page 510.

A young male, which had completed the autumn moult, was shot by Mr. E. Lort Phillips on October 25th, 1901, when snipe shooting with Mr. F. G. Gunnis in Rounach bog at the west end of Tiree, Inner Hebrides. The bird was very fat (*Bull. B.O.C.*, XII., p. 26; *Ann. S.N.H.*, 1902, p. 9). The species has been twice previously recorded in this country (Berkshire, 1864, Cardiff, 1888), and since it has been recorded several times in Greenland and breeds far north in North America, we think it should be admitted fully to the British list.

LITTLE CRAKE *Porzana parva* (Scop.). S. page 511.

SUSSEX.—One was caught near Rye in June, 1904 (N. F. T.).
SHROPSHIRE.—One was shot in November, 1898, at Petton Park, near Shrewsbury (H. E. Forrest, *Zool.*, 1900, p. 280).
IRELAND.—One was shot near Rathangan, co. Kildare, on November 12th, 1903 (Williams and Son, *Zool.*, 1903, p. 460). The bird has only once before occurred in Ireland.

(To be continued.)

PIED WAGTAIL REARING THREE BROODS.

IT may be of interest to record that a pair of Pied Wagtails have this year reared three broods from nests built in some ivy at one end of the house here (Stonewall Park, Edenbridge, Kent). The first brood of four left the nest on May 2nd. The second "brood" consisted of a Cuckoo, which left the nest on June 28th. The third nest contained three young, which fledged on August 1st. The Wagtails continued to feed the young Cuckoo until just before their third brood hatched.

E. G. B. MEADE-WALDO.

ON THE BRITISH BULLFINCH.

AT the meeting of the British Ornithologists' Club, held on June 17th, 1908, I exhibited a series of Bullfinches, clearly showing that the British race differed from its nearest ally, the *Pyrrhula pyrrhula europœa*, of Central Europe. The differences are, that the British race is slightly smaller, and that the female has the back darker brown, and the under-surface conspicuously darker and browner. The male, on the other hand, does not differ very appreciably in colour, though, if a series is compared, it is evident that the British form has the red of the underside as well as the grey of the upper-surface somewhat less brilliant.

At the meeting several members asked if I had compared the British Bullfinch with the great Northern Bullfinch *Pyrrhula pyrrhula pyrrhula*. To all who know these birds it is needless to remark that a comparison with the latter subspecies was unnecessary, as it is still larger and more brilliant than *P. pyrrhula europœa*, the grey of the upperside being purer, and the red of the under-surface brighter.

The somewhat darker and duller coloration of our British Bullfinch, and its slightly smaller size, again confirms the general inclination of British forms to be duller or darker, and often smaller than their continental representatives.

The name of the British Bullfinch must be—

PYRRHULA PYRRHULA PILEATA.

Under this name (*Pyrrhula pileata*) Macgillivray described it in 1837, in Vol. I., p. 407, of his "History of British Birds."

Of its distribution he says :—" The Bullfinch is generally distributed in Britain, occurring in most of our wooded and cultivated districts, but avoiding bare maritime tracts, as well as the northern islands, which are destitute of wood." Then, at the end of the article he adds :—" The Common Bullfinch is said by authors to be of general occurrence in the northern and temperate parts of Europe." It is thus quite clear that Macgillivray described the British Bullfinch, and that only, for he merely adds that it is " said by authors " to inhabit great parts of Europe besides. The author also says that he has " not observed any remarkable differences between individuals, indicating the existence of two species usually confounded, although I have heard it said that such have been met with." It follows that Macgillivray never came across the Northern Bullfinch, which occurs, though very rarely, as a straggler in England.

The case of the name of the British Bullfinch appears to me to be different from that of the Lesser Spotted Woodpecker. Macgillivray also bestowed a new name on this species in his work on " British Birds " (Vol. III., p. 86), calling it *Picus striolatus*. But then he says that he changed its name to *striolatus* because this species was by no means the smallest of even the Spotted Woodpeckers, and he regards this bird as "peculiar to Europe," saying that it is " said to be more abundant in the northern parts of Europe than in France and Germany," while it has not been found in Scotland, nor even in many parts of England. I therefore take it that Macgillivray re-named the " European " Linnean Lesser Woodpecker, and consequently I bestowed a new name on the British Lesser Spotted Woodpecker (*antea* Vol. I., p. 221). On the other hand, I think that we can safely adopt the name *pileata* for the British Bullfinch, and thus avoid creating a new name for this bird.

ERNST HARTERT.

HOW A CUCKOO DEPOSITED HER EGG.

MR. G. H. HIGSON kindly sends the following note from the huntsman of the Ynysfor Hounds, whom he describes as a very keen and accurate observer :—

" On the evening of May 24th I strolled down as far as the marsh to look for some nests, and found a Meadow Pipit's, with four eggs, quite cold. The little birds were following a Cuckoo close by, so I laid down in the rushes on the side of a ditch, within five yards of the nest, and watched. Presently the Cuckoo alighted near, and walking up to the nest, picked

up one of the Pipit's eggs in her beak.   This she put aside,
about two feet off, and then, walking back, she stooped with
her wings half raised, and laid her egg about three or four
inches from the side of the nest.   She then turned round
and pushed the egg most carefully with her beak into the nest.
Then she picked up the Pipit's egg in her beak and flew away,
dropping it about twenty yards further on.   The Meadow
Pipits were there, looking on as if they knew what she was
doing, for they stopped there and did not follow the Cuckoo."

"OWEN EPHRAIM."

TUFTED DUCK IN SCOTLAND.

IN the August number of BRITISH BIRDS, p. 84, some additional information is furnished regarding the Tufted Duck
as a Scottish bird.   Among other items, there is one upon
which I, and I am   sure   others, would   welcome   further
information.   I allude to Mr. Harvie-Brown's averment
that "Macgillivray states that it was *formerly* a common
bird in the Outer Hebrides."   This statement is not only of
considerable interest but has highly important bearings
on the history of this species as a British bird, and I would
ask Mr. Harvie-Brown to tell us *where* Macgillivray published
the information.   I have failed to find it in that distinguished
naturalist's writings with which I am acquainted, or in any
of Mr. Harvie-Brown's faunal works or papers, except in the
"Fauna of the N.W. Highlands " (from which you quote),
where, however, the desired reference is not afforded.

WM. EAGLE CLARKE.

STARK'S RECORD OF THE · BREEDING OF THE
SCAUP-DUCK AT LOCH LEVEN.

IN the August number of this magazine (p. 85) attention is
drawn to my old friend, the late Dr. A. C. Stark's, record of
the breeding of the Scaup at Loch Leven in 1880, under the
impression that it had been overlooked by Howard Saunders.
As a matter of fact, however, Saunders did not overlook the
record, with which he was perfectly familiar.   He specially
cited it, both in the fourth edition of " Yarrell," and in the
first edition of his own " Manual " (1889).   But a note in the
Appendix to the latter foreshadowed its suppression in the
second edition.   The note is as follows :—" As regards Mr.
A. C. Stark's very positive and detailed account (*Pr. R.
Phys. Soc. Edin.*, VII., p. 203) of the breeding of this species
on Loch Leven, Mr. W. Evans informs me that he subsequently
accompanied Mr. Stark to that spot several times and they

failed to identify a single Scaup, though *Tufted* Ducks were abundant, as they had been for years previously." In December, 1897, when working at the second edition of the "Manual," Saunders gave me to understand, in a letter now before me, that he was dropping the record, having made up his mind it was a case of "mistaken identification."

During part of the time when Stark was studying medicine in Edinburgh, he and I frequently took ornithological rambles together, and delightful outings they were, for Stark was a most interesting companion. It was in 1882 that he exhibited the "Scaup's" nest and eggs to the Royal Physical Society, and the following year I twice accompanied him to Loch Leven in the nesting season. Of course we looked out for Scaups, but could detect none. Tufted Ducks, however, were common, and we found several of their nests. The Tufted Duck, it should be noted, had been *proved* to breed there eight years before, and had probably done so for a much longer period (*cf.* my notes on the species in *Ann. S.N.H.*, 1896, pp. 148–155). It seemed strange that Scaups only, that is, as opposed to Tufted Ducks, were noted by Stark in 1880, and he frankly admitted the possibility of his having made a mistake in identification. I may here say that he frequently complained of injury to his eyesight through using the microscope. The opinion I then formed, and still hold, is that the nest in question was not a Scaup's but a Tufted Duck's. When the nest and eggs were on view in Stevens' auction rooms in June, 1902, I asked Saunders to tell me what he thought of them. His reply was : "I should say Tufted, decidedly." I do not know into whose hands this lot passed at the sale. Perhaps some reader of BRITISH BIRDS can tell me.

WILLIAM EVANS.

[Although we much regret having omitted to refer to the first edition of the "Manual," we are not altogether sorry to have been instrumental in resuscitating this erroneous record, since it has drawn forth these interesting details from Mr. W. Evans. The original record is a very important one and is very positively stated in the fourth edition of "Yarrell," and it is only right that all ornithologists should be put in possession of the exact facts with regard to it, so that they can judge for themselves. The details in the Appendix to the first edition of the "Manual" miss a very important point, viz., that Stark noted *only* Scaups in 1880, and the entire suppression of the record in the second edition, coupled with the comment that "assertions respecting the breeding of this species in Scotland lack confirmation," is somewhat misleading to those who do not possess the first edition. We fear that someone bought these eggs as veritable Scaup's, for they fetched £2 7s. 6d. at the Stark sale in 1902. We have to thank Messrs. A. H. Evans, J. A. Harvie-Brown, and Heatley Noble for having also drawn our attention to this error.— H. F. W. & N. F. T.]

THE DISTRIBUTION OF THE COMMON SCOTER IN SCOTLAND.

As I have elsewhere pointed out, the distribution of the Common Scoter in Scotland is peculiar, *e.g.*, Caithness and part of North Scotland, or " the Pentland area," low-lying lochs of the flow-lands ; the high-lying mountain lochs of certain remoter portions of Inverness-shire along the direction of the Great Fault of the Caledonian Canal ; [the Isle of Tiree, uncertain ?] ; and Ireland, as shown above on p. 86. J. A. HARVIE-BROWN.

PALLAS'S SAND-GROUSE IN YORKSHIRE AND KENT.

DURING the first week of June last three Sand-Grouse (*Syrrhaptes paradoxus*) were observed in a field of young corn in the eastern portion of Cleveland. Shortly afterwards one of them was picked up, dead ; and I have had an opportunity of examining this specimen, which is a male in excellent plumage. The other two birds were seen at intervals until the middle of June, when they both disappeared. T. H. NELSON.

ON July 4th last I obtained a satisfactory view of three Pallas's Sand-Grouse on the sand-hills north of Littlestone. H. G. ALEXANDER.

GREEN-BACKED GALLINULE IN NORFOLK.

ON June 19th, and for a fortnight previously, a Green-backed Gallinule '(*Porphyrio smaragdonotus*) was seen at Horsey by three different marshmen, one of whom recognised the bird from having seen a locally killed specimen some years previously, and the other two men's independent description was unmistakable. M. C. H. BIRD.

ABNORMAL EGGS OF THE RINGED PLOVER.

ON June 18th, 1908, I found on a lake island in Ireland a clutch of four abnormal eggs of (presumably) the Ringed Plover (*Ægialitis hiaticola*). In colour they were of a light greenish-blue, and without markings. The surface of the shells was somewhat rough, and with only one of them was it necessary to make full use of the drill ; in two of them a slight pressure of the drill only was required to penetrate a black and rotten spot on the shells. In the fourth there was a slight exudation of the contents through a small aperture with black edges. The site, too, of the nest was abnormal, for it was by the side of a small dead shrub, with a ragged robin and another plant close by it. The nest itself was a depression in damp moss. I find on reference to my notebook, that on May 10th, 1906, there were on the same island three Ringed Plover's eggs (normal) in a depression (lined with dead rushes) in mossy soil—though possibly on this island there is no spot that would give a normal Ringed Plover's nest, there is a considerable tract on the mainland, some two or three hundred yards off, where these birds nest in normal surroundings, and here in 1906 and 1907 I found very similar eggs to the faulty eggs of this year, it seems to me probable that they were all laid by one and the same bird.　　H. TREVELYAN.

[Major Trevelyan has kindly allowed us to submit the nest and eggs to the Rev. F. C. R. Jourdain, who reports upon them as follows :—" The eggs are certainly remarkable. The faint streaks at the end suggest those of Sandpipers, and so does the nest. However, I do not attach much importance to the latter, as from Major Trevelyan's letter it is clear that the bird only made the *depression* in the moss and placed one or two straws in it. Measurements in this case give little help, and weights are not of much use. White eggs are usually larger than normally-coloured ones, and these are obviously imperfect and prematurely laid, so that the weight would not be a safe test. I think, however, I have found a good criterion in the colour of the *inside* of the shell. Ringed Plovers' eggs, *when fresh*, show a distinct greenish tint, which fades somewhat, but is generally perceptible. Sandpipers' eggs I have always found yellowish inside. These eggs show a very distinct green when looked at against the light, and on that account I should ascribe them to the Ringed Plover rather than to the Common Sandpiper. A tendency ·towards the same aberration occurs also in the case of the Lapwing, Ruff, and Woodcock, but is rare among Limicoline birds on the whole.

Dr. Ottosson tells me (*in litt.*) that he has a clutch of Ringed Plover's eggs pale blue in colour, without any markings, and there is an abnormal set in Mr. P. F. Bunyard's collection with very pale bluish-green ground and a few fine jet-black spots and large underlying dark grey blotches ; a clutch of spotless bluish-green Curlew's eggs is recorded in the ' Zoologist ' (1903, p. 352) from Brecon."—F. C. R. J.]

PEBBLE NEST OF A RINGED PLOVER.

IN reference to Mr. Pycraft's article on the nest of the Ringed Plover (Vol. I., p. 373), it may be worth while to give details of a somewhat unusual nest which I found with four fresh eggs on July 2nd at Langston Harbour, near Portsmouth. The nest was formed of small pebbles, and a few little pieces of broken shell. It completely filled a rather deep hoof-mark of a cow in sun-baked mud. There were 2000 pebbles, weighing seven ounces, and they must have been collected from a distance of twenty yards.   H. LYNES.

LAPWING'S NEST WITH FIVE EGGS.

ON April 15th I was shown by a gamekeeper a Lapwing's (*Vanellus vulgaris*) nest with five eggs. I examined the eggs carefully, and found incubation had just begun. All the eggs were exactly similar, and looked as if laid by the same bird. The nest was in the middle of a large grass field, and no other Lapwings but the one pair were within half to three-quarters of a mile. The estate is strictly preserved, and no boys had been near to interfere, nor had the keeper any object in placing the fifth egg there. I think all were laid, without doubt, by the same bird, and as the case seems to be a perfectly authentic one I think it may be worth recording.
       R. H. RATTRAY.

[The Rev. F. C. R. Jourdain informs us that he has notes of the occurrence of five eggs in one nest in the case of the following species of *Limicolæ* :—Golden Plover, Lapwing (numerous instances), Redshank, Snipe, Common Sandpiper, Solitary Sandpiper and Curlew.—EDS.]

SOLITARY SANDPIPER AND OTHER WADERS IN KENT.

WHEN walking along the coast of Kent on July 18th last, I put up a Sandpiper which, from its very dark colour, I knew to be something out of the ordinary. I marked the bird down and stalked it behind a sandbank, getting to within seven or eight yards of it. Owing to its very dark greenish-brown plumage and pure white underparts, pale brown throat, and dark Sandpiper bill and legs, I concluded that I had obtained an exceptionally fortunate view of the Green Sandpiper. To make absolutely certain of its identity, I put it up, expecting to see the pure white rump, but, to my surprise, the rump was the same colour as the back, the white only coming up on either side, as in the Common Sandpiper.

I again stalked it, and put it up once more, and am now perfectly certain that it was a Solitary Sandpiper (*Totanus solitarius*). The first time I got near the bird I saw that there were two of the same kind, but I was only able to follow the one when they flew off.

I am very well acquainted with the Common Sandpiper (*Totanus hypoleucus*), having had ample opportunities of watching it on the Tay in Scotland, in Devonshire on the Tamar, occasionally at Woburn, and in innumerable other places. Not only does the very dark plumage of *Totanus solitarius* make it easy to distinguish from *Totanus hypoleucus*, but the wing bar, which is so conspicuous in the latter bird in flight, was absent. I may add that I have had several opportunities of watching Wood-Sandpipers at close quarters this summer, and on one occasion the Green Sandpiper.

On July 14th there was a small flock of eight or ten Sanderlings, also on the Kentish coast. On July 18th their numbers had greatly increased, as I first came across a party of thirty-one, then another of thirteen, and later on as the tide went out they were to be seen in small groups all along the shore. Nearly all of them still retained their red throats. On July 20th I saw a Curlew-Sandpiper with very red throat and breast, and small parties of Whimbrel were occasionally seen between the 13th and 21st. M. BEDFORD.

[On going to press we learn that a Solitary Sandpiper was shot at Littlestone on August 15th, and this seems confirmatory of the Duchess of Bedford's most careful observations. Her Grace is to be sincerely congratulated on having succeeded in an identification of such difficulty.—EDS.]

A HITHERTO UNRECORDED SPECIMEN OF THE LEVANTINE SHEARWATER FROM KENT.

DURING a visit to Canterbury in July last, in order to examine the bird collection there, I found in the Hammond Collection, which was bequeathed to the town in 1903, a specimen of a Petrel which at once attracted my attention. After comparing it with the Manx Shearwaters in the same case, and noting its points of difference, I consulted Dr. Godman's " Monograph of the Petrels " on my return home. I have not the slightest hesitation in pronouncing this bird to be a Levantine Shearwater (*Puffinus yelkouanus*).

The birds in the Hammond Collection are admirably housed and cared for, but, like those in almost every local museum that I have seen, urgently require proper labelling, and in this collection there are no labels at all except those which were attached by the original owner, and these are so small that it is extremely difficult to read many of them. With regard to the present bird, it has two hanging labels attached to its legs, in the late Mr. Oxenden Hammond's writing, which read as follows :—" Petrel undescribed, picked up dead at Wingham " [Kent] " about 1865. Considered a new species by Gould ; see his autograph attached. Mr. Howard Saunders has a closely resembling specimen from Gibraltar, but without the rosy breast ; he thinks it must be the Mediterranean form of *Puffinus anglorum*, but does not feel sure." " Breast rosy, like an adult Goosander."

I think we may take it that such a good ornithologist as the late Mr. Oxenden Hammond would not have stated that the bird was picked up at Wingham if he had any doubt on the point, and from his remark on the " rosy breast," which has, of course, now disappeared, the bird must have been very recently dead when it came into his hands.

I was not aware that any of the Petrels ever had this rosy tint in life, and I cannot find any mention of it with regard to the present species.

Mr. Hammond does not appear to have taken any further steps to have the identification of the bird made certain, and I presume thought that he was not justified in publishing the record, since Saunders expressed some uncertainty.

This bird is an example of the darker phase of the Levantine Shearwater in which the yellowish-brown wash on the flanks extends across the belly, and to a rather less extent up the breast. In other respects it exhibits the distinctive features of this species very clearly. It is a little larger than the Manx Shearwater, and the bill and wings are both slightly longer. The back is a deep brown instead of black, the under tail-coverts are brown instead of white, and as has been said above, there is no pure white on the breast and belly, which are everywhere washed with brown, and this is most intense on the flanks. The feathers of the breast also are mottled with dusky-grey. N. F. TICEHURST.

* * *

GOLDEN ORIOLE IN FIFESHIRE.—A female *Oriolus galbula* is reported by the Misses Rintoul and Baxter as having been obtained at Markinch on May 13th, 1908 (*Ann. S.N.H.*, 1908, p. 180).

GREAT GREY SHRIKES IN SCOTLAND.—Twenty-two occurrenees of *Lanius excubitor* are recorded in Mr. John Patterson's useful " Report on Scottish Ornithology for 1907 " (*Ann.*

S.N.H., p. 137). The following we have not previously referred to :—*Mull.*—One, March 9th. *Pentland Skerries.*—One, September 24th. *Shetlands.*—One flew on board a . boat twenty miles out on September 26th. *North Berwick.*—One, October 12th. *Gilston* (Fife).—One, November 4th ; one, November 28th (another is recorded at this place by the Misses Rintoul and Baxter (*t.c.*, p. 180) on April 22nd, 1908). *Colinsburgh.*—Two in November. *Auchnasheen.*—One, December 2nd. Another recorded in our pages (Vol. I., p. 263) by the Duchess of Bedford, is not referred to.

WOODCHAT SHRIKE IN SUSSEX.—Mr. J. A. Clark records that a male *Lanius pomeranus* was shot near Rye on September 15th, 1907 (*Zool.*, 1908, p. 269).

PIED FLYCATCHER NESTING IN AYRSHIRE.—We have omitted to refer to an interesting record of apparently the first breeding of *Muscicapa atricapilla* in Ayrshire, viz., at Glendoune, in 1907 (*cf.* M. Young, *Ann. S.N.H.*, 1907, p. 247).

CANARY SERIN IN SCOTLAND.—Mr. J. A. Harvie-Brown records (*Ann. S.N.H.*, 1908, p. 181) that a specimen of *Serinus canarius* was captured, in company with Linnets, at Springkerse, near Stirling, at the end of November, 1907. The bird had no appearance of previous confinement, and it was alive and still rather wild on May 29th, 1908. Mr. Harvie-Brown does not actually claim it as a truly wild wanderer, but we wonder if it would be so claimed by anyone less cautious were it to escape again !

ROSE-COLOURED STARLING IN SCOTLAND.—Major A. Hughes-Onslow writes that he had an excellent view on July 2nd last of a specimen of *Pastor roseus* on some sandy ground near Reay, in Caithness (*Field*, 11, VII., 08, p. 91).

DOMED NESTS OF JACKDAWS.—Mr. T. T. Mackeith records that he found in May, 1907, in West Renfrewshire, some Jackdaws' nests which were large structures, roofed over with sticks, with a hole large enough to admit the bird. They were built in spruce fir trees (*Zool.*, 1908, p. 232). This reminds us that Mr. W. Wells Bladen has for several years reported the occurrence of similar nests of the Jackdaw in Staffordshire (*cf. Trans. N. Staffs. F. Club*, 1901). Another domed nest of this bird was found by Mr. C. E. Wright near Kettering (*cf. Journ. Norths. N.H. Soc.*, 1899, p. 174). The Rev.

F. C. R. Jourdain found a colony in Shropshire in 1901 (*cf. Eggs of Europ. Birds*, p. 16), and other instances have been recorded.

SUPPOSED ALPINE SWIFT IN NORTH DEVON.—Mr. T. H. Briggs records (*Zool.*, 1908, p. 269) that he saw "recently" a Swift which he identifies as *Cypselus melba*, flying low at Lynmouth. His attention was directed to the "size" of the bird—presumably the large size which is, of course, a very striking characteristic of this species; but Mr. Briggs goes on to say that he distinctly "saw the grey underside" of the bird as it flew over his head. The Alpine Swift looks very *white* underneath when flying, and the use of the word "grey" in describing this distinctive characteristic makes us doubtful of the identification being correct. There was a sea-fog at the time.

SNOWY OWL IN THE OUTER HEBRIDES.—Mr. J. A. Harvie-Brown records a fine example of *Nyctea scandiaca* shot on South Uist in October, 1907 (*Ann. S.N.H.*, 1908, p. 182).

SCOPS OWL IN FIFESHIRE.—A female *Scops giu* was obtained near Largo. The Scops Owl has been recorded only eight times previously in Scotland. (W. Evans, *Ann. S.N.H.*, 1908, 183).

MONTAGU'S HARRIER IN SURREY.—Mr. Collingwood Ingram reports (*Zool.*, 1908, pp. 308–311) that a pair of Montagu's Harriers nested in Surrey this year in the same place as those recorded last year (*cf. antea*, Vol. I., pp. 237 and 351). Unfortunately the eggs failed to hatch. The nest was carefully protected by a Royal Society for the Protection of Birds' watcher, but possibly too much attention was paid to the nest by observers and photographers.

COMMON BITTERN IN HADDINGTONSHIRE.—The Rev. H. N. Bonar writes that a specimen of *Botaurus stellaris* (the third observed in East Lothian this year) was picked up dead on Gullane Links in April (*Ann. S.N.H.*, 1908, p. 183).

GADWALL IN ABERDEENSHIRE.—Messrs. L. N. G. Ramsay and A. L. Thomson satisfactorily identified two specimens of *Anas strepera* (a very rare bird in the district) in the estuary of the Don on September 1st, 1907 (*Ann. S.N.H.*, 1908, p. 184).

PINTAILS IN SHETLAND.—A pair of *Dafila acuta* was found breeding at Dunrossness in 1905 (*cf. antea*, p. 54). Mr. Harvie-Brown now announces that there are four or five pairs there this year (*Ann. S.N.H.*, 1908, p. 184).

TISH
RDS
TED·MAGAZINE
THE·BIRDS·ON
TISH·LIST

Vol. II.
No. 5.

NE·SHILLING·NET
LBORN·LONDON
RBY & C?

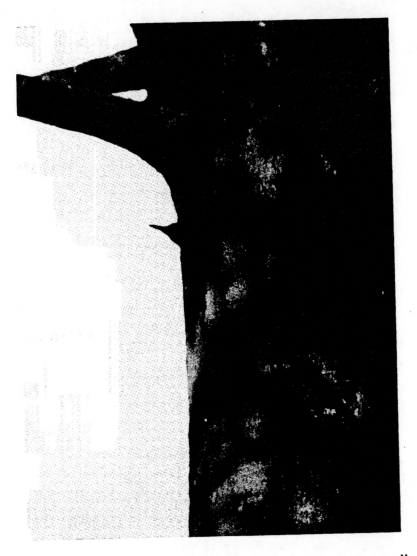

GREEN WOODPECKER. "KEEPING AN EYE ON THE ENEMY."

(Photographed by Miss Turner.)

BRITISH·BIRDS

EDITED BY H. F. WITHERBY, F.Z.S., M.B.O.U.

ASSISTED BY W. P. PYCRAFT, A.L.S., M.B.O.U.

CONTENTS OF NUMBER 5, VOL. II. OCTOBER 1, 1908.

GREEN WOODPECKER *versus* STARLING.

BY

EMMA L. TURNER, F.L.S.

(PLATE IV.)

WHILE wandering about soon after dawn on the morning
of May 8th, I came across one of the most amusing
incidents connected with bird life which I have ever
watched.

A pair of Green Woodpeckers, after having for some
years enjoyed undisputed possession of a nesting hole in
an oak tree, were engaged in a "tooth-and-nail" encounter
with a pair of Starlings which were maliciously endeavouring to obtain possession of their neighbour's

home.   The dispute raged continuously till May 14th,
when I settled it by putting up a nesting-box for the
Starlings; this they immediately annexed, and ultimately
both pairs brought off their respective broods in safety
and comparative peace.

During the campaign I saw much that was both
interesting and amusing in the tactics employed by the
opponents.   While the Woodpeckers were undoubtedly
the more powerful birds, the determination, readiness of
wit and general *finesse* of the active and irrepressible
Starlings commanded my respect.

The following is an account of what happened on May
9th between 6 a.m. and 3 p.m. and is typical of all the
after days of warfare, during which, however, I only
watched at odd times for two or three hours at a stretch.

When I arrived on the scene the Woodpeckers
flew away, being shy and easily alarmed, but soon
returned when I hid myself.   The moment the Starlings
were left to themselves they carried into the hole every
available bit of twig and rubbish they could seize,
working together with a feverish energy that made me
feel tired, so that in five minutes they seemed to have
collected ample material for several nests !   By-and-by,
however, a Woodpecker would return, then one Starling
carried on the fight while the other, when possible,
continued the nest building with more or less success ;
sometimes holding the entrance of the citadel while its
rightful owner clung to the outside (*see* page 144), looking
in and out and all around but not always daring to take
possession.   For although undoubtedly the stronger,
and able to hold her own when once inside the tree, the
Woodpecker seemed unable to cope with her smaller and
more active opponents at close quarters.   If she
ventured inside when either or both Starlings were in
possession a desperate scuffle could be heard which
generally ended in the defeat and ejection of the Woodpecker, but not always.   On one occasion I saw the
Woodpecker seize a Starling by the beak and drag it

forth, then slipping inside she soon ejected the other, but this was when her mate was near. The male Woodpecker did not take his fair share of the fight while I was watching, and often the hen bird would lean half out of the nest (*see* Plate 4) and call to him in soft complaining tones, but his answering cry generally came from a long distance off, and she was left for hours to continue the combat alone.

The Starlings, on the contrary, worked well together and sometimes a third came to their assistance. However, when once the Woodpecker gained possession of her home the Starlings literally had not a " look in," but sat disconsolately on a branch near at hand and watched, by no means without protest, while the Woodpecker slowly and daintily threw out each twig until the ground beneath the tree was strewn with *débris*. I wish it had been possible to obtain photographs of this part of the proceedings because the obvious enjoyment of the Woodpecker as she did this was worth recording. After watching every bit of rubbish till it reached the ground. she looked up at the discomfited pair of Starlings between each act and chuckled softly with her head on one side, while the lookers-on gave vent to sundry long-drawn-out screeches of disapproval. There was a particularly large and dry laurel leaf which one Starling had wrestled with and carried into the hole after great struggles, because its stiffness and length impeded the bird's flight. When this treasure was thrown out and fell to the ground with a dry rattle, both Starlings whistled so plaintively that I laughed aloud and frightened the Woodpecker so that she fled. Then the Starlings had another " innings," and for half an hour did what they pleased and threw out a large quantity of the wooden chips dear to the Woodpecker; but at noon the rightful owners again had full possession until 2.15, when something disturbed them, and the enemy held the citadel till 3 p.m., when I left, after seeing the Woodpeckers once more reinstated.

A favourite trick of the Starlings was to sit on a branch some little distance away and "yaffle." This at first always lured the Woodpecker from her hole, and during the week of fighting the Starlings became very proficient in "yaffling," but after a time the Wood-

Guarding the Entrance. (Photographed by Miss Turner.)

pecker learnt wisdom and was not deceived. So the fight alternated for a week until I began to fear for the ultimate success of the rightful owners of the nesting site, and even went so far as to harden my heart and consult with the powers that be as to the advisability of shooting the robbers. On one occasion, however, there had been three Starlings and one Woodpecker inside the tree together ; so it seemed more than probable that, if this marauding pair suffered the extreme penalty of the law for their sins, others might carry on the feud. Consequently, the nesting-box was tried first, with happy results ; for the Starlings occupied it the same evening (May 14th), and their young ones were fledged on June 19th, while the young Woodpeckers flew away a week later. Evidently even in wild nature the strongest does not always win, art and science hold their own. This the Starlings seemed to know well when they pitted their wits against mere physical strength ; for it seemed to me they would win finally by mere persistence and cunning.

When very agitated, the Green Woodpecker would rapidly ascend the bole of a neighbouring beech, and as rapidly descend *backwards* in a curiously jerky manner, as if she were climbing hand over hand down a rope. I have never before seen any of the Woodpeckers *descend* in this manner.* Owing to the height of the nesting hole I was obliged to use a telephoto lens, the slowness of which, added to the darkness of the wood, made it impossible to obtain photographs of the amusing attitudes and fighting tactics of these birds, and, when the young were hatched out, the surrounding foliage had become so dense as to make it practically impossible to get any picture with a slow lens. Out of some fifty plates exposed, the two accompanying photographs (from which part of the background has been taken out) were almost the only result worth mentioning ; but the enthusiastic photographer, even if unsuccessful in his art, sees so much of wonder and beauty that he has no cause to grumble if the pictorial results are not always just what he hoped for.

* This article was received before the publication of Col. Feilden's note on this subject (see p. 93).—EDS.

ON THE MORE IMPORTANT ADDITIONS TO OUR KNOWLEDGE OF BRITISH BIRDS SINCE 1899.

BY

H. F. WITHERBY AND N. F. TICEHURST.

PART XIV.

(*Continued from page* 129.)

BAILLON'S CRAKE *Porzana bailloni* (Vieill.).

ESSEX.—One was caught by a dog near Dagenham on October 3rd, 1874, and is now in the museum of the Essex Field Club (*Field*, 2, III., 04). ·

SURREY.—One was caught alive in Church Street, Godalming, in 1837, and is in the Charterhouse collection. An adult female was also caught alive between Mitcham and Carshalton about the end of May, 1847 (J. A. Bucknill, *B. of Surrey*, p. 274).

SUSSEX.—One was killed against the telegraph wires on Pett Level in June, 1907 (N.F.T., *cf. antea*, Vol. 1, p. 359).

KENT.—A female was shot by Captain R. Alexander near Lydd, on November 24th, 1906 (R.E.C., *Field*, 22, XI., 1906 ; *cf. antea*, Vol. 1, p. 359).

HAMPSHIRE.—Four occurrences are noted (Kelsall and Munn, *B. of Hants*, p. 271).

NORTH WALES.—A male was caught by a dog in a ditch near Colwyn Bay on November 6th, 1905 (H. E. Forrest, *Zool.*, 1905, p. 465).

CHESHIRE.—An adult male was captured alive near Stockport in May, 1905 (T. A. Coward, *t.c.*, 1906, p. 395).

SCOTLAND.—A female was shot at ·Thurso in September, 1898 (W. Arkwright, *Ann. S.N.H.*, 1899, p. 50).

[ALLEN'S GALLINULE *Porphyriola alleni* (Thompson).

An immature example of this African species alighted on a fishing-boat off Hopton, near Yarmouth, on January 1st, 1902, and was captured (J. H. Gurney, *Zool.*, 1902, p. 98). The species has occurred in the winter in Italy and Sicily, and it is possible that this bird was a genuine storm-driven migrant. The specimen is now in the possession of Mr.

J. B. Nichols. It may be noted that a bird of this species has been recorded as having been caught at sea 190 miles off the coast of Liberia.]

CRANE *Grus communis* Bechst. S. page 521.

NORFOLK.—On April 7th, 1898, four were seen by Mr. Pashley near Glaven, and they were afterwards reported at Weybourne, and again at Runton, after which they took their departure (J. H. Gurney, *Zool.*, 1899, p. 119). An immature bird was seen for about three weeks near Great Yarmouth in April, 1906 (J. E. Knights, *t.c.*, 1906, p. 194).

SCOTLAND.—A young bird appeared at the *Pentland Skerries* on May 1st, 1903, and was shot two days afterwards (J. Tomison, *Ann. S.N.H.*, 1903, p. 186). One was seen in *North Shetland* on May 16th, 1906 (T. E. Saxby, *t.c.*, 1907, p. 50). One was shot near *Stornoway* on May 14th, 1906. The species had not previously been recorded from the Outer Hebrides (N. B. Kinnear, *t.c.*, 1907, p. 84).

GREAT BUSTARD *Otis tarda* L. S. page 523.

INTRODUCTION.—In 1900 seventeen were imported from Spain and placed by Lord Walsingham on Lord Iveagh's estate at Elvedon, Norfolk. Fifteen survived the winter (they were feather pinioned), but they then left their secure retreat, where they had a run of some 800 acres, and appear to have dispersed over the country. Several were soon shot, and the whereabouts of four only were known at the end of 1901. By the end of the following year only two remained. They appear to have laid eggs, but no young hatched, and the experiment must be deemed an entire failure.

[IRELAND.—Two were seen near Thurles, co. Tipperary, and one of them was shot on December 20th, 1902 (Williams & Son, *Field*, 1903, p. 447). There is no previous authentic record of the occurrence of this bird in Ireland, and we think that these examples may have been " escapes." We believe that all the introduced birds mentioned above have not even yet been accounted for.]

[LINCOLNSHIRE.—Two females were killed, one at Weelsby, the other at Tetney, on December 15th and 29th, 1902 (G. H. Caton Haigh, *Zool.*, 1903, p. 368). These are looked upon by Mr. Gurney (*t.c.*, p. 125) as

genuine migrants, and not part of the Norfolk introduced
birds, but there seems no proof for this.]

JERSEY.—Two were shot on King's Meadow in
December, 1899 (H. Mackay, *t.c.*, 1904, p. 378).

LITTLE BUSTARD *Otis tetrax* L. S. page 525.

YORKS.—One was shot at Kilnsea on December 7th,
1902 (P. W. Loten, *Nat.*, 1903, p. 61).

NORFOLK.—One was shot at Feltwell on Jan. 25th,
1898 (J. H. Gurney, *Zool.*, 1899, p. 118). An adult
female was shot at Ludham on Nov. 26th, 1900 (*id.*, *t.c.*,
1900, p. 138). A male was shot at Caister-by-the-Sea
on Dec. 11th, 1902 (*id.*, *t.c.*, 1903, p. 137).

SUFFOLK.—A male in full summer plumage was shot
on May 3rd, 1898, at Kessingland, near Lowestoft.
This is the first instance of a bird in this plumage having
occurred in the Eastern counties (T. Southwell, *t.c.*,
1899, p. 31, and 1900, p. 115).

STAFFORD.—One was shot by a keeper at Warslow about
1899, but was not recorded at the time as it was killed during
the close season (F. C. R. Jourdain, *in litt.*)

DERBY.—A female was shot on Middleton Top, near
Youlgreave, on May 14th, 1901 (W. Storrs Fox, *Zool.*,
1901, p. 270).

KENT.—One was shot in Thanet on Dec. 20th, 1902
(C. Ingram, *t.c.*, 1908, p. 272).

SUSSEX.—One was shot at Ashburnham on Dec. 23rd,
1900 (G. W. Bradshaw, *t.c.*, 1900, p. 428)—[the date
should be Dec. 28th.—N.F.T.]. A female was shot
near Burpham on Dec. 16th, 1901 (W. Percival Westell,
t.c., 1902, p. 70). A female was killed against telegraph
wires at Hollington in February, 1902 (N.F.T.). A
female was shot at Westfield on Dec. 26th, 1905 (N.F.T.).

SOMERSET.—The Rev. W. Fox reported that a female,
previously unrecorded, was shot on Sedgemoor about 1872, and
was now in the possession of a small farmer (*Field*, 13, VII., 07.)

JERSEY.—A female was shot on Feb. 4th, 1902 (H.
Mackay, *t.c.*, 1904, p. 378).

STONE-CURLEW *Œdicnemus scolopax* (S. G. Gm.).
S. page 529.

KENT.—Owing to protection, the numbers annually
breeding in the county show a slight increase (N.F.T.).

SURREY.—There is at least one locality in which it
may be regularly seen (1898) (J. A. Bucknill, *Birds of
Surrey*, pp. 281-282).

MERIONETH.—One was obtained near Towyn on Jan. 6th, 1903 (H. E. Forrest, *Vert. F. N. Wales*, p. 323).

IRELAND.—One was reported to have been shot at Magheragollen, Gweedore, co. Donegal, on Oct. 12th, 1903 (D. C. Campbell, *Irish Nat.*, 1904, p. 119).

[It formerly bred in Oxfordshire (*Zool.*, 1903, p. 18; 1899, p. 437), Buckinghamshire (*t.c.*, 1903, p. 450), Cambridgeshire (*t.c.*, 1862, p. 8168), Bedfordshire (*Vict. Hist. Beds.*, p. 128), Nottinghamshire (*B. of Notts.*, p. 253)].

PRATINCOLE *Glareola pratincola* (L.).   S. page 531.

KENT.—A male was shot by Mr. Southerden at Jury Gap, Romney Marsh, on May 30th, 1903. The specimen, which was the first recorded example from Kent, is in Mr. Fleetwood Ashburnham's collection (N. F. Ticehurst, *Bull. B.O.C.*, XIII., p. 77). An adult pair were shot near the same place on July 19th, 1904 (N.F.T.).

SCOTLAND.—A young bird, only the second example of the species which has ever been obtained in Scotland, was shot on the Mill Burn, Rocksands, Montrose, by Mr. Stormond, on Nov. 4th, 1899 (J. A. Harvie-Brown, *Ann. S.N.H.*, 1900, p. 51).

CREAM-COLOURED COURSER *Cursorius gallicus* (J. F. Gm.).   S. page 533.

One was shot in Bouley Bay, Jersey, on Oct. 19th, 1896 (J. E. Harting, *Zool.*, 1896, p. 435). The Channel Islands are not included by Howard Saunders amongst the places where this bird has occurred.

DOTTEREL *Eudromias morinellus* (L.).   S. page 535.

NORTH WALES.—Four were seen on the top of one of the highest mountains in Merionethshire on May 10th, 1901, and on May 8th, 1902. A search was made in June, 1901, but none were seen (O. V. Aplin, *Ibis*, 1901, p. 517, and 1903, p. 133). It appears to occur sparingly on the mountains in spring (H. E. Forrest, *Vert. F. N. Wales*, p. 330).

IRELAND.—One was shot from a flock of more than a hundred birds (*thought* to be of the same species) in Donegal Bay on Nov. 29th, 1905 (A. R. Nichols, *Irish Nat.*, 1906, p. 45). Three females were shot at Athlone on Nov. 10th, 1906 (W. P. Williams, *t.c.*, 1907, p. 183).

OUTER HEBRIDES.—A bird-of-the-year was received from Eilean Mor, Flannan Isles, in September, 1906. The species had not been recorded previously from the Outer Hebrides (W. E. Clarke, *Ann. S.N.H.*, 1907, p. 53).

SCOTLAND.—Mr. Harvie-Brown notes an extension south, if not an actual increase in numbers and breeding area, in the Tay district, which began about 1900, suddenly, and was afterwards continued rapidly, especially about 1902-3 (*cf. Fauna of Tay Basin*, pp. 299-304).

RINGED PLOVER *Ægialitis hiaticola* (L.). S. page 539.

Inland Nesting.—In Worcestershire (D. R. Grubb, *Zool.*, 1902, p. 316) ; in Middlesex (R. B. Lodge, *t.c.*, 1901, p. 389).

KENTISH PLOVER *Ægialitis cantiana* (Lath.). S. page 543.

DURHAM.—An adult female was found dead near the North Gare breakwater (Teesmouth) at the end of May, 1904 (C. E. Milburn, *Nat.*, 1904, p. 283).

NOTTS.—One was seen on April 13th, 1904, near Mansfield (J. Whitaker, *Birds of Notts.*, p. 255).

KENT.—Owing to the rigid protection now in force in their breeding area, their numbers are steadily on the increase (N.F.T.).

KILLDEER PLOVER *Ægialitis vocifera* (L.). S. page 545.

A specimen shot at Peterhead by Mr. Andrew Murray, Jun., in 1867, was labelled " *Charadrius hiaticula*," and was discovered in the University Museum at Aberdeen and identified as an example of this species by Mr. W. P. Pycraft in July, 1904. This, therefore, is the first British-killed specimen (W. P. Pycraft, *Ann. S.N.H.*, 1904, p. 247).

LESSER GOLDEN PLOVER *Charadrius dominicus* P. L. S. Müller. S. page 549.

SURREY.—A specimen in the Charterhouse collection was shot on Epsom Racecourse on Nov. 12th, 1870 (J. A. Bucknill, *Birds of Surrey*, p. 283).

ESSEX.—One (which was afterwards identified at the British Museum) was shot by Mr. H. Nunn off Shell Haven Point, on the Thames, Aug. 6th, 1896 (H. Nunn, *Zool.*, 1897, p. 330).

SOCIABLE PLOVER *Vanellus gregarius* (Pall.). S. page 553.

IRELAND.—A female was shot on Aug. 1st, 1899, in a turnip field, by a farmer at Robinstown, near Navan, co. Meath (E. Williams, *Irish Nat.*, 1899, p. 233).

[It will be remembered that one was recorded from Kent in 1907, *vide antea*, Vol. I., p. 57.]

(*To be continued.*)

SOME EARLY BRITISH ORNITHOLOGISTS AND THEIR WORKS.

BY

W. H. MULLENS, M.A., LL.M., M.B.O.U.

III.—CHRISTOPHER MERRETT

(1614—1695).

(Continued from page 118.)

Rusticola minor, *the Snipe, or Snite*, I. 62. tab. 27. Scolopax, Gallinago minor Ald. tom. 3. 479. Gallinago sive Rusticola minor, G. 448. est altera Hujus species nuncupata, *the Jack Snipe.*

["Snite" is the old form of Snipe. Gesner (p. 483) gives "snyt" as an English name for the Sandpiper. "Jack Snipe," *cf.* Willughby (p. 25).]

Rusticola major, Scolopax. Gallinago, I. 88. tab. 31. *the Wood-cock*, Ald. tom. 3. 473. Rusticola vel Perdix Rustica major, G. 445. Utræq ; Hyeme huc migrant. ex Hibernia.

[Neither Turner nor Willughby mention the Woodcock as frequenting Ireland. Merrett's statement that it migrates hither from that country is derived from Giraldus Cambrensis' account (Chapter X.), "There are immense flights of Snipe (*acetæ*) . . . both the larger species of the woods and the smaller of the marshes."]

Ralla-Anglor, *the Rail, or King of the Quails*, Ald. 3. 455.

[*Cf.* Willughby (p. 23). Turner (p. 71) states, curiously enough, that he had not seen or heard the Corn Crake "anywhere in England, save in Northumberland alone."]

Upupa, *the Hoopee*, I. 62. tab. 27. Ald. 2. 704. G. 703. *In the New Forest in* Hampshire, & in Essexia, sed raro invenitur.

[The Hoopoe is still found occasionally in the New Forest. Merrett describes it as rare, *cf.* Charleton (p. 92), who calls it a Hoopoop, and states that it rarely visits this country, and that a friend of his killed one near London "Hyeme tamen Superiori." This bird, however, was not rare in

Norfolk.   Sir Thomas Browne, in his notes says : " Upupa,
or Hoopebird, so named from its note, a gallant marked
bird wch I have often seen and 'tis not hard to shoote them "
(*cf.* Southwell, p. 23).]

Pulveratrices domesticæ.

[p. 174] Gallus, *a Cock*, I. 82. tab. 29. Ald. tom. 2. 200.
. . . Hirsutis pedibus, ib.
. . . Palustris, *a Moor-hen*, G. 421, Morenna Angl. Ald.
2. 341.

[The Black Grouse. Turner (p. 87) calls it the Morhen
(*cf.* Willughby, p. 173). Aldrovandus (Lib. XIV., Cap. XV.)
treats " De Gallo Scotico Sylvestris & de Morhenna
Anglorum," and informs us that " Scoti in hoc genere marem
vocant Ane black cock, id est, Gallum nigram : fœminam
. . . ane grey hen, id est Gallinam fuscam."]

Gallina Rustica Turn. quam variis de causis Attagenem
esse conjicit.

[*Cf.* Turner (p. 87).]

Pulveratrices Lavatrices.

Fulica, *a Coot*, I. 88. tab. 31. Ald. 395. G. 344.
Ispida, *the Kings-fisher*, I. 88. tab. 31. Ald. 5. 520. G. 513.
Gallina Aq. I. 88. tab. 31.

[The figure in Jonstonus is that of the Water-Hen.]

Gallina serica, I. 88. tab. 31. sic dicta à splendore, Ald.
3. 470.

[The figure in Jonstonus is possibly that of the Godwit.]

Columba vulg. Livia, *the common House-Pidgeon, or Culver*,
I. 88. tab. 32. Ald. 2. 462. G. 245.

Columba Guttorosa perperam dicta *Cropper*, Ald. 2. 479.

Columbæ Cypriæ, *Jacobins*, I. 88. tab. 32. Ald. 2. 471.

Columbæ Turcicæ, Coloris sunt Betulini, cum oculis rubris,
Ald. 481.

Columb. Tabellariæ, *Carriers*.

Columb. Tremulæ, *Shakers*, suntq ; vel acuti vel lati
caudæ.

Columb. Hirsutis pedibus, *rough-footed Pidgeons*, I. 88.
tab. 32. Ald. 2. 466.

Columb. Angl. & Russica, G. 245. inter has majores vocantur,
Runts.

Columb. Galeatæ, *Helmets*.

[P. 175.] Columb. maculis nigris & aliæ rubris decoratæ,
Black and red Spots.

Columb. Percussores, *Smiters*.

Columb. Gyratrices, *Tumblers*, Omnes hæ Columbæ in Columbariis aluntur præterq ; has à curiosis educantur Turcicæ, Barbariçæ, *the Finikin, Cornew, Bastard bill, Light Horsman, Dragoon.*

Turtur, *the Turtle Dove*, I. 88. tab. 32. Ald. 2. 509. G. 277.

Palumbus major torquatus, *a Ring Dove, or Quist*, Ald. 2. 4. 7. In sylvis, Turn. *a Cowshot, a ringed Dove.*

Oenas seu vinago, *a Stock-Dove, or Wood-Pidgeon*, Ald. 2. 499. I. 88. tab. 32.

Passer domesticus, *the House-Sparrow*, I. 92. tab. 34. G. 581. Ald. 2. 534. idem quandoq ; albus invenitur, I. ibid. Ald. 556.

Passer pusillus in Juglandibus degens, I. 96. sine Icone, Ald. 2. 563.

[A variety of Sparrow, *cf.* Belon (pp. 363 and 364), who terms this bird " Moineau de Noyer," or " Friquet." Charleton (p. 78) calls it the " Wall-nut Sparrow."]

Junco, *the Reed Sparrow*, I. 166. tab. 53. Hujus datur minor species in Arundinetis prope Kingstoniam.

[The Reed-Bunting, *cf.* Turner (p. 103).]

Granivarœ Canorœ.

Carduelis, *a Gold-finch*, I. 69. tab. 36. Ald. 2. 801. G. 215. Aurivittis Turn.

Calandra Ald. 2. 847. est Alaudæ persimilis sed ipsa paulo major, Ramis arborum insidet an Passer torquatus G. *a Bunting.*

[*Cf.* Willughby (p. 208), and Belon (p. 271). Aldrovandus says it is called the " Challander " in England. Emberiza Callandra is the name applied to the Corn Bunting by Linnæus, *cf.* Syst. Nat., Ed. X., 1758.]

Coccothraustes, I. 98. tab. 37. Coccothraustes mas Ald. 2. 846.

[The Hawfinch. Though Merrett does not give the English name the figure in Aldrovandus is unmistakable (*cf.* Belon, p. 374).]

Fringilla, *the Common, or Chaffinch*, I. ib. Ald [Page 176.] 2. 817. G. 342. Turn. *a Sheld apl. a Spink.*

Monti-fringilla, *the Bramble, or Brambling*, I. 96. tab. 33. Ald. 2. 822. G. 343.

[P. 176.]

[The derivation of the word Brambling is obscure. Charleton's attempt is : " The Brambling, or Brier-finch (*utpote rubris sœpe insidens, eorumg' fructibus victitans*) "; " rubris " = on brambles.]

Chloris, *the Green-finch*, I. ib. Ald. 2. 851. G. 226. Turn.
Acanthis, Spinus, Ligurinus.

Citrinella, *the Yellow-hammer*, I. 96. tab. 36. Ald. 2. 859.
Emberiza flava Turn. *a Youlring*, G. 591.

[*Cf.* Swainson (p. 69).]

Linaria, *the Linet*, I. 96. tab. 36. and 98. tab. 37. Ald. 2.
824. G. 550.

Luteola, *a Siskin*, rara apud Anglos avis nec uspiam fere
alibi quàm in caveis cernitur semel in agris Cantabrigianis
se vidisse recordatur, Turn.

[Merrett's account of this bird is taken *verbatim* from Turner
(p. 109), " Caveis " being Turner's word for cages.]

Alauda, *the Lark*, I. 98. tab. 37. 38. Ald. 2. 845. G. 67. Turn.
Galerita.

Alauda pratensis, *the tit-Lark*, I. ib. Ald. 2. 849.

Alauda cristata, *the wood-Lark*, G. 72. Ald. 2. 841. I. 98.
tab. 37.

Rubicilla, *a Bull finch, a Hoop, and Bul Spink, a Nope,*
I. 120. tab. 43. Pyrrhula sive Rubicilla Ald. 2. 745. G. 662.

Baccivoræ.

Turdus vulg. *the Song-Thrush*, I. 98. tab. 37. Ald. 2. 600.
& Turn. *a Thrussel.*

Turdus Viscovorus, *the Misletoe Thrush, or Saith*, I. 102.
tab. 39. Ald. 2. 583. G. 688. & Turn. simpliciter, *a Thrush.*

[Willughby (p. 187) calls it the Missel-bird, or Shrite ; and
Charleton (p. 83) the Shreitch.]

Turdus Illas, *the Wind Thrush*, I. 102. tab. 39. Turdus
minor Ald. 2. 598. Turdus minor Illas vel Tylas, G. 689.
Turn. *a Wind Thrush.*

[The Redwing. Turner does not call it a Wind Thrush, as
Merrett states, but a Wyngthrush (p. 173). The name Wind
Thrush is applied to the Redwing to this day in Somersetshire
(*cf.* Swainson, p. 5).]

Trichas, *the Feldefare*, Turdus pilaris Ald. 596. [Page 177.]
tantum hyeme apud nos reperitur.

[The Fieldfare, *cf.* Willughby (p. 24).]

. Merula, Collyrion Turn. *the Black-bird, or black Ousle*, I.
104. 140. Ald. 2. 604. G. 543.

Sturnus vulg *the Stare, or Starling*, I. 104. t. 40. Ald. 2. 632.
G. 677. circa turres & altiora ædificiorum culmina.

Sturnus Cinereus, Ald. 2. 638. I. 104. t. 40.

[Probably the immature Starling.]

Cæruleo, *a Clot Bird, a Smatch, or Arling, a Stone-check,*

nidulatur in Cuniculorum foveis, & sub lapide in Anglia Turn.

[The Wheatear—also on p. 178 as Œnanthe. *Cf.* Turner (p. 53), Clot (bird) = Clod. Clod bird for clodhopper (*cf.* Swainson, p. 10). The account of the Wheatear's nesting in rabbit burrows is derived from Turner (p. 53).]

Insectivoræ.

Picus viridis, *the Green Wood pecker, or Hickwall*, I. 110. t. 41. Ald. 1. 849. G. 642. Chlorion, Virio, *a Witwoll*, Turn.

Picus varius major, I. 110. t. 41. Ald. 846.

Picus varius minor & mas, I. ib. Ald. 1. 847.

Picus murarius, *the Creeper, or Wall-Creeper*, I. ib. Ald. 1. 852. G. 644.

[Merrett distinguishes this bird from the Tree-Creeper, which he refers to as " Certhia, the Ox-eye Creeper." Willughby (p. 143) says : " They say it is found in England ; but we have not as yet had the hap to meet with it." Turner does not mention the Wall-Creeper, but there is no reason why Merrett should not be correct in including it in his list of British birds (*cf.* Saunders' *Manual of British Birds*, p. 119, and Gilbert White's VIIIth letter to Marsham, in Harting's second edition of the *Natural History of Selborne*). Charleton (p. 86) calls this bird the " Creeper, or Spider Catcher," which latter name Willughby also adopts.]

Picus Cinereus, I. 110. t. 41. Sitta seu Picus Cinereus Ald. 1. 853.

[The Nuthatch, *cf.* Willughby (p. 143). " Gaza* retains the same name, calling it ' in Latina, Sitta.' Later writers style it *Picus cinereus*, *i.e.*, the ash-coloured Woodpecker." Charleton calls it the " Nut-breaker," or " Nut-jobber."].

Juynx, seu Torquilla, *the Wryneck*, I. 114. t. 42. Ald. 1. 866. G. 515.

[Charleton calls this bird the " Wrynecken," or " Emmet-hunter."]

Certhia, *the Ox-eye Creeper*, I. 114. t. 42. Ald. 2. 870. Certhia Turn. G. 223.

[For explanation of name Creeper *cf.* Turner (p. 53).]

Passer Troglodytes, *a Wren*, I. ib. Ald. 2. 651. G. 588. Trochilus, Senator, Regulus Turn.

Curruca, *the Hedge Sparrow*, G. 326. Ald. 2. 753. Hypolais seu Curruca, I. 122. t. 45.

---

* Theodorus Gaza, ob. 1480. The translator of Aristotle's " History of Animals " into Latin from the original Greek.

Hirundo, *the House Swallow*, I. 114. t. 42. [Page 178.] Ald.
2. 662. G. 492. vivit per Hyemem in mineris stanneis Cor-
nubiensibus & in Rupibus marinis.

["It lives during the winter in the tin mines of Cornwall,"
cf. Carew (Fol. 25).]

Hirundo Riparia, *the Sand Martin, or Shore-bird*, I. ib.
Ald. 2. 695. *a bank Martnet*, G. 508.

Hirundo agrestis sive Rustica Plinii, *a Martin*, Ald. 2. 693.
I. 114. t. 42.

Hirundo apus, *a black Martin, or Martlet*, Ald. 2. 699. I.
114. t. 42. *a Rock or Church Martnet*, G. 507.

[The Swift, *cf.* Willughby (p. 214.)]

Parus major, *the Common Titmouse*, I. 120. t. 43. Ald. 2.
713. G. 578. *the great Titmouse*, vel *the great Ox-eye* Turn.

[*Cf.* Turner (p. 131), and Swainson (p. 32).]

Parus Cæruleus minor, I. 122. t. 44. Ald. 721. G. 579. *the
less Titmouse*, Turn.

Parus ater, seu Carbonarius, *the Coalmouse*, I. ib. Ald. 2.
723. G. 579.

[*Cf.* Willughby (p. 241) "Cole-mouse."]

Parus Caudatus, *the least, or long taild Titmouse*, I. ib.
Ald. 2. 716. G. 580.

Motacilla, *a Water Wagtail*, I. 122. t. 44. Ald. 2. 727. G. 557.
Culicilega, *a Wag tail*, Turn.

Motacilla flava rostro longiusculo nigricante, I. 122. t. 44.
Ald. 2. 859. G. 559.

[Presumably the Yellow Wagtail (*M. raii*). The epithet
Longiusculus—somewhat lengthy, as applied to the beak is,
however, more descriptive of the Grey Wagtail (*M.
melanope*).]

Rubetra, *the Stone-Chatter, or Blackberry-eater*, & Turn.
mortetter, I. 122. t. 45. Ald. 2. 740. *a Moortiting* Aquilonari-
bus.

[*Cf.* Turner (p. 159). Charleton also calls the Stone-Chat,
the "Blackberry-eater, Morteller, or Black-cap." For this,
and Moor-titing, or Moor-titling, *cf.* Swainson (p. 12).]

Rubecula, *the Ruddock, Red-breast, and Robin Red-breast*,
I. ib. Ald. 2. 742. G. 661.

Ruticilla, Phænicurus, *the Red-start*, I. 120. 43. Ald. 2. 747.
a Redtail, G. 663. ex Turn.

[Gesner has "Angli a redetale" (p. 699). According to
Charleton (p. 91), this bird possibly hibernated in England.]

Oenanthe, *the Wheat ear, or White tail*, I. 122. [Page 179.] t. 45. Ald. 2. 763. G. 567. in agro Warwicensi *Fallow Smiters*.

[The Wheatear as Cæruleo (p. 177). Fallow Smiters— Swainson (p. 9) has Fallow Smich ; and in Wiltshire it is known as the Horse Smatch, or Snatcher (A. C. Smith, *Birds of Wilts*). Smiter possibly from Smit, SW. Smet, grease or fat.]

Luscinia, Lusciniola, *the Nightingale*, I. ib. Ald. 2. 777. G. 532.

Morinellus, *the Dotterel*, Ald. 3. 540. G. 554. in agro Lincolniensi certo anni tempore capitur jocose, vide *Camden*.

[The account of the taking of this bird given by Camden (who apparently derived it from Caius, *cf.* Evans' Turner, p. 203), is as follows :—

" Dotterells, so named from their dotish foolishnesse, which being a kind of birds as it were of an apish kind, ready to imitate what they see done, are caught by candle light according to foulers gesture : if he puts forth an arme, they also stretch out a wing : sets he forward his legge, or holdeth up his head, they likewise doe their : in briefe, whatever the fouler doth, the same also doth this foolish bird, untill it be hidden within the net " (Camden, Philemon Holland's edition, 1610 (p. 543) ; *cf.* also Willughby (pp. 309, 310).]

Aquaticæ Palmipides.

Cygnus, *the Swan*, I. 136. t. 48. Ald. 3. 8. G. 321.

Anser Domesticus, *the Goose*, mas vocatur, *the Gander*, I. 136, t. 48. Ald. 3. 102. G. 125.

Chenalopex, vulpanser, *a Bergander*, nusquam alias vidi nisi in Thamesi fluvio aiunt tamen esse frequentem in insula Tenia (Thanet.) vocatâ & illic in scrobibus cuniculorum nidulari, Turn.

[The Shelldrake. Turner (p. 25) says " our people nowadays name it Bergander " (*i.e.*, Burrow-gander). Caius, however, suggests quite another derivation for Berg, which he thinks may be from Brend, or Bernd, meaning variegated (*cf.* Evans' Turner, p. 195). Ray, apparently using the same idea, calls the Goosander, a Bergander (p. 94).]

Anser ferus, I. 136. t. 48. Ald. 3. 150. G. 140.

Capricalca, Capricalze Scotis, Ald. 3. 164. I. 136. t. 48. G. 146.

[The Capercailzie, also Urogallus (p. 173).]

Anseris speciem vidi in Cimelio Tradescanti sub nomine squeed una cum Ovo ex Insula Scotica Bass dicta, in qua

quam plurima avium genera stato anni tempore nidificant uti
etiam in insula Vecti.

["In Cimelio Tradescanti," *i.e.*, the Tradescant Museum,
which was the origin and basis of the Ashmolean Museum at
Oxford. ."Squeed" (?) not the Gannet which is mentioned
by Merrett on the same page.   Possibly the Eider Duck is
meant, a few still breed on the·Bass.]

Bernicla Brenta, *the Brant Goose*, I. 136. t. 48 Ald. 3. 167.
Brenthus, G. 95.

Gustarda Avis Scotica Ald. 3. 163. G. 145.

[The Great Bustard ; also p. 173.   The name Gustard was
applied to the Great Bustard by Hector Boethius, or Boece
(1465-1536), the author of " Scotorum Historiæ," *cf.* Willughby
(p. 178), and Gray's " Birds of the West of Scotland " (p. 248) :
" Besides these, we have another foule in Mers, more strange
and uncouth than all these aforementioned, called a Gustard,
fullie so great as a swan, but in colour of feathers and tast of
flesh little differing from a Partridge."]

Anser Bassanus, sive Scoticus, *a Soland Goose*, G. 145. ex
insula Bass non procul Edinburgo.

[*Cf.* Evans' Turner (p. 197).   Soland Goose = the Gannet,
cf. Willughby (328).]

Anas Domesticus, *the Duck*, mas, *the Drake*, I. 142. t. 49.
Ald. 3. 188.   Anas cicur, G. 83.

[Page 180.]   Harle, *the black Diver*, I. 148. t. 49. *a Shell
Drake* in Norfolcia.

[For " Harle," as applied to the Red-breasted Merganser,
cf. Swainson (p. 164) : " Shell Drake in Norfolk."   Swainson
(p. 163) gives Shell Duck as a name of the Goosander.   The
figure in Jonstonus is possibly meant for the Merganser.]

Anas fera, I. ib. Ald. 3. 222. G. 101. *the Wild Duck.*

Anas fera fusca, I. 142. t. 49. Ald. 3. 221. in Paludibus
Lincolniensibus.

[The figure in Jonstonus is possibly meant for the Scoter.
Merrett may here have meant the Pochard.   Ray gives the
English equivalent as ." Pochard " (p. 96), and Charleton,
" the Red-headed Widgeon " (p. 99).]

Anas Platyrhincus Ald. 231. in paludibus Crowlandiensibus,
I. 142. t. 49.

[The Shoveller.   The figure in Aldrovandus is distinct.]

Querquedula, *the Teal*, I. 142. t. 49. Ald. 3. 549. G. 91.

A Gaddel, Ornithopolis nostris sic dictus est magnitudine
Anatis, rostrum simillimum rostro Querquedulæ, sed ali-
quanto magis cærulescit.

[The Gadwall, *cf.* Willughby (p. 374), where the name Gadwall is seemingly used for the first time. The derivation of Gadwall is obscure, *cf.* Newton " Dict. Birds " (p. 297).]

Penelope major, *the Widgeon*, Ald. 3. 219. I. 142. t. 49.

Penelope fæm, Ald. 3. 220.

Colymbus major, *the great Ducker*, I. 136. t. 48. Ald. 3. 252.

[The Great Northern Diver, *cf.* Willughby (p. 342), and Swainson (p. 213).]

Colymbus a Norwegis Lumme, a nostratibus, *Razor bill*, *Worm*, 304. ex Auctario Clus. pag. 367. Mr. *Willoughby*.

[*Cf.* "Museum Wormianum " (p. 304), and. Willughby (p. 342). The word "Loom," or "Loon," is applied to the Divers in general (*cf.* Swainson, p. 213).]

Colymbus Cristatus seu Auritus, *Worm*, ib. sine Icone, idem.

[Possibly the Great Crested Grebe.]

Colymbus medius, *the Dive-dapper*, or *Arsfoot*, I. 136, t. 45. Ald. 3. 258.

[Arsfoot, a name given to the Grebes on account of the position of their legs (*cf.* Swainson, pp. 215, 216). So also the Razorbill and Guillemot are known in Yorkshire as "fèet in Ass."]

Colymbus minimus, *the Dab Chick*.

[The Little Grebe (*cf.* Willughby, p. 340, and Swainson, p. 216).]

Mergorum serrati-Rostratorum species major & minor, in fluvio *Tame* in agro Warwicensi an. 1664. cum rigidissima fuerit hyems, Mr. *Willoughby*.

[Probably the Goosander and Merganser, or the male and female of one of these species (*cf.* Willughby, p. 27, and Charleton, p. 95).]

Mergus Turn. (ut sentio) qui vidit in rupibus marinis nidificantes, juxta Ostium Tinæ fluvii [Page 181.] in Norfolcia, Hoc me ditavit Doctissimus affinis meus Ds. *Jenner* Sclopeto transfosso in agro Wiltoniensi.

[Turner's Mergus is the Cormorant (*cf.* pp. 111, 113).]

Corvus aquat. *the Cormorant*, Ald. 3. 263. I. t. 27. Carbo aquat. G. 121. in Cornubia *Shags*, Turn. mergus.

[The same as the above.]

Onocrotalus, sive Pelicanus, *the Pelicane*, I. 128. t. 46. Ald. 3. 47.

[Pelicans were (1660-1670) kept in captivity in the Royal

Aviary in St. James's Park (cf. Willughby, p. 327, and Charleton, p. 94). Sir Thomas Browne, writing to Merrett under date September 13th, 1668, says :—" In your Pinax I find Onocratalus, or Pellican, whether you meane those at St. James or others brought over or such as have been taken or killed heere I knowe not. I have one hangd up in my howse wch was shott in a fenn ten miles of about 4 yeares ago and because it was so rare some conjectured it might bee one of those which belonged unto the King & flewe away " (cf. Southwell, pp. 64 and 16).]

Pelicanus sive Platæa, a *Shovelard*, I. 128. t. 46. ex agro Lincoln. Turn. a *Spoon bill*.

[The Spoonbill, cf. Turner (pp. 151 and 41), and Willughby (p. 288).]

Larus major & minor albus, *the Sea Mew*, I. 126. t. 46. Ald. 3. 65. & simpliciter, *Gul, Sea Gul, or Sea Cob*.

[For " Sea Mew," as applied to the Common Gull (*L. canus*), cf. Swainson (p. 207). For " Sea Cob," cf. Turner (p. 79).]

Larus, quem Cornubienses indigitant, a *Ganet*, forsan detorto nomine a Gavia vel a Gallicorum Gavian quod idem sonat, est par Anseri, palmipes, rostrum rotundum cæruleum, corpus grisei coloris, alte volat alausasq ; minores solas captat.

[No doubt one of the Skuas, probably the Great Skua (cf. Willughby, p. 348). Merrett states that it catches " alausas," by which he means pilchards, cf. p. 185, where the word is spelt " alosa."]

Puphinus Anglicus, *the Puphin*, G. 657. ex Insula Anglesey, & Cornubia, Anas Artica Clusii, & Fraterculus Ald. 3. 230. lin. 13, 14.

[Cf. Turner, p. 205, and Carew, Fol. 35.]

Fissipides Aquaticæ.

Ciconia, *the Stork*, I. 148. t. 50. Ald. 3. 311. G. 230. raro huc advolat.

[Cf. Willughby (p. 286) for an account of the specimen " taken on the coast of Norfolk " which he received from Sir Thomas Browne. Writing to Merrett September 13th, 1668' Browne states that he had seen two, " one in a watery marsh 8 miles of, another shott whose case is yet to bee seen."]

. Ardea Cineria, *the Ash coloured Heron, or Hern, Hernshaw*, I. 148. t. 50. Ald. 3. 378. Ardea pulla sive Cinerea, G. 187.

[Cf. Swainson, p. 144.]

Ardea alba, G. I. 152. t. 51. Ald. 390. G. 189, *a Mire Drumble.*

[This may either mean a white (albino) Heron, or the Spoonbill, which Merrett has already mentioned. (*Cf.* Turner, p. 39.) Merrett gives it the name " Mire Drumble," which was in the form of Mire Drum, or Mire Drumble, commonly applied to the Bittern (*cf.* Willughby, p. 25, and *Century Dict.*, Vol. V.). Charleton, apparently following Merrett, calls the Spoonbill a " Mire-drumbel."]

Ardea stellaris, *the Bittourn*, I. ib. Ald. 405. G. 100.

[Bittern, *cf.* Turner (p. 41), and Willughby (p. 25), where we are told that its common name was " Night-Raven." (*Cf.* Swainson, p. 146.)]

[p. 182] Ardea minor, I. tab. 56. quam ad me transmisit Ds. *Jenner*, ex agro Wiltoniensi.

[Possibly the Little Bittern.]

Avis pugnax, I. 154. t. 52. *a Rough*, est tertia in Tab.

Avis pugnax, quarta in dicta Tab. *a Reev.* utræq ; ex agro Lincoln. est fæmina superioris.

Hæmantopus mas & fæm. *Red shanks*, I. 154. t. 52.

Arquata, seu numenius, *the Curliew*, Ald. 3. 426. I. 152. t. 51. G. 197.

Arquata congener, *a Stone Curliew*, huic rostrum breve, accipitrinum, pennæ milvi, Phasiano par magnitudine, Dilicatissimæ avis ex agro Hantoniensi, Ds. *Hutchinson* Ornithopola Londinensis.

[The Stone-Curlew was found in Hampshire (*cf.* Gilbert White, XVth letter to Pennant), and still breeds in that county.)]

Vannellus, *the Lapwing, bastard Plover, or Pewit*, insula quædam ab iis nomen fortitur in Essexia : Huc enim migrant præcise ad diem Divo Georgio sacrum, vide *Fuller*, 318. I. 166. t. 53. Ald. 3. 526. G. 692.

[*Cf.* Turner (pp. 77 and 175). For accounts of this migration of the Pewet Gull on St. George's Day to the promontory of the Ness, or Naze, in Essex, *cf.* Charlton (p. 108), and Fuller, " Worthies of England " (Vol. 1, p. 494) :—

" There is an island of some two hundred acres, near Harwich, in the parish of Little Oakley, in the Manor of Matthew Gilby, esquire, called the Puet island, from Puets [Fuller was, of course, referring to the Black-headed, or Pewet Gull, *Larus ridibundus*] in effect the sole inhabitants thereof. Some affirm them called in Latin *Upupæ*, whilst others maintain that the Roman language doth not reach the name, nor land afford the bird. On Saint George's Day precisely, they pitch on the island, seldom laying fewer than four or more than six eggs."

This mention of migration on St. George's Day, April 23rd, coincides with the fact that this is the day on which cattle in Eastern Europe are shifted from winter to summer pasture.]

Vannello congener capella vulgari procerior, at Turdo minor, pennis Cæruleis & crista longa, ex Cornubia Ds. *Gunthorp.*

[Smaller than a Thrush . . . with a long crest ? If *minor* be a misprint for *major*, this is the Green Plover or Lapwing.]

Pluvialis cinerea, *the Grey Plover*, I. 166. t. 53. Ald. 3. 531. G. 647.

Pluvialis flavescens hujus meminit, I. pag. 165. sine Icone.

Pluvialis vulg. *the Whistling Plover, or green Plover,* in Ericeto Lincoln.

[This is the Golden Plover (*cf.* Willughby, p. 308 ; *cf.* Southwell, p. 20).]

Rallus Itallorum, I. p. 147. Ald. 3. 98. utribique sine icone.

[One of the Rails.]

[Page 183.] Trynga Ald. 3. 814. I. 166. t. 53. juxta stag. na Ichthyophylatica, & rivulorum margines, in agro Warwicensi, Mr. *Willoughby.*

[The Sandpiper (*cf.* Willughby, p. 301).]

Trynga paulo minor, Ald. 3. 482. I. 166. t. 53. idem.

Merulam aquat. vidit volantem in Cumberlandia Ds. *Willoughby,* Ald. 3. 486. I. 166. t. 53.

[The Dipper, also on p. 171 as Cornix Aquaticus *cf.* Willughby (p. 149).]

Charadrios ab incolis, *Sea Lark,* in littoribus Cambrobritannicis, præsertim in statione Belli Mauritii, I. 166. t. 53. Charadrios sive Hiatula, Ald. 3. 537. Ds. *Willoughby.*

[" Sea Lark," *cf.* Willughby, p. 310. Probably the Dunlin, or the Ringed Plover—possibly both (*cf.* Swainson, pp. 182 and 193).]

Grus, *the Crane,* I. 166. t. 45. Ald. 3. 329. G. 474.

[*Cf.* Turner, p. 97.]

Crex, *a Daker Hen,* est avis longis cruribus, cætera coturnici, (nisi quod major sit) similis, quæ in segete & Lino, vere, & in Principio æstatis non aliam habet vocem quam crex, hanc enim vocem semper ingeminat, quam ego Arist. Crecem esse puto, nusquam in Anglia nisi in sola Northumbria, vidi, &

audivi, & an sit eadem cum Ortygometra superius memorata nescio, suspicor tamen esse Turn, me talem vidisse & audivisse ad *Wheatley* quinq ; Oxonio Milliaribus memini.

[Also on p. 73 as Ralla—a Daker Hen. The Landrail, or Corncrake (*cf.* Turner, p. 71, and Willughby, p. 316). For Daker Hen, *cf.* Swainson (p. 177). It is still called the " Daker Hen " in Yorkshire and Lincolnshire.]

Tres sequentes· aves cum nominibus & breviusculis descriptionibus mihi communicavit Ds. *Hutchinson* Ornithopola Lond. quas se vidisse ait in agro Lincoln.

[" Ornithopola," = a dealer in birds.]

Non est avis aquatica querquedula paulo minor Rostrum ei rotundum, tenue & argastum, superius paululum incurvatum, toto ventre albes- [Page 184.] cit, dorsum nigrum, caput cristatum unde forsan ei nomen, sc. a monacha velata.

[" Non est," a misprint for " Nun est," as it appears in the 1666 edition. Merrett here, no doubt, refers to the Smew (*cf.* Ray, *Collection of English Words*, p. 95). It is called the Nun to this day (*cf.* Yarrell, Vol. IV., p. 499 ; Swainson, p. 165).]

Crickaleel, est priori æqualis, Cærulea in alarum supernis, caput collumq ; maculata, ad ingluviem coloris grisei inde deorsum albescit vel contra quoad ventris colorem.

[This may be the Garganey. " Crickaleel " may be onomatopæic. The Garganey is known as the " Cricket Teal," *cf.* Swainson (p. 158), who states that from its cry it is known as Cric Cric (Jura), Criquet (Savoy), Kriechentlein (Germany). Sir T. Browne says :—" We have a kind of teale which some fowlers call Crackling Teale " (*cf.* Southwell, p. 83).]

Gossander, palmipes & cristata ventre aureo, rostro longo & angusto, caro flavescit & cocta tota facessit in oleum, non est edulis, ex agro Lincoln. videtur esse Puphini species.

[Sir Thomas Browne's comment on this note is:—"Gossander videtur esse puphini species—worthy Sr that which we call a gossander & is no rare foule among us is a large well colourd & marked diving fowle most answering the Merganser. It may bee like the puffin in fattnesse and Ranknesse butt no foule is I think like the puffin differenced from all others by a peculiar kind of bill " (Letter to Merrett, December 29th, 1668 ; Southwell, p. 72).]

OLD ENGLISH NESTING BOTTLES.

EARTHENWARE bottles of the form shown in the accompanying photograph were in common use on the barns and other farm buildings in Kent and Sussex a century ago. They were put up in rows under the eaves, and their object was to facilitate the collection of the eggs and young of the House-Sparrows,

which were then universally destroyed, rewards being given for them in every parish. Probably this custom was equally common in other counties. Nesting bottles of this form may be seen depicted on the ends of cottages in some of Morland's pictures. Some forty years ago they were comparatively common in the part of Kent in which I live, but lately I had some difficulty in finding any that were entire.

Only half of the back of the bottle was earthenware, half

being left open to admit the hand, while a small nick was made in the upper half for the nail on which the bottle hung.

E. G. B. MEADE-WALDO.

BLACK REDSTARTS IN MERIONETH.

ON August 23rd a pair of Black Redstarts (*Ruticilla titys*) were seen at Tal-y-llyn, near Towyn, by Mr. H. N. Kirkby, who watched them at close quarters for some time. The hen bird was seen there again on the 30th of the same month. The species is rare in North Wales, but has been recorded three times previously in the same neighbourhood—in each case a single bird (*cf. Vert. Fauna N. Wales*, p. 82).

H. E. FORREST.

BLACK-HEADED WAGTAIL IN KENT.

A FINE male Black-headed Wagtail was shot at Fairfield, Lydd, Kent, on June 3rd, 1908. It was taken by me to the Natural History Museum, and identified as *Motacilla flava melanocephala*.

J. B. NICHOLS.

GREAT GREY SHRIKE IN SCOTLAND.

IN your last issue you mention twenty-two occurrences of the Great Grey Shrike (*Lanius excubitor*) in Scotland during the year 1907. Yet another specimen, unrecorded in that list, was shot at Long Hope, in Orkney, in the November of that year.

H. W. ROBINSON.

[The Duchess of Bedford informs us that a female was shot early in April, 1908, in the valley of the Palnure, two miles from the place where one was seen in the previous autumn, as recorded in Vol. I., p. 263.—EDS.]

TWO-BARRED CROSSBILL IN SUSSEX.

IT may be interesting to record that a fine pair of Two-Barred Crossbills (*Loxia bifasciata*) were shot together at Penhurst, near Ashburnham, Sussex, on March 10th, 1908. The cock is in fine red plumage, the hen in yellow. They were seen in the flesh by Mr. W. R. Butterfield after being sent to Mr. Bristow of St. Leonards. They are now in my collection.

J. B. NICHOLS.

TUFTED DUCKS NESTING IN THE OUTER HEBRIDES.

IN the references by MS. Marginal Notes in a copy of MacGillivray's " British Birds," the statement by Dr. C.

Gordon is clear enough in itself.   Dr. C. Gordon was a personal
friend of MacGillivray.   The repetition in my "Fauna of
the N.W. Highlands and Skye," as regards their almost complete disappearance for some years, is also correct, and was
upon the authority of personal investigations, and also upon
the authority of Mr. D. Guthrie, who, by the date of the
notes in " The Annals," 1896, pp. 3-22, had been some seventeen or eighteen years head-keeper to Sir Reginald Gordon
Cathcart, in South Uist.   I quoted in the first instance from
the annotated copy of MacGillivray's, which was lent to me,
but the second time from memory of the passage.

There appears to be little mystery—or none at all—in the
sequence of the accounts of the Tufted Ducks in the Outer
Hebrides.   Dr. C. Gordon as early as 1851, when he dates his
marginal notes in the fifth volume of MacGillivray's " British
Birds," spoke of the Tufted Duck as " common and plentiful "
in South Uist during the winter (vide Annals S.N. Hist., 1896,
pp. 3-22).

Mr. D. Guthrie, however, a most careful and capable
observer, reported this species as much scarcer in years
subsequent to 1892 ; and he had been resident in South
Uist at that time since about 1874.

In 1893 actual record of nesting took place in South Uist, and
Mr. Guthrie verified some of his previous statements of its
doing so, and sent me an egg taken from a nest by himself.
Four pairs were known to breed in 1906, and one pair in 1907,
by Bahr and Kinnear.   Mr. Guthrie also had spoken of the
Tufted Duck having been in unusual numbers in South Uist in
the winter of 1902-3.

J. A. HARVIE-BROWN.

DISTRIBUTION OF THE COMMON SCOTER IN
SCOTLAND.

WITH reference to Mr. Harvie-Brown's note (antea, p. 134)
on the distribution of the Common Scoter in Scotland, it may
be of interest to state that a large flock of Common Scoters
was seen off the south end of the island of Graemsay, in
Orkney, during the first week in March this year.

With one exception, viz., a single adult male seen in company with an old Goldeneye drake on the Loch of Harray,
among a large and widely scattered flock of Pochard drakes,
on the last day of February, 1905, this is the only time I,
personally, have come across the Common Scoter in Orkney
in winter.   Whether they occur on and around the island of
Tiree in the Inner Hebrides I cannot say, but I spent the

greater part of three consecutive winters on this particular island without seeing the species. However, they might easily have been there for all that, as, owing to the local lochmen being unwilling to go more than a quarter of a mile from land, and then only in very calm weather, I was rather handicapped as far as my observations of the sea Ducks were concerned.

H. W. Robinson.

PALLAS'S SAND-GROUSE IN CHESHIRE.

On or about the 11th of June, 1908, two Sand-Grouse (*Syrrhaptes paradoxus*) were observed in a field of roots at Wythenshawe, Cheshire, by Mr. H. V. MacMaster. Their plumage and "pigeon-like" heads at once attracted his attention, and he stood for some time at a distance of about thirty-five yards from them watching them feeding. When he approached a little nearer, one of the birds got up and called "chack, chack," and then both flew away with remarkably rapid and strong flight, which reminded him of the flight of the Golden Plover, a bird which is common on the Withenshawe fields in winter. Mr. MacMaster, though he was struck with the long wings and tails of the birds when they rose, is not prepared to say whether they were a male and female.

T. A. Coward.

SUPPOSED BLACK GROUSE AND PTARMIGAN FROM IRISH CAVES.

The mistake over this subject in the "Irish Naturalist" (1899, pp. 17 and 37) has unfortunately been adopted in BRITISH BIRDS (*antea*, p. 127). As I pointed out in "The Birds of Ireland" (p. 231), I had the able assistance of Mr. E. T. Newton and Dr. Forsyth Major, as well as of Dr. Scharff to determine the humerus from the Ballynamintra Cave, in co. Waterford, of which I was the finder. The conclusion arrived at was that this bone agrees far more closely with that of a common fowl, and as it was found in the superficial stratum, I have no doubt it was brought in by a fox in recent times. It can be seen in the Dublin Museum, where it is labelled *Gallus*. Among the numerous bones of birds found by me during the past eight years in the caves of Sligo, Clare and Cork, and which Mr. E. T. Newton has kindly determined for us, the Black Grouse is not represented, and I know of no evidence that it was indigenous in Ireland.

As regards the supposed bones of Ptarmigan, these also

were compared by Dr. Forsyth Major and Dr. Scharff with bones of the several species of Grouse, and they were found to agree better in some respects with those of Red Grouse than with those of Ptarmigan. Some bones from Kish Cave, co. Sligo, have been referred by Mr. E. T. Newton to Red Grouse (?) or Ptarmigan (?). The former is common in Ireland while the latter is unknown.

R. J. USSHER.

THE OYSTER-CATCHER'S METHOD OF FEEDING ON THE EDIBLE MUSSEL.

THE systematic methods adopted by Oyster-catchers in abstracting mussels from their shells are admirably detailed by Mr. J. M. Dewar in the " Zoologist " for June.

It is somewhat surprising to find that no shells larger than $1\frac{5}{8}$ inches by $\frac{7}{8}$ inch were found opened, while shells less than 1 inch by $\frac{1}{2}$ inch were swallowed whole.

The larger shells are dealt with in a most methodical manner. No attempt is made to attack them when their valves are closed : hence, those left high and dry by the tide, or in rain-water pools, are always passed over unmolested. The shell must be more or less gaping to arrest attention. As everyone knows who is familiar with mussel-scalps, these molluscs assume varied positions, sometimes presenting the ventral, sometimes the dorsal border, uppermost, and sometimes one end of the shell. And of these positions, shells with the dorsal borders uppermost are most sought for, no less than 78 per cent. of the empty shells left by Oyster-catchers having occupied this position at the time of attack.

Each mussel is approached in the line of its long axis, and generally, for some inexplicable reason, this approach is made " from the front." Should the shell be slightly gaping a tentative tap is given, as if to ascertain whether the slit is large enough for the beak to enter. If the experiment is favourable, the beak is thrust home by a series of jerks, forcible and rapid. When the blow is delivered a little to one side, so as to force inwards a portion of one side of a valve, more deliberation is displayed, which suggests that the abstraction of the animal from its case is a matter of certainty, the body being dragged out through the hole made, in spite of the closing of the valves.

The author describes, in great detail, a number of methods in the use of the beak as a lever, after it has once been thrust down between the valves. These we cannot repeat, but it should suffice to say that the simplest method employed is to shake the beak violently from side to side till the valves are laid open by the fracture of the adductor muscles. Another method is to turn the beak through a quarter of a circle, either by walking round the victim, or turning the head in the neck. In yet another, the head is lowered almost to the ground, and the point of the bill is thrust between the valves ; the bird then moves its head to the left whereby the two valves are forced apart.

Only about 9 per cent. of shells are opened through the ventral borders, which may be accounted for by the fact that this border is generally undermost. It is a noteworthy fact indeed that these buried shellfish are found at all ; often they are discovered when buried by a layer of sand or mud as much as an inch in depth. In their search for this buried treasure, the bill is used as a sort of divining rod, the ground being tapped here and there, until a victim is found.

Some 13 per cent. of shells are opened through their posterior ends. Many buried shells are opened in this fashion. Indeed, the author declares that this can be " the only route to the interior of the buried shells, the long axes of which are vertical." This statement, however, requires some qualification, since he also contends that " more mussels are opened by way of the ventral borders when buried than when exposed to view."

Finally, Mr. Dewar contends that he has " brought forward observations which seem to prove that the Oyster-catcher, far from being actuated by blind impulse, on the contrary proceeds deliberately to remove certain structures (the adductor muscles) which hinder the achievement of their desires." It may be questioned whether this is not placing to the credit of the Oyster-catcher a degree of intelligence which it does not possess.

KILLDEER PLOVER IN KENT.

On April 21st, 1908, Mr. Bristow informed me that he had seen the previous day three strange Waders on one of the " fleets " in Romney Marsh, not far from Lydd. He was not sure of their identity, but was struck by their unusually long tails. The following day all doubt was set at rest by the receipt of one of them, which he at once brought to me in the flesh. It was an adult specimen of the Killdeer Plover (*Ægialitis vocifera*), and had been shot by a shepherd at the place where the three were seen. The second was shot on

April 21st, and the third on the 22nd, these I did not see until after they had been mounted.

These are the first of this common American species that have been killed in Kent, and bring up the British-

Killdeer Plover, shot near Lydd, Kent, on April 21st, 1908.

taken examples to six. One is now in the collection of Mr. J. B. Nichols, and a second in that of Mr. C. J. Carroll.

N. F. TICEHURST.

[We are much indebted to Mr. J. B. Nichols for the loan of his specimen, and for the permission to reproduce the accompanying photograph of it.—EDS.]

SOLITARY SANDPIPER IN KENT.

WITH reference to the Duchess of Bedford's note in the last number of BRITISH BIRDS (p. 136), the Solitary Sandpiper was shot by a visitor at Littlestone, and therefore at no great distance from where she saw it on August 15th. Mr. Bristow received it two days later, and kindly brought it to me in the flesh. From its condition it had evidently passed unrecognised, and it looked as though it had been shaken up in a " game " bag with other birds : it was soddened with blood and melted fat, sand and sea-water, and so was a very sorry-looking object, but its tail-feathers and axillaries gave unmistakable proof of its identity.

I may, perhaps, draw attention here to the great immigration of Waders that took place on the Kent coast at the end of July, of which the Duchess of Bedford's notes give evidence. Mr. M. J. Nicoll informs me of Ruffs and other species seen by him about the same time, and on the night of July 23rd, when at Folkestone, I heard large numbers of Waders, chiefly Sanderlings, passing over the town for two hours or more.

N. F. TICEHURST.

LATE NESTS OF THE GREAT CRESTED AND LITTLE GREBES.

ON August 12th I visited, with a friend, one of the Surrey breeding haunts of the Great Crested Grebe, and we were fortunate in discovering a nest containing four eggs, which were only very slightly stained, and certainly had not been incubated more than a week. I think it is somewhat exceptional to find eggs in the nest during August, as this species does not appear to be double-brooded. That the Little Grebe is so is well-known, but I think it worth recording that on July 27th we found a nest with a fresh egg (another was laid the next day) belonging to a pair of birds, which were accompanied by chicks not more than a day or two old.

A. G. LEIGH.

* * *

MARKING BIRDS.—Dr. Otto Herman, Director of the Hungarian Central Bureau for Ornithology, informs us that he has begun marking young Storks, Herons, Gulls and Swallows, by means of an aluminium ring which is fastened around the leg of the bird and bears the inscription "Budapest," followed by a number which corresponds to the entry in a register book. Should anyone capture a bird so marked he is requested to send the ring to the Hungarian Central Bureau for Ornithology, Jozsef-korut, 65, Budapest VIII., Hungary, accompanied by a notice stating the locality, time and particulars of capture.

BIRD-LIFE IN DUBLIN BAY. — Under this title Mr. Alexander Williams gives an interesting account of the changes in the sea and shore bird-life of the vicinity of Dublin during the last twenty-five years (*Irish Nat.*, 1908, pp. 165-170).

How to Attract and Protect Wild Birds. By Martin
Hiesemann. Translated by Emma S. Buchheim, with an
introduction by Her Grace the Duchess of Bedford.
(Witherby & Co.) Illustrated. 1s. 6d. net.

THE purpose of this little book is to set forth the methods
employed by the Baron von Berlepsch to provide suitable
nesting-places and food for various birds, and to protect them
from their enemies. Wonderful success has attended these
methods at Seebach, where exhaustive experiments have been
made for many years by Baron von Berlepsch. The
statement that "we can only preserve and increase our
birds by restoring . . . . . the opportunities for nesting
of which we have robbed them " is perhaps more applicable in Germany, where high forestry has robbed many
birds of nesting-places by the cutting down of decaying
trees and undergrowth, than it is in England. At the same
time the fact that the number of birds can be actually
increased by providing them with suitable nesting-
places is a most interesting one, and is sufficiently sub-
stantiated by the experiments here described. All our
readers are probably well aware of the value of nesting-boxes
as means of attracting such birds as Tits, Nuthatches and
Wrynecks, but we have never heard of Woodpeckers nesting
in boxes in England as they do in Germany. This may be
due to the fact that old timber is much more plentiful
in this country, but we are inclined to think that if the
Berlepsch box were adopted under the conditions so carefully
described in this little book, even Woodpeckers would
be induced to nest in them. This nesting-box has been
designed and is manufactured with elaborate care. After
exhaustive experiments, the Baron made the most interesting discovery that all the holes made by the various
species of Woodpeckers are formed on a uniform plan.
Special machines have at length been constructed to produce
"boxes" which are faithful imitations of the Woodpecker's
nesting hole down to the smallest detail, and the use of these
has met with remarkable success. Equally interesting are
the methods here described of pruning and growing bushes in
various ways to make them attractive to birds for nesting
purposes, and also of feeding birds in winter in the most
effective way at a minimum of cost. We may hope that the
methods here described will be adopted so universally that
people will compete as to how many nests they have in their
gardens rather than as to how many birds they have caught
or killed.

BRITISH BIRDS

AN·ILLUSTRATED·MAGAZINE
DEVOTED·TO·THE·BIRDS·ON
THE·BRITISH·LIST

NOVEMBER 2,
1908.

Vol. II.
No. 6.

MONTHLY·ONE·SHILLING·NET
326·HIGH·HOLBORN·LONDON
WITHERBY & C?

BRITISH·BIRDS

EDITED BY H. F. WITHERBY, F.Z.S., M.B.O.U.
ASSISTED BY W. P. PYCRAFT, A.L.S., M.B.O.U.

SOME EARLY BRITISH ORNITHOLOGISTS AND THEIR WORKS.

BY

W. H. MULLENS, M.A., LL.M., M.B.O.U.

IV.—MARTIN MARTIN

(*Ob.* 1719).

THE islands of the Outer Hebrides have from an early
period attracted the attention of the traveller and the

naturalist. The romantic wildness of their situation, their difficulty of access, and the strange manners and customs of their sequestered population, have all appealed strongly to the curious inquirer, and we thus have a considerable mass of information concerning them and, incidentally, their natural history, compiled at a time when the fauna of far more accessible and perhaps important districts remained neglected and unrecorded.

Far out in the wild Atlantic, over one hundred miles from the mainland of Scotland, lies the lonely island of St. Kilda, the " Hirta " of the ancients. Although mentioned briefly by Joh. de Fordun (*ob. circa*, 1380) in his " Scoti-chronicon," and by Boethius (1465-1536) in the " Scotorum Historia," published in 1527, the first detailed account we have of the Island of St. Kilda, and certainly the first made from personal observation, is that dealt with in the present article. It was prepared by Martin Martin, a factor of the Clan Macleod, who in the year 1697, in the summer season and " to the almost manifest hazard of the author's life," visited the island in company with Mr. John Campbell, minister of Hawis.*

During Martin's stay in St. Kilda, which extended over three weeks, he devoted a certain amount of time to the observation of the birds of the island, and amongst them to the Garefowl, or Great Auk, and it is chiefly owing to his description of this extinct and famous bird that Martin's book—curious and entertaining as it otherwise is—is of such interest to the naturalist of the present day.

Of Martin Martin we know but little. He was born, as we are told in the preface to his book, " A late Voyage to St. Kilda," " in one of the most spacious and fertile isles in the west of Scotland* ; and besides his liberal education at the University, had the advantage of seeing foreign places, and the honour of conversing with some

---

* For further particulars as to the early history of St. Kilda, *vide* Seton's "St. Kilda, past and present." Edinburgh, 1 vol., 8vo, MDCCCLXXXIII.

† Possibly the Isle of Skye.

of the Royal Society, who raised his natural curiosity to survey the isles of Scotland more exactly than any other " ; . . . . Martin took his degree of M.A. at the University of Edinburgh in 1681, and subscribed his name to the customary oath as " Martinus Martin," and he seems to have died in 1719. In addition to his voyage to St. Kilda, Martin also published a more extensive work entitled " A Description of the Western Islands of Scotland," London, 1703, 1 vol., 8vo, which contains several short notices of the birds of the different islands. This book the great Dr. Johnson had studied before he made his tour to the Hebrides with the faithful Boswell in 1773. There is a copy of this work in the Advocates' Library,* on the title page of which is endorsed the following :—

" This very book accompanied Mr. Samuel Johnson and me in our Tour to the Hebrides in Autumn, 1773. Mr. Johnson told me that he had read Martin when he was very young. Martin was a native of the Isle of Sky, where a number of his relations still remain. His book is a very imperfect performance ; and he is erroneous as to many particulars, even some concerning his own island. Yet as it is the only Book upon the subject it is very generally known. I have seen a second edition of it. I cannot but have a kindness for him notwithstanding his defects.

16 April, 1774. JAMES BOSWELL."

In Boswell's " Life of Johnson " we are told that the " great lexicographer " was at first pleased to approve of Martin's work, but that afterwards he changed his opinion and hurled at the unfortunate author one of his ponderous bolts : " No man now writes so ill as Martin's ' Account of the Hebrides ' is written. A man could not write so ill, if he should try." Though surely poor Martin had done his best to disarm hostile criticism by informing us in his Preface that :—

" This (i.e., " The Natural History of 'em ") I had a

* Cf. Seton's " St. Kilda," p. 18.

particular regard to in the following description, and have everywhere taken notice of the Nature of the Climate and soil, of the Produce of the places by sea and Land . . . . and that in such variety as I hope will make amends for what Defects may be found in my stile and way of Writing ; for there's a Wantonness in Language as well as in other things . . . ."

A second edition of this book was published in London. 1716, " very much corrected." To come, however, to his more important work, the full title is as follows :—

A late / Voyage / to / St. Kilda, / The Remotest of all the / Hebrides, / or / Western Isles of Scotland. / With / A History of the Island, Natural, Moral, / and Topo- graphical. Wherein is an Account of their / Customs, Religion, Fish, Fowl, &c. As also a Rela- / tion of a late Impostor there, pretended to be / Sent by St. John Baptist. / By M. Martin, Gent. / London : / Printed for D. Brown, and T. Goodwin : At the Black Swan and / Bible without Temple-Bar ; and at the Queen's Head against / St. Dunstan's Church in Fleet Street. MDCXCVIII.

1 Vol. 8vo.

Collation : 1 p. Short Title + 1 p. Title, reverse of both blank, + pp. 2, Address, + pp. 4, Preface, + pp. 4, Contents, all unnumbered, + pp. 158, map,* and plate of two birds to face p. 53.

This, the first edition, of which a facsimile title page is given opposite, is rare.

The second edition is said to have been published in 1716.

The fourth and best edition, which was reprinted in Pinkerton's " Collection of Voyages and Travels," was published in London in 1753. 1 Vol., 8vo.

This latter work, according to " The History of the Works of the Learned," Vol. V., was " very agreeable to the curious, especially to such as have any true taste for natural and experimental philosophy."

* A fac-simile of the map is reproduced on page 179.

A LATE

VOYAGE

TO

St. KILDA,

The **Remoteſt** of all the

H E B R I D E S,

O R

Weſtern Iſles of SCOTLAND.

W I T H

A Hiſtory of the Iſland , Natural , Moral, and Topographical. Wherein is an Account of their Cuſtoms , Religion , Fiſh , Fowl, &c. As alſo a Relation of a late I M P O S T O R there, pretended to be Sent by St. *John Baptiſt.*

By M. M A R T I N, *Gent.*

L O N D O N:

Printed for *D Brown,* and *T. Goodwin*. At the *Black Swan* and *Bible* without *Temple-Bar* ; and at the *Queen's Head* againſt St *Dunſtan's* Church in *Fleetſtreet*. M DC XC VIII.

The collation of the fourth edition is as follows :—
pp. 4 unnumbered + pp. 79. The last two wrongly
numbered 70 and 63 respectively. Frontispiece (a map
and figure of two birds).

Martin treats of the birds of St. Kilda in pp. 46-67
of the first edition, and in pp. 26-36 of the fourth, the
accounts in both editions being nearly identical.

Amongst the land birds he enumerates :—

" Hawks extraordinary good, Eagles, Plovers, Crows,
Wrens,* Stone-Chaker, Craker, Cuckoo."

Of the sea fowl, however, as may be expected, he gives
us a fuller description ; and he thus commences it with
his historic description of the Great Auk :—

" The Sea-Fowl are, first, Gairfowl, being the stateliest
as well as the largest Sort, and above the size of a Solan
Goose of a black colour, red about the Eyes, a large white
spot under each, a long broad Bill ; it stands stately,
its whole Body erected, its wings short, flies not at all ;
lays its egg upon the bare Rock, which if taken away,
she lays no more for that Year ; she is whole-footed,
and has the hatching Spot upon her Breast, *i.e.* a bare
spot from which the Feathers have fallen off with the
Heat in hatching ; its Egg is twice as big as that of a
Solan Goose, and is variously spotted Black, Green,
and Dark ; it comes without regard to any Wind,
appears the first of May, and goes away the middle of
June." †

Martin further records the fact that the inhabitants
of St. Kilda made use of " the Bones, Wings, and
Entrails of their sea-fowls " to add to the composts of
straw and ashes with which they manured their lands,
and this and the fact that they consumed the eggs and
flesh of the Garefowl may have contributed to its

---

* The list of Land birds is given as it stands, and it will be noted
that though Martin mentions the Wren, he does not describe it.

† This quotation is from the fourth edition, the description of the
Gair-fowl in the first edition is almost word for word the same, but a
trifle more obscure, and has the amplification, " he is Palmipes, or
whole-footed."

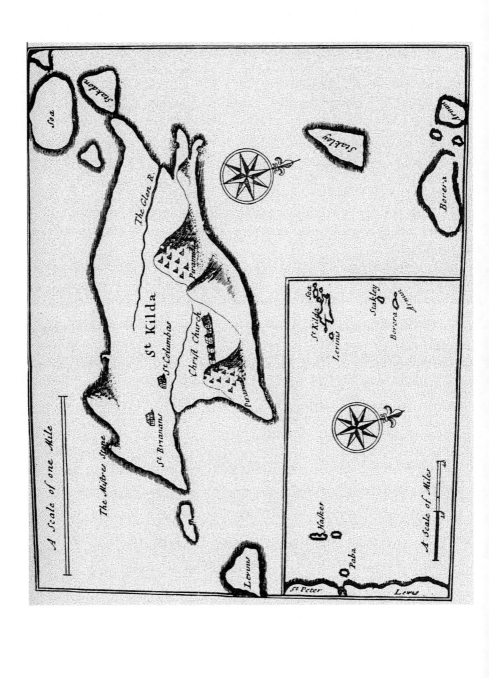

extermination (cf. Symington Grieve, *The Great Auk*, London, 1885, 1 Vol., 4to, pp. 76 and 119).

Robert Gray, in his invaluable work, " The Birds of the West of Scotland " (Glasgow, MDCCCLXXI., 1 Vol., 8vo), p. 442, says, " It is, I think, doubtful whether Martin ever saw the bird, as in another and larger work entitled, ' A Description of the Western Islands of Scotland,' published five years afterwards, and in which he gives a full account of St. Kilda and its birds, he does not even mention it, but it should be noted that the description of the birds in this book is not in any way so complete as that in the same author's ' A late Voyage to St. Kilda.' "

In this work Martin gives a considerable account of the Solan Goose, and amongst other curious statements, tells us :—

" The Solan Geese are always the surest sign of Herrings, for where-ever the one is seen, the other is always not far off.   There is a Tribe of Barren Solan Geese which have no Nests, and sit upon the bare Rock ; these are not the Young Fowls of an Year old, whose dark colour would soon distinguish them, but old ones, in all things like the rest ; these have a Province, as it were, allotted to them, and are in a separated state from the others, having a Rock two hundred Paces distant from all other ; neither do they meddle with, or approach to those Hatching, or any other Fowls ; they sympathize and Fish together ; this being told me by the Inhabitants, was afterwards confirmed to me several times by my own observation " (1st ed., p. 52).

And of the Fulmar :—

" And when the young Fulmar is ready to take Wing, he being approached, ejects a quantity of pure Oyl out at his Bill, and will make sure to hit any that attacks him, in the Face, though seven Paces distant . . . . but the Inhabitants take care to prevent this by surprizing the Fowl behind, having for this purpose a wooden dish fixed to the end of their Rods, which they hold before his Bill as he spouts out the Oyl ; they surprize him also from behind by taking hold of his Bill, which they tie with a thread, and upon their return home they untie it with a Dish under to receive the Oil . . . ." (p. 56, *op. cit.*).

Among his observations on the nesting habits of the sea birds the following passage may be quoted :—

" Every Fowl lays an Egg three different times (except the Gair-fowl and Fulmar, which lay but one) ; if the First or Second Egg be taken away, every Fowl lays but one other Egg that Year, except the Sea-Malls, and they ordinarily lay the Third Egg, whether the First and second Eggs be taken away or no " (p. 64, *op. cit.*).

The supply of sea-fowl was of course a most important factor in the life of the island, and Martin computed the consumption of Gannets alone as follows :—

" We made particular Enquiry after the number of Solan Geese consumed by each Family the Year before we came there, and it amounted to Twenty two thousand five hundred in the whole Island, which they said was less than they ordinarily did, a great many being lost by the badness of the season, and the great Current into which they must be thrown when they take them, the Rock being of such an extraordinary Height, that they cannot reach the boat " (p. 115, *op. cit.*).

Only the briefest notice can here be made of the manners and customs of the inhabitants of St. Kilda, " of their dexterity in climbing," in which " custom had perfected them, so that it is become familiar to them almost from their cradles ; the young boys of three years old begin to climb the walls of their Houses "—" of the beauty of their voices and the soundness of their lungs," to which " the Solan Goose Egg supp'd Raw doth not a little contribute." How they possessed but one steel and tinder-box among a population of one hundred and eighty souls ; and how their native ignorance alone prevented them from being the most fortunate of mankind.

" There is this only wanting to make them the Happiest

People in this Habitable globe, viz., That they themselves
do not know how happy they are, and how much they
are above the Avarice and Slavery of the rest of
mankind."

Enough perhaps has been quoted to show the nature
of this somewhat rare and curious book, the precursor
of many others dealing with St. Kilda and the Scottish
Islands. Among which may be mentioned the Rev. A.
Buchan's "Description of St. Kilda, the most remote
Western Isle of Scotland," published in Edinburgh,
1741 ; the Rev. Kenneth Macaulay's " A Voyage to and
History of St. Kilda," London, 1764 ; an anonymous
work entitled " A Voyage to Scotland, the Orkneys and
the Western Isles of Scotland," London, 1751 ; and the
" Travels in the Western Hebrides : from 1782 to 1790,"
London, 1793, by the Rev. John Lane Buchanan [in no
way to be confounded with George Buchanan (1506-1582),
the Scotch historian], which last work affords us the
pleasing statement that :—

" The Gare Fowl is four feet long, and supposed to
be the pigeon of South America."

And so farewell to Martin Martin ; would that he had
noted more of what he saw in St. Kilda when he set out
for that almost " unknown land," having, as he tells us,
" embark'd at the Isle of Esay in Hawies the 29th of
May, at six in the Afternoon, 1697.   The Wind at S.E."

NESTING HABITS OF THE MARSH-WARBLER.

BY

PERCY F. BUNYARD, F.Z.S., M.B.O.U.

MR. WARDE FOWLER has described* the nesting habits
of the Marsh-Warbler (*Acrocephalus palustris*) so admirably that perhaps little that is new can be added
to his observations. Nevertheless, the bird is so rare
as a breeding species in this country that a short account
of my experiences with two pairs this summer may be
of interest. On June 26th I received a telegram from
a friend in Somersetshire to say that he had found a
nest with eggs, and had also another pair of birds under
observation. In the hope of hearing the birds singing
we were on the spot where the nest had been located
(elevation 500 ft. above sea-level) just after sunrise, but
only heard a few notes during a wait of some few minutes.
My companion then took me to the nest, which was
situated in a broad, rough, overgrown hedge (composed
of whitethorn and elder) on the side of a lane, and
bordering a field of wheat. The nest was on the field side
of the hedge, about three feet from the ground, and was
beautifully concealed (a good deal of herbage was removed
for the purpose of photographing). The nest was supported by two stems of bracken, and one of stinging nettle,
round which the nest had been built. This nest resembled
in general appearance that of a Whitethroat much more
than that of a Reed-Warbler. It was loosely constructed
on the exterior, and the interior was well and evenly lined.
The materials used in the exterior were rather coarse grass,
two pieces of frayed-out cotton, and one small feather
(possibly from one of the birds). Finer grass was used
as the lining was approached, and this was composed of

* " Zoologist," 1906, pp. 401-9.

fine fibrous roots, a single flowering head of grass (still green), and a very little horsehair.   The outside measurements of the nest were : depth, 3½ in. ; diameter, 4 in., extending to 5 in. where the nest was built round the stems of the supports, tapering downwards to the centre almost

Nest of Marsh-Warbler in Somerset, June 28th. 1908.

to a point ; interior diameter, 2 in. ; depth, 1½ in.   This nest was originally located on June 10th by watching the birds building, and at the time of my visit, June 28th, it contained four eggs in an advanced state of incubation. They are quite typical, and cannot be confused with the eggs of the Reed-Warbler, and indeed I have not yet seen well-authenticated eggs of the Marsh-Warbler which could easily be mistaken for the eggs of the Reed-Warbler. The British-taken eggs of *A. palustris* appear to me rather larger and less pointed than Continental eggs, and it would be interesting to know if others have noticed this. I re-visited this nest about 11 a.m. for the purpose of taking the photograph here reproduced. The bird was sitting, but slid off quietly on my approach, and although I remained in the vicinity of the nest (in the hope of getting a photograph of the bird itself) for nearly two hours, the alarm-note was only uttered once, and the birds were nowhere to be seen. I was disappointed in not seeing and hearing more of the birds, and their extreme shyness and quietness struck me more than anything in connection with this interesting experience.

On the evening of the same date we visited a small osier bed (150 ft. above sea-level) in which some other Marsh-Warblers had been previously located, and after watching for some time we saw the birds continually diving down among the rough growth near a large plant of the cow-parsnip, in which we afterwards found a nest containing five newly-hatched young. This nest was supported by three stems of the plant, and was similar to the one just described. The mouths of the young were of a beautiful rich lemon-yellow, and on the back of the tongue were two conspicuous black spots, placed horizontally.

ON THE DOWN-PLUMAGE AND MOUTH-COLORATION OF SOME NESTLING BIRDS.

BY

C. B. TICEHURST, m.a., m.r.c.s., l.r.c.p., m.b.o.u.

OF all the books which have been written on British birds not one, as yet, has dealt satisfactorily with the question of the sequence of plumages and, so far as I know, none give even the barest description of the down or natal plumage of even our commonest birds.

Mr. Pycraft deplored this fact, and in the course of two excellent articles (*vide antea*, Vol. 1, pp. 102 and 162) gave a brief outline of the different kinds of down-plumage recognisable, and made some remarks upon their significance, at the same time suggesting that further investigation into the matter would be valuable.

The sequence of plumages is a study which has long interested me; and I am certain that the collection of a large amount of material in reference to this subject, as well as on the coloration of the mouths of nestlings, as suggested by Mr. Pycraft (*cf.* Vol. I., p. 129) would, when worked out on comparative lines, yield some important results relating to the question of morphological ornithology.

As Mr. Pycraft has already explained (*cf.* Vol. I., p. 162) the different types of down, I shall here only state that in all the Passerine birds which I have examined, the type of down present is that of the pre-penna, and belongs to the mesoptyle generation. These pre-pennæ, I need hardly remark, are not distributed all over the body, but are arranged in definite tracts. Further, the development of the pre-pennæ in these tracts varies considerably in different genera, and even in the different species of the same genus.

The *inner supra-orbital tracts* consist of few pre-pennæ which are situated above the eye on each side, and from thence pass backwards, each tract forming, in most species, a line or crescent of down.   It will be noticed that the inner supra-orbital tracts are present in every species which I have examined which has down at all.

The *outer supra-orbital tracts* consist of two or three small, short pre-pennæ on each side, situated between the edge of the upper eyelid and the inner supra-orbital tract.   They are present in the Mistle-Thrush, Meadow and Red-throated Pipits, Chaffinch, and Brambling.

The *occipital* tracts consist of two or three fairly large, well-developed, pre-pennæ situated on each side of the occiput.

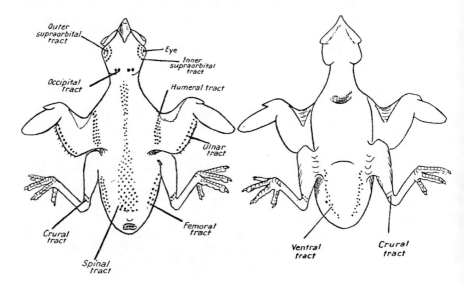

Diagram showing the Down-tracts of Nestling Birds.

The two tracts usually form a line, or crescent, of down when well developed.   They are present in every species examined which has any down.

The *humeral tracts* are usually well developed, and run obliquely downwards and outwards from the base of the neck across each humerus just in front of the shoulder joint.   They are replaced by the " scapular " feathers of the juvenile plumage.   They are present in all the birds that I have examined which have down, except the Wren (see note under this species).

The *spinal tract* runs down the centre of the dorsum from about the level of the shoulder joint to the end of the sacrum in those species in which it is well developed.   The length varies in different species, in some the anterior part being slightly developed or absent, in others the posterior part. The breadth is greatest in the lumbar region. This tract is present in all species examined except the Blue Tit (see note under this species).

The *ulnar tracts* consist of small pre-pennæ on the ulnar margin of each wing. Each pre-penna is replaced later by the secondaries, and in some cases by the secondary coverts. These tracts are absent in the Wheatear, Robin, all four Tits, and Wren.

The *femoral tracts* are situated laterally on each side just beneath the femur. The pre-pennæ never seem to be long, and are more closely approximated to each other than in some of the other tracts. This tract is absent in the Thrushes and Robin, the Tits, Swallow, and Sand-Martin.

The *ventral tract* is situated laterally on each side of the abdomen, and runs obliquely from the middle line towards the upper end of the femoral tract. In character it resembles the femoral tract, and is widest posteriorily. This tract is present in the Meadow-Pipit, Starling, and in all the *Fringillinæ* examined which have down.

The *crural tract* consists of a few small inconspicuous pre-pennæ forming a circle round the lower end of the crus, just above the ankle joint. It was noted in the Red-throated Pipit, all the *Fringillinæ* examined which have down, and in the Snow-Bunting.

I am indebted to my friend, Mr. J. L. Bonhote, for notes or specimens of the Bearded Tit, Red-throated Pipit, Tree-Sparrow, Brambling, Lesser Redpoll, Snow-Bunting, and Kingfisher.

MISTLE-THRUSH *Turdus viscivorus* L.

Down. *Colour.*—Greyish white, some pre-pennæ having buffish white tips.

Distribution.—Inner and outer supra-orbital, occipital, humeral, spinal and ulnar. In some there is a pre-penna on the bastard wing. The outer supra-orbital tract is not found in the Blackbird or Song-Thrush.

Coloration of Mouth. Inside, orange; no spots; externally, flanges lemon-yellow.

SONG-THRUSH *Turdus musicus* L.

Down. *Colour.*—Buffish white.

Distribution.—Inner supra-orbital, occipital, humeral, spinal and ulnar.

Coloration of Mouth. Inside, orange; no spots; flanges lemon-yellow.

BLACKBIRD *Turdus merula* L.

DOWN. *Colour.*—Greyish white.
Distribution.—Inner supra-orbital, occipital, humeral, spinal and ulnar.
COLORATION OF MOUTH. Inside, orange ; no spots.

WHEATEAR *Saxicola œnanthe* (L.).

DOWN. *Colour.*—Dark grey.
Character.—Moderate length.
Distribution.—Inner supra-orbital, occipital, humeral, spinal and femoral. It will be noted that there is no ulnar tract, and the spinal tract is a very short one, confined to the middle of the dorsum.
COLORATION OF MOUTH. Inside, orange ; no spots.

REDBREAST *Erithacus rubecula* (L.).

DOWN. *Colour.*—Dull jet-black.
Distribution.—Inner supra-orbital, occipital, humeral, and spinal. Here also there is no ulnar tract, and the short spinal tract does not reach the sacrum.
COLORATION OF MOUTH. Inside, orange ; no spots.

WHITETHROAT *Sylvia cinerea* Bechst.

DOWN. Absent.
COLORATION OF MOUTH. Inside, yellowish orange ; one brownish black spot at the base of the tongue on each side.

LESSER WHITETHROAT *Sylvia curruca* (L.).

DOWN. Absent.
COLORATION OF MOUTH. Inside, orange. Tongue spots as in *S. cinerea.*

GARDEN-WARBLER *Sylvia hortensis* (Bechst.).

DOWN. Absent.
COLORATION OF MOUTH. Inside, deep pink with a violet tinge ; one brownish oval spot on each side of the base of the tongue.
N.B.—The absence of down in these three species of the genus *Sylvia* is noteworthy.

HEDGE-SPARROW *Accentor modularis* (L.).

DOWN. *Colour.*—Greyish black.
Character.—Fairly long, and well developed.
Distribution.—Inner supra-orbital, occipital, humeral, spinal, ulnar and femoral.

COLORATION OF MOUTH. Inside, orange. A black spot on each spur of the base of the tongue, and another, brown and more diffuse, situated subterminally. The latter disappears completely 4–5 days after the chick is hatched, which accounts for the fact that it was not noted by Mr. Pycraft (*cf. antea*, Vol. I., p. 130).

BEARDED TITMOUSE *Panurus biarmicus* (L.).

DOWN. Absent. (N.B.—Spirit specimen.)

GREAT TITMOUSE *Parus major* L.

DOWN. *Colour.*—Whitish grey.

Character.—Moderate in length but scanty, a few pre-pennæ only in each tract.

Distribution.—Inner supra-orbital, occipital, humeral and spinal.

COLORATION OF MOUTH. Inside, lemon-yellow; no tongue spots.

COAL-TITMOUSE *Parus ater* L.

DOWN. *Colour.*—Greyish.

Character.—Moderate in length but very scanty, consisting of a few pre-pennæ only in each tract.

Distribution.—Inner supra-orbital, occipital, humeral, and spinal.

COLORATION OF MOUTH.—Inside, orange; no tongue spots; externally, flanges lemon-yellow.

MARSH-TITMOUSE *Parus palustris* L.

DOWN. *Colour.*—Grey.

Character.—Rather short in length, and very scanty, consisting of a few pre-pennæ only in each tract.

Distribution.—Inner supra-orbital, occipital, humeral, and spinal.

COLORATION OF MOUTH. Inside, orange; no tongue spots.

BLUE TITMOUSE *Parus cœruleus* L.

DOWN. *Colour.*—White.

Character.—Moderate in length but very scanty, consisting of a few pre-pennæ only in each tract.

Distribution.—Inner supra-orbital, occipital, and humeral.

COLORATION OF MOUTH. Inside, lemon-yellow; no tongue spots; externally, flanges lighter yellow.

N.B.—It is possible that the scanty pre-pennæ which form the spinal tract may have been rubbed off in those individuals which I examined; if not, the absence of that tract in this species is worthy of note.

WREN *Troglodytes parvulus* K. L. Koch.

DOWN.   *Colour.*—Greyish black.

Character.—Scanty.

Distribution.—Inner supra-orbital, occipital and spinal.

COLORATION OF MOUTH.   Inside, yellow ; no tongue spots ; externally, flanges lemon-yellow.

N.B.—Since this is the only species in which I have noted the absence of the humeral tract, it is possible that it is slightly developed but had become rubbed off in the nest.

PIED WAGTAIL *Motacilla lugubris* Temm.

DOWN.   *Colour.*—Grey.

Character.—Moderate in length.

Distribution.—Inner supra-orbital, occipital, humeral, spinal, ulnar and femoral.

COLORATION OF MOUTH.   Inside, yellow ; no tongue spots ; externally, flanges very pale yellow.

MEADOW-PIPIT *Anthus pratensis* (L.).

DOWN.   *Colour.*—Whitish grey.

Character.—Moderate in length.

Distribution.—Inner and outer supra-orbital, occipital, humeral, spinal, ulnar, femoral and ventral, the last being very scanty.

COLORATION OF MOUTH.   Inside, deep pink ; no tongue spots ; externally, flanges orange.

RED-THROATED PIPIT *Anthus cervinus* (Pall.).

DOWN.   *Colour.*—Greyish black.

Character.—Long ; femoral and crural tracts scanty.   '

Distribution.—Inner and outer supra-orbital, occipital, humeral, spinal, ulnar, femoral and crural.

N.B.—As I only had a spirit specimen to examine it is possible that the ventral tract, which in the Meadow-Pipit is only slightly developed, may have been overlooked.

SWALLOW *Hirundo rustica* L.

DOWN.   *Colour.*—Grey.

Character.—Fairly long. Tracts scanty.

Distribution.—Inner supra-orbital, occipital, humeral and spinal.

COLORATION OF MOUTH.   Inside, lemon-yellow; no tongue spots ; externally, flanges whitish.

SAND-MARTIN *Cotile riparia* (L.).

DOWN.   *Colour.*—Gray, rather darker on the humeral tract.
Character.—Rather short, scanty.
Distribution.—Inner supra-orbital, occipital, humeral, spinal
and ulnar.   On the last two very scanty.   Prepennæ of
ulnar tract present on the secondary coverts.
COLORATION OF MOUTH.   Inside, lemon-yellow ; no tongue
spots ; externally, flanges lemon-yellow.

GREENFINCH *Ligurinus chloris* (L.).

DOWN.   *Colour.*—Greyish white.
Character.—Medium length, sparse on the crural and ventral
tracts.
Distribution.—Inner supra-orbital, occipital, humeral, spinal,
ulnar, femoral, ventral and crural.
COLORATION OF MOUTH.   Inside, deep crimson, no tongue
spots ; externally, gape white, beak horn colour with a
yellowish tint.

HAWFINCH *Coccothraustes vulgaris* Pall.

DOWN.   *Colour.*—Snow-white.
Character.—Long and plentiful.
Distribution.—Inner supra-orbital, occipital, humeral, spinal,
ulnar, femoral, ventral and crural.
COLORATION OF MOUTH.   Inside, violet pink ; no tongue
spots ; externally, flanges yellowish orange and whitish
at the angles.   The bill during the first two days is not
markedly large, but it rapidly grows in size.

HOUSE-SPARROW *Passer domesticus* (L.).

DOWN.   Absent.
COLORATION OF MOUTH.   Inside, yellow ; no tongue spots ;
externally, flanges lighter yellow.

TREE-SPARROW *Passer montanus* (L.).

DOWN.   Absent.
N.B.—Taking into consideration the development of the down
in the other *Fringillinæ* the absence of it in the genus
Passer is a most remarkable fact.

CHAFFINCH *Fringilla cœlebs* L.

DOWN.   *Colour.*—Greyish.
Character.—Moderate in length and quantity.
Distribution. — Inner and outer supra-orbital, occipital,
humeral, spinal, ulnar, femoral, ventral and crural.

COLORATION OF MOUTH. Inside, violet red, but the hard palate is orange ; no tongue spots ; externally flanges white.

BRAMBLING *Fringilla montifringilla* L.

DOWN. *Colour.*—White.

Character.—Length moderate, well developed, except on the crural tract.

Distribution. — Inner and outer supra-orbital, occipital, humeral, spinal, ulnar, ventral, femoral and crural. Pre-pennæ of the ulnar tract are attached to the secondaries only. Ventral tract well marked at the posterior end.

(N.B.—From spirit specimen.)

LESSER REDPOLL *Linota rufescens* (Vieill.).

DOWN. *Colour.*—Greyish.

Character.—Long ; ventral and crural tracts scanty, pre-pennæ on the secondaries only make up the ulnar tract.

Distribution. — Inner supra-orbital, occipital, humeral, spinal, ulnar, femoral, ventral and crural.

COLORATION OF MOUTH. Inside, carmine ; no tongue spots ; externally, gape white, a carmine spot at angle of gape due to colour of inside showing through.

BULLFINCH *Pyrrhula europœa* Vieill.

DOWN. *Colour.*—Blackish grey.

Character.—Abundant and long.

Distribution. — Inner supra-orbital, occipital, humeral, spinal, ulnar, femoral, ventral and crural. Pre-pennæ of the ulnar tract are present on the secondaries and their coverts.

COLORATION OF MOUTH. Inside, violet red, no tongue spots ; externally, flanges whitish.

SNOW-BUNTING *Plectrophenax nivalis* (L.).

DOWN. *Colour.*—Dark grey.

Character.—Fairly long, spinal tract thicker anteriorly than posteriorly, crural very scanty and minute, the other tracts well marked.

Distribution. — Inner supra-orbital, occipital, humeral, spinal, ulnar, femoral and crural.

COLORATION OF MOUTH. Inside, not noted ; externally, gape yellow ; beak yellow.

STARLING *Sturnus vulgaris* L.

DOWN. *Colour.*—Greyish white, a shade darker on the head,

Character.—Fairly long and plentiful.

Distribution. — Inner supra-orbital, occipital, humeral, spinal, ulnar, femoral, and ventral. Spinal tract long and well marked. Ventral tract scanty and not well marked.

COLORATION OF MOUTH. Inside, orange; flanges very broad and lemon-yellow in colour; no tongue spots; gape huge; externally, flanges lemon-yellow.

SKYLARK *Alauda arvensis* L.

DOWN. *Colour.*—Light sandy, dark at the base of the pre-pennæ giving the whole a peculiar "leveret" appearance.
Character.—Fairly long and abundant.
Distribution. — Inner supra-orbital, occipital, humeral, spinal, ulnar, and femoral; spinal tract confined to small area over and just above sacrum.

COLORATION OF MOUTH. Inside, orange-yellow; two oval black spots at base of tongue situated bilaterally, another triangular spot forms the tip of the tongue; externally, flanges whitish.

KINGFISHER *Alcedo ispida* L.

DOWN. Absent. (N.B.—Spirit specimen.)

Falco subbuteo Vit.

ON THE MOUTH-COLORATION OF SOME NESTLING BIRDS.

BY

ANNIE C. JACKSON.

HAVING read, with interest, Mr. Pycraft's article on the colouring of the inside of the mouths of nestling birds in BRITISH BIRDS for October, 1907, I determined, if possible, to make some observations during the following spring; the more so, as but little material seemed to have been collected on the subject. I now give the results of my observations, which seem to prove that the spotted type of mouth in nestlings is far from common.

Unfortunately, I was not able to examine any of the Tit family; for nesting as they do in crevices and holes it is difficult to reach the young birds. I was struck with the fact that all the downy chicks of the Order *Limicolæ* which I have had the opportunity of examining, had very inconspicuously coloured mouths. But before one can draw any conclusions as to the significance of this, it will be necessary to have a complete list of the colouring of the inside of the mouths in nestling birds of the helpless type.

MISTLE-THRUSH. — *Mouth* (inside), yellow, unspotted. *Flanges* (outside), pale yellow. Nest well lighted.

SONG-THRUSH. — *Mouth* (inside), orange-yellow. *Flanges* not nearly so large and conspicuous as those of the young Starlings. Nest well lighted.

WHITETHROAT.—*Mouth* (inside), yellow, with a dark semicircular band stretching from one spur of the tongue to the other. The band should rather be called "dusky," than black.

WILLOW-WREN.—*Mouth* (inside), unspotted yellow. Nest well lighted.

WREN.—*Mouth* (inside) and *flanges* very pale lemon-yellow, unspotted. Nest badly lighted.

TREE-CREEPER.—*Mouth* (inside), yellow, unspotted. The

nest was situated in the split trunk of a pine tree and was fairly well lighted.

GREY WAGTAIL.—*Mouth* (inside), yellow, unspotted.   Nest placed under a ledge of rock, fairly lighted.

MEADOW-PIPIT.—*Mouth* (inside), flesh-coloured, rather paler on the spurs of the tongue ; unspotted.

SPOTTED FLYCATCHER.—*Mouth* (inside), yellow, unspotted. Nest well lighted.

GREENFINCH.—*Mouth* (inside).   The tongue red, spurs white ; palate red, shading into purple.   *Flanges* (inside), purple ; (outside), deep red.   Nest moderately lighted.

CHAFFINCH.—*Mouth* (inside) purplish red.   *Flanges* (outside), pale yellow.   Nest well lighted.

LESSER REDPOLL. — *Mouth* (inside).   The tongue and flanges purplish-red, spurs of tongue white, palate white and horny.   Edges of upper mandible, inside, blackish.   *Flanges* (outside), pink.   Nest well lighted.

BULLFINCH.   *Mouth* (inside), red.   *Flanges* (inside), purple ; (outside), pale yellow.   Nest moderately lighted.

YELLOW BUNTING.—*Mouth* (inside), purple-red, unspotted. Nest well lighted.

STARLING.—*Mouth* (inside), yellow, palate bristly.   *Flanges* (outside), light yellow and very large and conspicuous.   Nest in a thick yew tree and badly lighted.

SKYLARK.—*Mouth* (inside), yellow, with three black spots on the tongue forming the angles of a triangle, the base of the triangle corresponding to the base of the tongue ; also a blackish spot at the tip of the lower and upper mandibles. Nest well lighted.

WOOD-PIGEON.—*Mouth* (inside), dirty white.   *Flanges* inconspicuous dark grey.   Nest moderately lighted.

LAPWING.—*Mouth* (inside), tongue and palate white, palate having peculiar little roughnesses like tiny teeth running back towards the throat.

OYSTER-CATCHER.—*Mouth* (inside), fleshy-pink.

COMMON SANDPIPER.—*Mouth* (inside), bluish-grey.

COMMON REDSHANK.—*Mouth* (inside), pale bluish-grey.

CURLEW.—*Mouth* (inside), pale fleshy-pink.

ARCTIC TERN.—*Mouth* (inside), fleshy-pink.

NOTES ON THE COMMON CUCKOO IN INDIA.

BY

MAJOR H. A. F. MAGRATH, M.B.O.U.

So much has been written on the subject of the Common Cuckoo (*Cuculus canorus*) that one might suppose nothing more remains to be recorded; yet the following notes, based on observations of this species in the North-Western Himalayas, may be of interest to the readers of BRITISH BIRDS.

In the North-Western Himalayas this bird arrives at its breeding grounds about the middle of April, and from the middle of this month to the middle of June its familiar call is a common sound on the hill-sides; but once the middle of the month is past it gradually decreases. The latest record I have is July 13th. During the time that the call is uttered, I have noticed that the body is by no means invariably held in the horizontal position with which we are most of us familiar. On the contrary, it sometimes assumes a semi-upright attitude. Further, I have noticed that while the call is being made the body is swayed slightly from side to side, and this swaying motion is especially marked in the tail.

In my experience the Cuckoo's notes do not alter as the season advances, though the contrary is usually held to be the case. The bird is probably more vigorous at the beginning of the season, and the call may then be more prolonged. The typical tri-syllabic call is, I believe, entirely connected with the proximity of the female. The well-known variations of the ordinary call are as likely to be heard at the beginning as at the end of the season.

It is strange that no observer seems to have noticed that the Cuckoo, like many, if not all, song-birds, acquires his full song only by degrees. I have met with it early in April in the plains, when in the spring passage, and heard the ludicrous attempts to produce the call result, for the first two hours at any rate, in nothing more than a croaking sound ! The full call is, however, acquired in a day or two, but is very feeble : probably the full compass is not attained till the breeding grounds are reached, that is to say, when the bird has become sexually ripe.

As regards the eggs of the Cuckoo in India I can say but little from personal observation. But in the summer of 1907, when in the Thandiani-Hazara district, at an elevation of 9,000 feet, I found three blue Cuckoos' eggs in nests of the Dark Grey Bush-Chat (*Oreicola ferrea*) and the Indian Blue-Chat (*Larvivora brunnea*), and as *Cuculus canorus* was the only Cuckoo, to my knowledge, frequenting the vicinity of these nests, I could only attribute the eggs to this species. In order to make sure, however, I shot a female Cuckoo, and with great luck took from the oviduct fragments of shell (the egg having been broken) of a beautiful pure blue, which tallied with the egg found in the nest of *Larvivora brunnea*. The eggs of this bird, it may be remarked, were of a delicate, spotless, blue colour, while those of the Dark Grey Bush-Chat were spotted with a few tiny specks of darker greenish-blue. But the eggs of this last species present some variation, showing different shades of pale bluish-green, speckled more or less densely with chestnut and pale rufous.

WOOD-PIGEON "DIPHTHERIA."

NOT the least of the aims of BRITISH BIRDS is the advancement of the Study of Economic Ornithology, and the great interest which was taken in our endeavour to penetrate the mystery surrounding the so-called "diphtheria" in Wood-Pigeons shows that our readers are in entire sympathy with this most important object.

The appeal which we made for material met with a most hearty response ; and in the able hands of Dr. C. B. Ticehurst this material was made to yield some most interesting and valuable results. But, as may be seen from his Report, published in our issue for August last, many points require further elucidation ; and we feel that, having gone so far it is our bounden duty to go further, till all possible facts have been ascertained. We, therefore, turn again to our readers for help in providing material which Dr. Ticehurst, once more, has kindly promised to deal with.

It has been contended that Wood-Pigeon diphtheria is communicable to man ; but, so far, Dr. Ticehurst's investigations do not lend much support to this view. It is certainly significant that it appears to be by no means so readily spread among other birds—and notably game-birds—as was supposed. Having regard to the importance of this aspect of the disease, and to the statements which have been made thereon by other workers with regard to the spread of the disease among domesticated animals and man, further research is emphatically to be desired.

The importance of this enquiry must be perfectly obvious to everyone ; and we may remark that its significance is fully appreciated—as might be expected—by medical men. The "Lancet," September 5th, 1908, in commenting on Dr. Ticehurst's paper in our Magazine, expressed a hope that we might be induced to continue what we had begun. And as Dr. Ticehurst is again willing to place his skill at our disposal we appeal to our readers for help during the coming winter in filling up the schedules issued with this number. And it may be remarked that we shall be as grateful for negative, as for positive evidence. Further copies of the schedule will be sent to any of our readers who may desire to enlist the sympathy of others who, as yet, do not happen to be among our subscribers.

THE EDITORS.

BARRED WARBLERS IN NORFOLK.

On September 11th, Mr. H. A. V. Maynard, shooting with me in the Cley bushes, secured an immature specimen of the Barred Warbler (*Sylvia nisoria*). Its appearance in the bushes was very light, and it showed no inclination to skulk. The wind was N.W., and it had been raining all the morning, the bird making its appearance just after the clearing shower.

E. C. ARNOLD.

On September 12th T. Cringle, one of Lord Leicester's keepers, shot a young female Barred Warbler on the Wells Marshes. Unfortunately it was very badly damaged by the shot. There were only a few birds in the bushes on that day, one Common Whitethroat being the only other warbler recognised, but there was a distinct increase of Meadow-Pipits on the marsh, and I think there were some Rock-Pipits also.

F. G. PENROSE.

YELLOW - BROWED WARBLERS, RED - BREASTED FLYCATCHERS, BLUETHROATS AND OTHER BIRDS IN NORFOLK.

THE following notes from the neighbourhood of Blakeney of the chief movements of migrants observed during September, 1908, may be of interest. From September 7th to 20th the wind was chiefly westerly, south-westerly and southerly, and practically no migrating small birds were seen until September 18th, when a large number of Pied Flycatchers and a good many Common Redstarts appeared, but by September 21st they had nearly all left. After a wet day with a south-west wind on the 22nd, the weather cleared and the wind veered to the north-east on the 23rd. On the morning of this day I shot an immature Red-breasted Flycatcher (*Muscicapa parva*). A few Redstarts and Blackcaps and one Ring-Ouzel were the only other migrants seen in the morning, but during the afternoon a large migration set in. My son (W. R. G. Richards) shot a female Yellow-browed Warbler (*Phylloscopus superciliosus*) and Pinchen (a well-known local fowler) shot a male of the same species. Both birds were very tame. We saw also numbers of Redstarts and a few Pied Flycatchers, Black-caps, Garden-Warblers and Ring-Ouzels, while one Bluethroat (*Cyanecula suecica*) was also seen and shot. On September 24th Ramm (another well-known local fowler) shot a mature male Red-breasted Flycatcher in very fine plumage, and several more Bluethroats were seen and shot. On the 25th the wind went back to the west and the migration considerably decreased, but Ramm shot another immature male Red-breasted Flycatcher, and several more Bluethroats were obtained. On the 26th, the wind being south-westerly, the birds had nearly all gone, while on the 27th we saw only one Redstart.

Since leaving Norfolk I have had word from Ramm that he shot another Yellow-browed Warbler (a mature male) on October 2nd.

F. I. RICHARDS.

YELLOW-BROWED WARBLERS IN YORKSHIRE.

ON September 23rd, 1908, I shot in Holderness, Yorkshire, on the sea coast, a male (apparently adult) of the Yellow-browed Warbler (*Phylloscopus superciliosus*). The yellow bars on the wings attracted my attention, as the bird fluttered up from some buckthorn bushes, the flight much resembling that of the Willow-Wren. A thick sea-fog prevailed, following a night of heavy rain, the wind being slight, and from the south-east. The bird was identified in the flesh by Mr. H. F. Witherby, who kindly prepared the skin for me. The gizzard was full of small flies and other minute insects.

ARTHUR R. GALE.

ON September 30th I had the good fortune to obtain a Yellow-browed Warbler near the same place as the one recorded above. The weather was (and had been) clear and hot, with a light southerly breeze. There was very little movement of birds apparent, and the Yellow-browed Warbler was quite alone, and was very lively. Its gizzard was full of small flies, and the bird was fat, so that it may well have been travelling down the coast in a leisurely fashion. It was a male and, judging by the texture of the skull, which I have always found an infallible test, an adult.

H. F. WITHERBY.

A SUSSEX RUFOUS WARBLER.

Aëdon galactodes or *A. familiaris?*

IN Borrer's "Birds of Sussex" (pp. 63-64), there is an account of the first example of the Rufous Warbler shot in the British Islands. Mr. A. L. Butler has recently called my attention to the fact that the figure of this specimen is undoubtedly drawn from a specimen of *Aëdon familiaris*, the brown central pair of rectrices, which is one of the chief characteristics of this form, being well shown in the plate.

Can any of your readers inform me where the original specimen is ? I do not recollect seeing it in the Booth Museum.

If this example should prove to be referable to *Aëdon familiaris*—which I strongly suspect—the specimen recorded by Mr. J. B. Nichols in your January number (Vol. I., p. 257) is the second recorded example of this form in the British Islands.

M. J. NICOLL.

[Borrer states that the bird was moulting, and that the feathers on the back and tail, " especially the central ones of the latter, are much worn " (*Birds of Sussex*, p. 64), which may account for the colouring of these feathers. If correctly coloured the bird in the plate appears too dark on the back for *A. familiaris*.—EDS.]

WHITE WAGTAIL INTERBREEDING WITH PIED WAGTAIL IN DEVONSHIRE.

ON April 8th last I noticed a White Wagtail on my lawn (near Sidmouth). It only stayed a short time, though I was able to get a good view of it. As it did not put in an appearance again I imagined it to be only a traveller, but early in June I met with a bird, which may have been the same one, at the other end of the village. I watched it for some time feeding in a roadside ditch outside some farm buildings, after which I lost it. It was back at the same place about an hour later, this time accompanied by a male Pied Wagtail, with which pairing took place. It was not till June 13th that I was able to find the nest, which was situated in the stump of an old straw rick, and contained six eggs. I took these on the 14th, as the rick was to be thrown down on the following day. They only differ from Pied Wagtails' eggs with which I have been able to compare them, in having the surface markings brown without any shade of grey, and bolder in character. The bird appeared to me to be less suspicious than the Pied Wagtail usually is, and did not hesitate to go back to its nest while under observation.

It may be well to add that I have had opportunities for watching White Wagtails at close quarters in Scotland, and that a pair of Pied Wagtails were nesting in the ivy of my house at the same time as the pair above recorded were nesting in the rick, so that I had good opportunities for comparing the hen , Pied with the hen White Wagtail. The sharply defined black hood of the latter and the pure grey colour of its back and upper tail-coverts were most distinctive.

AMYAS W. CHAMPERNOWNE.

LESSER REDPOLL NESTING IN ESSEX.

I DO not know whether there are any records of the Lesser Redpoll's nesting in Essex, but probably, in any case, instances are sufficiently uncommon to be worth noticing. A pair built a nest this year at the very top of a standard pear tree in my garden at Chelmsford. On July 28th the pair of old birds were accompanied by two young ones, and this little family party, a rather noisy one, remained about here for two or three weeks off and on, but have now apparently quite disappeared. The nest, on examination, proved to contain one much decomposed young one, so that apparently the clutch consisted of three eggs.

LEONARD GRAY.

BREEDING OF THE CROSSBILL IN COUNTY DUBLIN.

ALTHOUGH the common Crossbill (*Loxia curvirostra*) has been noticed on several occasions in the Scalp, and elsewhere, in co. Dublin, there is no note of its having bred in the county, and all records of its appearances have been, I think, in June, July, or August, when small flocks usually wander over the country. The following notes of its breeding in co. Dublin this year may therefore be of interest.

About mid-June, 1907, Mr. C. V. Stoney and myself saw a flock of fifteen Crossbills in the Scalp. This flock had increased to about twenty birds in August. They never left the neighbourhood during September, October, November, December, and in January we commenced to search most carefully for a nest. By March 7th the flock had been reduced to three or four pairs, and still there was no sign of a nest. On March 16th Mr. Stoney heard a Crossbill singing in the fir woods, and while trying to locate the bird he saw another Crossbill a few feet from him in a Scotch fir. Watching it, he saw it run along a dead branch of the tree with head down, and nip off with its bill a twig, and fly with it into a Scotch fir close by. The nest, about 35 feet up, could be distinctly seen with the aid of glasses, and was just commenced, being a mere platform of twigs, with daylight showing through. On March 28th I climbed to the nest and watched the sitting bird from a distance of about 18 in. for a long while. I touched its back with my fingers before it left the nest, and then it stayed quite near me all the time I was in the tree. The nest, which was very compact, was lined with dead grass—no feathers or fur—and it had the usual platform of larch and fir twigs. It contained three eggs, quite different

to any Crossbill's eggs I have ever seen, the ground colour
being blue, almost as blue as in the egg of a Bullfinch,
sparingly spotted with dark brown ; one egg had a lilac
streak. R. HAMILTON HUNTER.

CIRL BUNTING SINGING IN OCTOBER.

AT noon on October 18th—a dull, muggy morning—I heard
a Cirl Bunting in full song at Heath, near Leighton Buzzard.
During the quarter of an hour that I waited at the spot,
the bird, which was perched on the top of a thorn hedge,
sang persistently at intervals of a few seconds. Is not mid-
October a late date for this species to be in song ?

CHAS. OLDHAM.

LATE NEST OF THE KINGFISHER.

ON October 10th, 1908, I was informed that there was a
Kingfisher's nest in the banks of the Wenning, near Bentham,
Yorkshire. I went and inspected the nest and found it to
contain four young nearly ready to fly. The late date is
remarkable, and the very warm weather we have been having
may partly account for it. GRAHAM W. MURDOCH.

SCOPS-OWL OFF ABERDEENSHIRE.

IT may be worth while to put on record that I have in my
possession a male Scops-Owl (*Scops giu*), which was captured
on a trawler about twenty-five miles off the coast of
Aberdeenshire in October, 1900. This bird was in an
exhausted state, and although the plumage was in fair
condition it was much faded. From this arises another
question : What is the nautical limit within which a bird
may be called " British " ? E. R. PATON.

HONEY-BUZZARD IN SHROPSHIRE.

I RECENTLY examined a fine example of the Honey-Buzzard
which had been shot in North Shropshire, about the last day
of September, 1908. It appears to be a male in its second
year, and belongs to the dark-brown form. The last prior
record in the county was in August, 1881, when three are
said to have been seen near Ludlow, one of which was caught.
H. E. FORREST.

GREY PHALAROPE IN SUMMER IN DEVONSHIRE.

ON the morning of May 14th last I was surprised to find a
Grey Phalarope (*Phalaropus fulicarius*) fluttering on a path
in my garden (near Sidmouth). The bird was hopelessly

crippled, having evidently fallen a victim to the heavy gale which had raged during the night. It proved to be a female in summer plumage, the tips of not more than five or six grey feathers showing among the chestnut of the lower breast. I find only two previous occurrences recorded from this county of this species in summer plumage.

AMYAS W. CHAMPERNOWNE.

NESTING OF THE COMMON SNIPE IN KENT.

In 1896 my brother and I found a single pair of Snipe nesting in Kent (cf. Zoologist, 1897, p. 271), but since then I have no certain record of any having bred. However, on April 21st of this year, I was walking with a friend along one of the many "levels" which connect up with Romney Marsh, and he told me that there had been several Snipe there for some time, and on that day we saw three or four pairs flying round and uttering their summer note, but we did not hear them "drumming." I had no time on that day to search for a nest. On June 16th I was again in the same spot, and saw at least two pairs flying round and "drumming," and from their behaviour they evidently had young about, but the state of the grass made a search for them impossible. My friend told me that the Snipe were "drumming" nearly every day between my two visits, so that I do not think that there can be any doubt that they had bred there. The "levels" were unusually wet all through the summer, which probably accounted for Snipe breeding there this year, and I have noticed before that these birds are particularly influenced by the state of a prospective breeding ground, a place which is wet and marshy one year and holding several pairs, will be perhaps too dry another year and the birds will be absent; the obvious inference being that under one condition the food supply will suffice, and under the other it will not.

C. B. TICEHURST.

PECTORAL SANDPIPER AND BARTRAM'S SANDPIPER IN KENT.

A PECTORAL SANDPIPER (Tringa maculata) frequented a piece of marshy ground in Kent for several days during July, 1908. This bird was first noticed by the Duchess of Bedford and myself on July 14th. It was very shy, but by careful stalking I obtained a very good view of it through binoculars at about twenty yards' distance. Owing to the somewhat worn appearance of the plumage I take it to have been an adult bird. Its flight was somewhat peculiar, and reminded one

of the "soaring" breeding flight of a male Redshank. Her
Grace informs me that this bird was still in the same place
on July 21st.

On July 18th a Bartram's Sandpiper (*Bartramia longicaudata*) was shot on Romney Marsh, and I examined it in the
flesh two days later in Mr. Bristow's shop at St. Leonards.
It was an adult male in good condition, but in somewhat
worn breeding plumage. On July 23rd Mr. Bristow informed
me that on the previous day (the 22nd) he saw a bird on
Pevensey Level which he believes to have been a Bartram's
Sandpiper.

The interesting note by the Duchess of Bedford on the
Solitary Sandpiper in Kent, in the August number of
BRITISH BIRDS, coupled with the present records, seem to
point to an immigration of American sandpipers to England.
It would be interesting to know if any of your correspondents
have noticed similar arrivals of American species in Britain.

With the possible exception of the Scilly Islands, Sussex
and Kent can claim to have produced more records of American
waders than any other part of Great Britain. Possibly this
is owing to the fact that there are more observers in these
counties than elsewhere on the south coast. At any rate,
there can be no doubt that the tendency of these waders is
to follow a west to east line of flight.

M. J. NICOLL.

PECTORAL SANDPIPER IN NORFOLK.

AT Cley, between September 1st and 17th, 1908, I repeatedly
saw a bird which I judged to be the Pectoral Sandpiper
(*Tringa maculata*). The first time it got up it uttered the
note which I remembered hearing at Aldeburgh, some years
ago—a double chirp. I watched it once through glasses at
about twenty yards, and thought I made out the pectoral
band. It was often with Dunlins, and I could always pick
it out by its superior size, but for many days it escaped the
notice of the other frequenters of the estuary, mainly, I think,
because it uttered its note very seldom, and the note when
uttered was so low. It was the last wader I saw before I
left the place.

E. C. ARNOLD.

THE LEVANTINE SHEARWATER IN BRITISH
WATERS.

I HAVE received some very interesting information from
Mr. W. J. Clarke (the Scarborough wildfowler) with regard
to the occurrence of the Levantine Shearwater (*Puffinus*

yelkouanus) off the Yorkshire coast. In his " Monograph of the
Petrels," now in course of publication (p. 107), Dr. Godman
gives the range of this species as practically confined to the
Mediterranean, although its disposition to wander northwards
occasionally was evidenced by the fact that it had been recorded several times from the seas to the south and east of
Great Britain. If we exclude the Yorkshire records, these
occurrences appear to be as follow : Devon, three ; Hampshire, one ; Kent, one ; Northumberland, one. The
Yorkshire records up to the date of Mr. Clarke's most recent
observations are as follow :—

1. 1877, autumn, near Redcar (T. H. Nelson, *B. of Yorks*, p. 760).
2. 1880 (about), Flamborough   Do.   Do.
3. 1890, Aug. 16th. Flamborough   Do.   Do.
4. 1898, Oct., Bridlington (R. B. Sharpe, *Bull. B.O.C.*, X., p. 48).
5. 1899, Feb. 4th, ♀ adult, Scarborough (T. H. Nelson, *B. of Yorks*, p. 761).
6. 1900, Sept. 13th, ♀ jun., Scarborough   Do.   Do.
7. 1900, autumn, Scarborough   Do.   Do.
8. 1902, Sept. 1st, ♂ adult, Scarborough   Do.   Do.
9. 1904, Sept. 17th, ♀, Scarborough (W. J. Clarke, *Zool.*, 1905, p. 74).
10. 1904, Sept. 27th, Scarborough   Do.   Do.
11. 1907, Sept. 9th, ♀, Scarborough (W. J. Clarke, *in litt.*).
12. 1907, Sept. 19th, ♀, Scarborough   Do.
13. 1907, Sept. 19th, ♂, Scarborough   Do.
14. 1907, Sept. 28th, ♀, Scarborough   Do.
15. 1908, Sept. 4th, ♂, Scarborough   Do.
16. 1908, Sept. 21st, Scarborough   Do.
17. 1908, Sept. 24th, Scarborough   Do.

Mr. Clarke writes that out of twenty-two Shearwaters
which he has had through his hands since 1890, twelve have
been specimens of the Levantine species. With one exception, all these were shot from a boat from four to eight
miles from land, and most of them in the dusk of the
evening. Mr. Clarke, who has himself obtained several of
these birds, considers the Levantine to be the commonest Shearwater off the coast of Yorkshire in the autumn,
but in his experience it never approaches near the shore, and
must be sought in the dusk. It looks on the wing, he says,
distinctly larger and darker than the Manx Shearwater.

It would certainly seem by Mr. Clarke's valuable observations
that the Levantine Shearwater migrates regularly northward
in autumn, and if this be the case not only is our knowledge
of the distribution of the bird affected, but we have the
anomaly of a species migrating north in autumn. Shearwaters
are difficult birds to observe, and the Levantine has for many
years been confused with the Manx Shearwater, but for those
who like to repeat that there is nothing more to be learnt
about British birds, and that there is nothing to be learnt from

the occurrence of " stragglers," here is an occasion to think again. We hope that Mr. Clarke's observations will induce some of our readers, who have opportunities for doing so, to go out in boats in the dusk of the evening. and study Shearwaters.

With reference to the rosy tint of the breast referred to by Dr. N. F. Ticehurst (*antea*, p. 138) Mr. Clarke writes as follows :—" I have examined a good many freshly-killed, as well as a couple of living, specimens, and none of them showed the slightest sign of any rosy tint on the breast."

* * *

H. F. WITHERBY.

GOLDEN ORIOLE IN LINCOLNSHIRE.—A specimen of *Oriolus galbula*, which is a somewhat rare visitor so far north as Lincolnshire, flew against a telegraph wire " recently " (? August, 1908) at Gainsborough (F. M. Burton, *Nat.*, 1908, p. 359).

SWALLOW'S NEST ON A LAMP SHADE.—The nest of a Swallow on the shade of an electric-lamp is recorded and a photograph given, with a summary of previously-recorded curious nesting sites for this bird (*Field*, 12, IX., 1908, p. 514).

LESSER REDPOLL NESTING IN SUSSEX.—Mr. R. Morris reports that at least one pair of *Linota rufescens* bred again (*cf. antea*, Vol. I., p. 183) this year at Maresfield (*Zool.*, 1908, p. 350).

CHOUGH IN LANCASHIRE.—Mr. E. Bell reports that a specimen of *Pyrrhocorax graculus* was shot near Wigan in the middle of September last. The Chough has previously occasionally wandered to Lancashire (*Field*, 26, IX., 1908, p. 590).

HOOPOE IN ROSS-SHIRE.—Colonel W. H. E. Murray records that an example of *Upupa epops* was caught at Geanies on September 9th. The Hoopoe is not often recorded from Scotland (*Field*, 19, IX., 1908, p. 547).

SAND-GROUSE IN ESSEX. — " R. M." reports that an example of *Syrrhaptes paradoxus* was shot on Great Mollands Farm, South Ockenden, on September 1st, 1908 (*Field*, 12, IX., 1908, p. 514).

RUFF IN CO. CLARE.—A pair of *Machetes pugnax*, a rare casual visitor to Ireland, were shot on September 4th in co. Clare (H. V. Macnamara, *Field*, 12, IX., 1908, p. 514).

SABINE'S GULL IN NORFOLK.—On September 1st, 1908, a *Xema sabinii* was shot on Breydon (F. A. Arnold, *Zool.*, 1908, p. 352). The bird was an adult in full summer plumage (J. H. Gurney, *in litt.*).

WOOD-PIGEON ENQUIRY.

1d.

STAM

THE EDITORS OF

"BRITISH BIRDS,"

326, HIGH HOLBORN,

LONDON, W.

WOOD-PIGEON ENQUIRY No. 2.

You are urgently requested to answer as many of the questions detailed below as possible, and to forward the Schedule to the Editors of *British Birds* by March 1st, 1909. It is hoped that the readers of the Magazine will thus co-operate and collect a large number of facts. The result of the enquiry will be announced in a future number, when all the observations have been collated and compared.

Observer's Name and Address ....

....

District (state County) in which observations were made ....

MIGRATORY MOVEMENTS.

1. Have they been plentiful this winter compared to other years, especially to last year?

2. When did the flocks arrive?

3. When did they depart?

4. Has the food supply been scarce or plentiful, and of what has it consisted?

DISEASES.

THROAT DISEASE.   FEATHER DISEASE.

1. Have you noticed either disease, and at what time of year? If no disease has been met with, *please note the fact.*

WOOD-PIGEON ENQUIRY.

THE EDITORS OF

"BRITISH BIRDS,"

326, HIGH HOLBORN,

LONDON, W.C.

BRITISH BIRDS

BIRDS

AN·ILLUSTRATED·MAGAZINE
DEVOTED·TO·THE·BIRDS·ON
THE·BRITISH·LIST

DECEMBER 1,
1908.

Vol. II.
No. 7.

MONTHLY·ONE·SHILLING·NET
326·HIGH·HOLBORN·LONDON
WITHERBY & C⁰

BRITISH·BIRDS

EDITED BY H. F. WITHERBY, F.Z.S., M.B.O.U.
ASSISTED BY W. P. PYCRAFT, A.L.S., M.B.O.U.

ON THE NESTING OF THE SCAUP-DUCK IN SCOTLAND.

By P. H. BAHR, M.A., M.B., B.C., F.Z.S., M.B.O.U.

PERHAPS there is no more interesting fact in the history
of the modern ornithology of these islands than the

remarkable spread of the breeding area of many of the duck-tribe during recent years.

The causes which made its allies, the Tufted Duck* and Shoveler,† common resident species in Scotland, have also affected the Scaup-Duck. There are several early records of its supposed breeding in Scotland. Under the head of "Scaup Pochard" Selby writes ‡ : " a single female was shot by Sir William Jardine in a small loch between Loch Hope and Eriboll in 1834 ; she was attended by a young one, which unfortunately escaped among the reeds. This is the first instance of its breeding in Great Britain that I am aware of." In June, 1868,§ Mr. J. A. Harvie-Brown shot an adult male Scaup in Sutherlandshire, " which had been frequenting the loch for some days ; and from its unwillingness to leave the locality, though repeatedly disturbed and fired at, I am fully persuaded that the female was sitting on her eggs at no great distance. With my friend, Mr. W. Jesse, I also in June, 1867, obtained a laying of duck's eggs, and though failing to identify them, they closely resembled eggs of this species from Lapland."

In 1880 the late Dr. A. C. Stark recorded a nest and eggs found on Loch Leven which he considered to be those of a Scaup.‖ Full details have recently (antea, p. 132) been given by Mr. W. Evans, which show that these were no doubt the eggs of a Tufted Duck.

The records cited above were not accepted by Professor Newton¶ as authentic.

The first authentic nest appears to have been discovered by Mr. Heatley Noble, and was recorded in the " Ibis " for

* J. A. Harvie-Brown, " Ann. S.N.H.," 1896, pp. 3-22; " Proc. Roy. Phys. Soc. Edin.," Vol. XIII., pp. 144-160.

† Id., "Fauna N.W. Highlands," p. 232; "Ann. S.N.H.," 1902, p. 212.

‡ " Edin. New Philosph. Journal," XX., p. 293, 1836; vide also Yarrell, 4th ed., Vol. IV., p. 425.

§ " Proc. N.H. Soc., Glasgow," 1875, II., pp. 120 et 121.

‖ A. C. Stark, " Proc. Roy. Phys. Soc. Ed.," 1881-83, VII., p. 203.

¶ " Dict. Birds," p. 815.

1899,* and the " Annals of Scottish Natural History."†
The nest was found in Speyside, and was placed in rushes
on an island five feet from the water's edge ; it contained
three eggs when found. It was revisited in a week's
time, but the duck was not sitting. Mr. Noble got within
ten feet of her and watched her for some time close to the
nest. Two days after 'he watched her leave the nest,
which contained nine eggs, and was deep, cup-shaped, and
better made than most ducks' nests. In a recent
number of BRITISH BIRDS‡ he has described the down
and feathers taken from this nest.

The Scaup-Duck has also been recorded§ as nesting
in the islands south of the Sound of Harris. Mr. Harvie-
Brown's correspondent writes : " They are numerous,
and have bred for the last four years. Two pairs to my
knowledge in 1897, 1898, and 1899, and three pairs in
last season, 1900." It was believed to have bred again
in 1901, and in June, 1902, a young bird still in down,
ten days old, was sent to Mr. Harvie-Brown by Guthrie,
a keeper in these islands.

Finally, the nest figured overleaf (Fig. 1) was discovered
on a journey made to these islands with Mr. N. B.
Kinnear for this express purpose.|| Professor Newton,
with his critical acumen, would not at first accept the
record because of the similarity of the eggs to those of the
Tufted Duck, but did so after a careful comparison of
the down and feathers with those taken from the nest
found in 1899 by Mr. Heatley Noble.¶

Mr. Kinnear and I spent a fortnight in June, 1906,
searching innumerable lochs for signs of nesting Scaup-

---

* H. Saunders, "Ibis," 1899, p. 648 ; "Bull, B.O.C.," VIII., p. 5.

† J. A. Harvie-Brown, " Ann. S.N.H.," 1899, p. 215.

‡ BRITISH BIRDS (Mag.), Vol. II., p. 38.

§ J. A. Harvie-Brown, "Ann. S.N.H.," 1902, p. 211 ; vide also
Guthrie, "Ann. S.N.H.," 1903, p. 76.

|| Kinnear, "Ann. S.N.H.," 1907, p. 82.

¶ "Ootheca Wolleyana," Vol. II., pp. 591 and 2.

Ducks.   For this   purpose   we   had   the   owner's   kind
permission, and the assistance of his keepers.

On June 4th we discovered a solitary old male Scaup

Fig. 1.—Nest of Scaup-Duck.   June 11, 1906.   (Photographed by
P. H. Bahr.)

riding asleep on a peculiarly desolate and unpromising-
looking loch.   He was accompanied by a solitary Tufted
drake, and when seen from a distance, even with the aid
of glasses, they seemed, when asleep with head tucked

FIG. 2.—Duck and Drake Scaup and Nesting Site. (Drawn by P. H. Bahr.)

under wing, to be indistinguishable. In certain lights the grey feathers on the back of the male Scaup do not show up well, so that these birds can easily be missed when in company with Tufted Ducks. We waded to all the islands, but to no purpose. On the next day, June 5th, we were rewarded by the sight of two pairs, the drakes in each instance floating lazily about, fast asleep, with one leg cocked up in the air, as I have depicted in Fig. 2. No more evidence was forthcoming till the 9th, when I explored a loch famous for its trout and the variety of its bird-life. I was lying for no less than four hours at a stretch on a small islet some twenty feet in diameter, when I became aware that I was being watched by a pair of Scaups, every bit as carefully as I myself was watching some Black-headed Gulls. The duck appeared to be very anxious, and was swimming about, evidently on the *qui vive*, in the neighbourhood of another rocky islet, while her mate varied his vigil with an occasional "forty winks," the temptation for which he seemed unable to resist. A search was made, and to my surprise I found that for four hours I had unconsciously been lying within six feet of a duck's nest, to all appearance the very one I was in quest of. It was placed under a boulder, was lined with dark brown down, and contained nine olive-green eggs, which were not covered up. Owing to the trampled-down condition of the surrounding vegetation I had grave doubts as to whether the bird would return. Consequently I retired to another island some quarter-mile away and kept watch. Soon I saw the Scaup duck swim behind the island and disappear, so I resolved not to disturb her, but to return on the following Monday, the 11th, with Kinnear.

When the day arrived we rowed up to the island with very anxious hearts. A brown duck flew off, scuttled along the water, and settled in a distant part of the loch, where she was soon joined by an undoubted Tufted drake ! ! ! So a further search was made ; on the next island, only some three hundred yards away, a Shoveler's

nest with eleven eggs was found. While I busied myself
taking some photographs, Kinnear explored the remaining
island of the series. This was also very small, had a rocky
base, with large tufts of long grass on top. A brown
duck was flushed off a nest, and settled on the water but
twenty yards from the shore, and was watched for a
quarter of an hour at a stretch through glasses. At
the same time a drake Scaup was seen riding out on the
loch at some distance from the island, and he, afterwards,

Fig. 3.—The Duck Scaup coming off the Nest.   (Drawn by P. H. Bahr.)

in company with a drake Tufted, flew past me in my
hiding-place. Kinnear refrained from investigating any
further, but noted that the nest was situated in a deep
hollow some nine inches below the level of the ground, and
well guarded by large tufts of grass (Fig. 1) ; a trampled
pathway led up from the edge of the water to the nest.

An hour after he returned with me. On our approach
the duck was seen leaving the nest, threading her way,
with neck stretched in front of her, through the matted
grass. She scuttled into the water (Fig. 3) and remained
within thirty yards of the bank. The white patch and

broad bill were plainly visible, and set aside all doubts
of identification, so that we did not deem it necessary
to take any further steps.   She showed her uneasiness
by giving vent to a sort of guttural grunt, splashed the
water with her wings, and finally dived out of sight.
The nest was almost invisible, so cleverly were the grasses
pulled over the top.   The eggs were covered with down,

FIG. 4.—Nest of Tufted Duck in the same hollow as the Scaup's nest in
Fig. 1.   May 28, 1907.   (Photographed by P. H. Bahr.)

so she had evidently heard us approaching.   A quantity
of blackish down, amongst which were greyish-white
feathers, mixed with much fine grass, lined the nest.
The eggs were nine in number, of an olive-green hue,
and of the same size as those in the Tufted Duck's nest
we had just discovered.   In the bottom of the nest we
found a quantity of old eggshells, and it seems that this
hollow had done service on several other occasions.   The

eggs were hard set, and the young were within three days of hatching.

That same evening we saw five more Scaup-Ducks, two drakes and three ducks, on a sea-loch. Towards the end of the same month another nest was found by the keeper on the same loch. In 1907 I returned, and though I searched every likely situation, no trace of a Scaup was seen. The same hollow contained the nest of a Tufted Duck* (Fig. 4), from which the old bird was disturbed on three occasions and identified ; it contained nine eggs, and was to all intents and purposes exactly similar to the Scaup's of the year before. There was also a Shoveler's nest in exactly the same position. It is significant that the nests of these three species should be found on contiguous islands, where, not so many years ago, they were unknown.

Though essentially a circumpolar species, the Scaup-Duck has been recorded as nesting in north Germany, once by Baldamus in Anhalt, and twice by Blasius in ponds near Brunswick.†

---

* *Cf.* "Ann. Scot. Nat. Hist.," 1907, p. 213.

† Rudolf Blasius, "Naumann, Naturgesch. V. Mitteleurop.," new ed., 1896-1904, Vol. X., p. 153; Howard Saunders, "Man. B.B.." 1899, p 449.

SOME EARLY BRITISH ORNITHOLOGISTS AND THEIR WORKS.

BY

W. H. MULLENS, M.A., LL.M., M.B.O.U.

V.—ROBERT PLOT (1641—1696)

AND SOME EARLY COUNTY NATURAL HISTORIES.

IN the year 1661, Joshua Childrey (1623—1670), antiquary, schoolmaster, and divine, published in London a small duodecimo work entitled "Britannia Baconia : / or, the Natural / Rarities / of / England, Scotland, & Wales." This book, although of no particular value in itself, being merely a brief and somewhat imperfect compilation, was nevertheless destined to be of some considerable influence on the literature of natural history in this country. For, according to Wood's "Athenæ Oxonienses " (p. 339), it inspired Robert Plot (1641—1696), the first keeper of the Ashmolean Museum at Oxford, with the idea of writing the "Natural History of Oxfordshire," which appeared in 1677, and which was followed in 1686 by a "Natural History of Staffordshire," the work of the same author ; who is said to have also contemplated similar histories of Middlesex and Kent. These two works of Robert Plot's also proved in their turn to be the forerunners of a numerous series of county natural histories by different writers.

The full title of the "Natural History of Oxfordshire" is as follows :—

"The / Natural History / of / Oxford-shire, / being an Essay toward the Natural History / of / England. / By R.P., LL.D. / [quotation from Arat. in Phænom.] / [engraving] Printed at the Theater in Oxford, and are to be had there : / And in London at Mr. S. Millers, at the Star near the / West-end of

St. Pauls Church-yard. 1677. / The price in sheets at the Press, nine shillings. / To Subscribers, eight shillings.

1 Vol. folio.

Collation : pp. 4, Imprimatur & Title. + pp. 8. un. + pp. 358. + pp. 12, Errata & Index. Map & XVI. Plates.

A second edition of " The Natural History of Oxfordshire " appeared in 1705 " with large additions and corrections : also a short account of the Author, &c."

It cannot be said that Plot's observations on the birds of Oxfordshire contain anything of much interest or value ; he was a somewhat credulous writer, and seems to have been a better authority on plants than on birds, and, indeed, is mentioned by the eminent John Ray, in the latter's " Synopsis Methodica Stirpium Britannicarum " as " Robertus Plot LL. Doctor, e cujus Historiis Naturalibus lectu sane dignissimis territorii tum Oxoniensis, tum Stafford- iensis, non pauca in Historiam & Synopsin hanc nostram transtuli." It must not, however, be forgotten that Plot's book was written at a time when but little was known of British birds, in fact, the " Natural History of Oxfordshire " was published a year before the appearance of Willughby's famous " Ornithology " (English translation).*

Robert Plot dealt with the birds of Oxfordshire on pp. 175—180 of his book, under the head " of Brutes." It will here suffice to state that, having informed his readers that there was but little that he could mention in the way of new matter " since the feathered kingdom has been so lately and so carefully surveyed by the learned and industrious Francis Willughby," he proceeds to describe, amongst other birds, one " about the bigness of a sparrow, with a blue back, and a reddish breast, a wide mouth and a long bill . . . . from the noise that it makes commonly called the Wood-cracker," † although this bird, which was undoubtedly the Nuthatch,

* The Latin edition of Willughby's work entitled " Ornithologiæ Libri tres " appeared in 1676.

† *Cf.* Swainson, p. 35.

had been duly noticed in Willughby's Latin edition of the " Ornithology " (pp. 19 and 95).

Plot's other work, " The Natural History of Staffordshire," was published in 1686, and is altogether a more important and far rarer book than the one above mentioned. Its full title is as follows :—

The / Natural History / of Stafford-shire / by / Robert Plot, LL.D. / Keeper of the / Ashmolean Museum / and / Professor of Chymistry / in the / University / of / Oxford. / Ye shall describe the Land, and bring the Description hither to me. Joshua 8. v. 6. / [Engraving] Oxford / Printed at the Theater, Anno M. DC. LXXXVI.

1 Vol. folio.

Collation : pp. 16 un. + pp. 450 + pp. 14, Index, " Proposalls of the Author," and list of Subscribers. Map, XXXVII. Plates, and extra Plate of " Armes omitted." (This last plate is very seldom found in the original state.)

Birds are treated of in Chapter VII., pp. 228—236, and though the observations are somewhat fuller than in the same author's " Natural History of Oxfordshire," their principal interest lies in the curious account of the nesting of the Pewit (*i.e.*, the Black-headed Gull, *Larus ridibundus*). A small portion of this account is given in the fourth edition of Yarrell's " British Birds " (Vol. III., p. 599).* But as it is of considerable interest we here give it in full, together with a facsimile of the original plate, showing the taking of the young Pewits.†

" But the strangest whole-footed water fowle that frequents this county is the Larus Cinereus Ornithologi, the Larus Cinerus tertius Aldrovandi, and the Cepphus of Gesner and Turner ; in some Counties called the black-Cap, in others the Sea or Mire-Crow, here the Pewit ; which being of the migratory kind, come annually to certain pooles in the Estate of the right Worshipfull Sr. Charles Skrymsher Knight to

* The quotation in " Yarrell " is by no means word perfect; it did not appear in the first three editions of that work.

† Plot uses the spelling, Pewit or Pewet, indifferently.

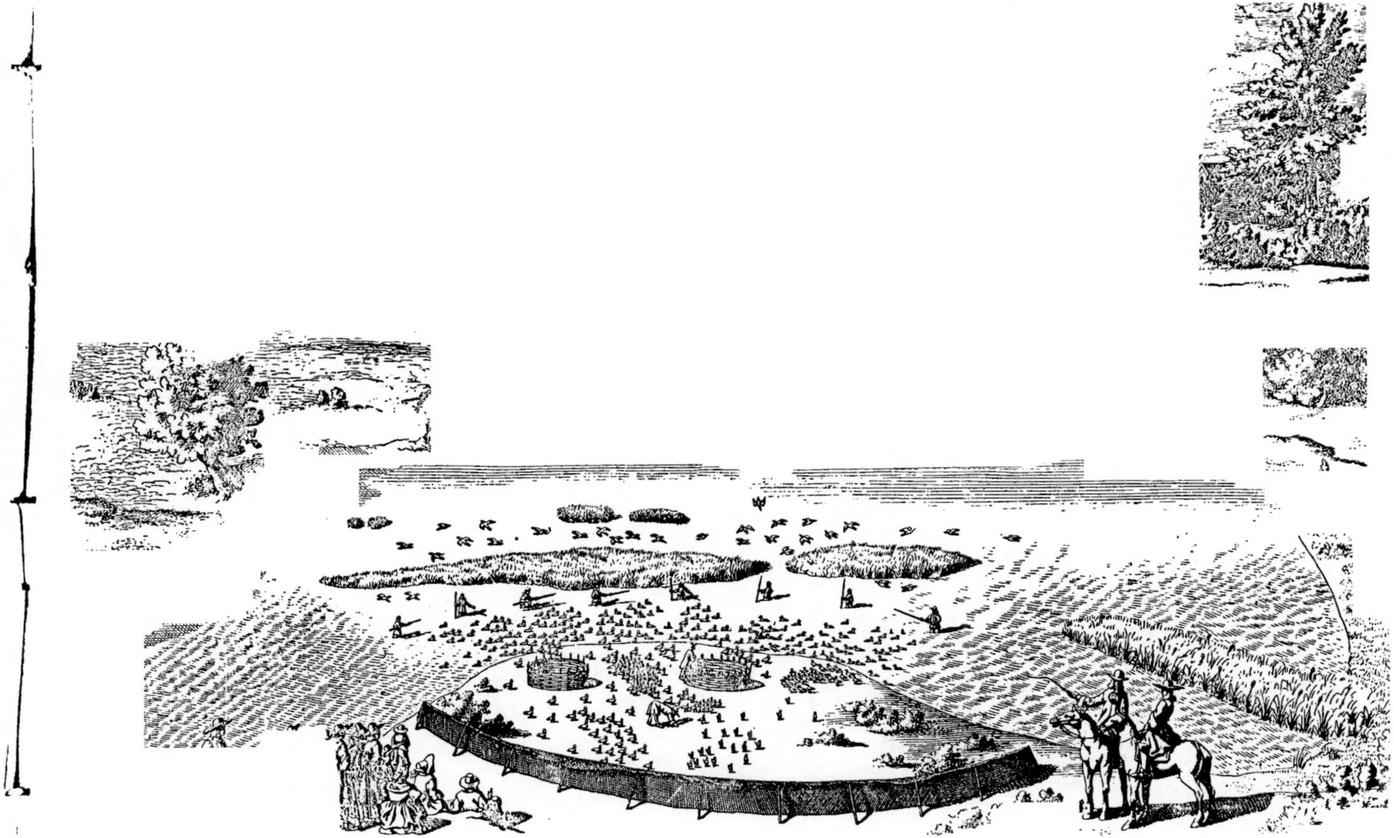

[*BRITISH BIRDS, Vol. II., Pl. 5.*]

build and breed, and to no other estate in or neer the County, but of this Family, to which they have belong'd ultra hominum memoriam, and never moved from it, though they have changed their station often. They anciently came to the old Pewit poole above mentioned, [chap. 6. §§. 36, 40, 42] about ½ a mile S.W. of Norbury Church, but it being their strange quality (as the whole Family will tell you, to whom I refer the Reader for the following relation) to be disturb'd and remove upon the death of the head of it, as they did with-in memory, upon the death of James Skrymsher, Esq., to Offley-Moss near Woods-Eves, which Moss though containing two gentlemans land, yet (which is very remarkable) the Pewits did discern betwixt the one and the other, and build only on the land of the next heir John Skrymsher, Esq., so wholy are they addicted to this family. At which Moss they continued about three years, and then removed to the old pewit poole again, where they continued to the death of the said John Skrymsher, Esq.; which happening on the Eve to our Lady-day, the very time when they are laying their Eggs, yet so concerned were they at this gentleman's death, that notwithstanding this tye of the Law of Nature, which has ever been held to be universal and perpetual, they left their nest and Eggs ; and though.they made some attempts of laying again at Offley-Moss, yet they were still so disturbed that they bred not at all that year. The next year after they went to Aqualat, to another Gentleman's Estate of the same family (where though tempted to stay with all the care imaginable) yet continued there but two years, and then returned again to another poole of the next heir of John Skrymsher deceased, called Shebben poole in the parish of high Offley where they continue to this day, and seem to be the propriety as I may say (though a wild-fowle) of the right worshipfull Sr. Charles Skrymsher Knight, their present Lord and master. But being of the migratory kind their first appearance is not till about the latter end of February and then in number scare above six, which come as it were as harbingers to the rest, to see whether the Hasts or Islands in the pooles (upon which they build their neasts) be prepared for them ; but these never so much as lighten, but fly over the poole scarce staying an hour : about the sixth of March following, there comes a pretty considerable flight, of a hundred or more, and then they alight on the hasts, and stay all day, but are gon again at night. About our Lady-Day, or sooner in a forward Spring, they come to stay for good, otherwise not till the beginning of April, when they build their nests, which they make not of sticks, but heath and rushes, making them but shallow, and laying generally but 4 eggs, 3 and 5 more rarely, which are about the bignes of a small Hen-egg. The Hasts or Islands are prepared for them between Michaelmass and Christmass, by cutting down the reeds and rushes and putting them aside in the nooks and corners of the hasts, and in the valleys to make them level ; for should they be permitted to rot on the Islands, the Pewits would not endure them.

" After three weeks sitting the young ones are hatch't, and about a month after are almost ready to flye, which usually happens on the third of June, when the Proprietor of the poole orders them to be driven and catch'd, the Gentry comeing in from all parts to see the sport ; the manner thus. They pitch a Rabbit-net on the bank-side, in the most convenient place over against the hasts, the Net in the middle being about ten yards from the side but close at the ends in the manner of a bow ; then six or seven men wade into the poole beyond the Pewits, over against the Net, with long staves and drive them from the hasts, whence they all swim to the bank side, and landing run like Lapwings into the Net, where people standing ready, take them up, and put them into two penns made within the bow of the Net, which are built round, about 3 yards Diameter, and a yard high or somewhat better, with small stakes driven into the ground in a circle, and interwoven with broom and other raddles, as in Tab. 19, at the bottom whereof is represented in Sculpture, the poole, and whole method of taking these Pewits ; and Norbury Manor at the top, the seat of the Proprietor, a most generous Encourager of this work. In which manner they have taken off them in one morning 50 dozens at a driving, which at 5s. per dozen (the ancient price of them) comes to twelve pounds ten shillings : but at several drifts that have been anciently made in the same morning, there have been as many taken as have been sold for thirty pounds, so that some years the profit of them has amounted to fifty or three score pounds, besides what the generous Proprietor usually presents his Relations, and the Nobility and Gentry of the County withall, which he constantly does in a plentifull manner, sending them to their houses in Crates alive, so that feeding them with livers, and other entrals of beasts, they may kill them at what distance of time they please, according as occasions present themselves, they being accounted a good dish at the most plentiful tables. But they commonly appoint 3 days of driving them, within fourteen days or thereabout, of the second or third of June ; which while they are doing, some have observed a certain old one that seems to be somewhat more concerned than the rest, being clamorous, and striking down upon the very heads of the Men ; which has given ground of suspicion that they have some Government amongst them, and that this is their Prince, that is so much concern'd for its Subjects. And 'tis further observed that when there is great plenty of them, the Lent-Corn of the Country is so much the better, and so the Corn-pastures too, by reason they pick up all the worms, and the Fern-flyes, which though bred in the Fern, yet nip and feed on' the young corn and grass, and hinder their growth."

Robert Plot, as we are informed in " a short account of the Author " appended to the Second Edition of the " Natural History of Oxfordshire," was the son of Captain Robert Plot, of Borden in Kent, and was born in 1641 at Sutton-Barn in the said parish. He was educated at Magdalene Hall in the University of Oxford, and afterwards at University College there. In the year 1683 he was appointed first Keeper of the Ashmolean Museum, and about the same time was made Professor of Chemistry to the University. In 1694 he was nominated Mowbray Herald, by Henry, Duke of Norfolk, and died at his house, Sutton-Barn, April 30th, 1696. A monument to his memory stands in the Parish Church at Borden.

In the year 1700* Charles Leigh (1662—1701), " Doctor of Physick," published in imitation of Plot's works a worthless " Natural History of Lancashire, Cheshire, and the Peak in Derbyshire," the full title of the book being as follows :—

The / Natural History of / Lancashire, Cheshire, / and the / Peak, / in Derbyshire : / with an / Account / of the / British, Phœnician, Armenian, Gr. and Rom. Antiquities / in those / Parts. / By Charles Leigh, / Doctor of Physick. / Oxford : / Printed for the Author, and to be had at Mr. George West's,/ and Mr. Henry Clement's, Booksellers there : Mr. Edward Evet's / at the Green-Dragon, in St. Paul's Church-yard ; and Mr. John / Nicholson, at the King's Arms in Little-Britain, London. MDCC.

1 Vol. folio.

* In 1684 appeared the " Scotia illustrata sive Prodomus Historiæ Naturalis," Edinburgh, 1 vol. folio, of Robert Sibald (1641—1722) which contained some short notes on the birds of Scotland.

Collation : pp. 20 un. + pp. 4. + pp 2 un. + pp. 196. (Book II.) pp. 99. (Book III.) pp. 112 + pp. 33 Index. Portrait, Map & XXIV. Plates. (Numerous mistakes in pagination.)

Birds are treated of, pp. 157—164, Book I., but Leigh's ornithological observations are useless and trivial, though he could not well complain of any lack of material, since he informs us on p. 157 that " These Counties afford us great variety of Birds, and in some places, even, clog the Inhabitants with their Plenty."

County Natural Histories now began to appear at frequent intervals, and contained more or less useful notices of the local birds, but it is here only possible to mention some of the rarer or more important of them in their chronological order :—

1709. Robinson (Thomas)—

An / Essay / towards a / Natural History / of / Westmoreland / and / Cumberland. / . . . . By Tho. Robinson, Rector of / Ousby in Cumberland. / London . . . . 1709.

1 Vol. 8vo. (Contains some worthless remarks on birds, pp. 64—68, and pp. 94—98 of the " Moral Conclusions," which form the latter part of the work.)

1712. Morton (John)—

The / Natural History / of / Northamptonshire : / . . . . by John Morton, M.A., / Rector of Oxenden in the same County . . . . / London . . . . MDCCXII.

1 Vol. folio. (Birds, pp. 423—438.)

1758. Borlase (William)—

The / Natural History / of / Cornwall. / . . . . By William Borlase, A.M.F.R.S. / Rector of Ludgvan, and Author of the Antiquities of Cornwall. / Oxford . . . . MDCCLVIII.

1 Vol. folio. (Birds, pp. 242—248, the information being chiefly derived from Carew's Survey of Cornwall, 1602.)

1769. Wallis (John)—

The / Natural History / and / Antiquities / of Northumberland : / and so much of the County of / Durham / as lies between the Rivers Tyne and Tweed ; / . . . . By John Wallis, A.M. / London MDCCLXIX.

2 Vols. 4to. (Birds, pp. 309—346, a considerable account.)

1772. Rutty (John)—

An / Essay / towards a / Natural History / of the / County of Dublin, / . . . . By John Rutty, M.D. / . . . . 1772.

2 Vols. 8vo. (Birds, Vol. I., pp. 295—343, and IV. plates of birds.)

1789. Pilkington (James)—

A / view / of / the / Present State / of / Derbyshire ; / . . . . by James Pilkington. / . . . . London. / MDCCLXXXIX.

2 Vols. 8vo. (Birds, Vol. I., pp. 480—496.)

We will conclude our short list with the earliest of the local ornithologies, *i.e.* :—:

1809. Tucker (Andrew)—

Ornithologia Danmoniensis ; / or, / an history of the habits and Economy / of / Devonshire Birds. / Embellished with coloured plates, engraven from accurate and / Beautiful Drawings from Nature : / . . . . By Andrew G. C. Tucker. / . . . . London : / Printed for the Author, and published by T. Cadell and / W. Davies, Strand. / 1809.

1 Vol. 4to.

" An ambitious work of which not even the whole of the somewhat turgid Introduction was published, but the two parts printed show the author to have been a physiologist, anatomist and outdoor-observer far beyond most men of his time " (*cf.* Newton's *Dictionary of Birds ;* introduction, p. 45).

ON THE SONG OF THE WOOD-WARBLER.

BY

H. W. MAPLETON, B.A., M.B.O.U.

In May last, while availing myself of a very good opportunity of observing a Wood-Warbler in full song my attention was called to the fact that this bird has two distinct songs. As I do not remember to have seen this fact recorded in works on British birds, I thought it possible that a few notes on the subject might prove of interest. Of course, everyone knows that in the songs of accomplished vocalists, such as the Nightingale and the Song-Thrush, many distinct phrases are utilized in a variety of combinations, but in the case of the Wood-Warbler there are two distinct songs, which bear no resemblance to each other, either in tone or phrasing, and which, when the bird is singing well, are very rarely mixed. The first of these is the ordinary song, which needs no description here.

The second song, which is much rarer than the first, varies considerably in different individuals as regards the number of syllables, though the tone is constant. In the case of the first bird I had under observation, on May 16th, it consisted of from 9–12 syllables—the average number in this case being 10. It is sweet, and rather plaintive in tone, falling gradually from F sharp to E flat, or possibly D. [This interval I am not certain about, as I verified it on the pianoforte from memory only.] In character it resembles to a certain extent the ecstatic " tail-end " of the full song of the Tree-Pipit. The last and lowest note of this song seems to be the same as that used as a call-note when the young are fledged and flying about in family parties.

On May 17th I came across another Wood-Warbler, and timed the bird, roughly, for ten minutes, during which it sang No. 1—the ordinary song—over fifty times, and No. 2 only about five times. Neither on this, nor on the previous occasion, did I hear these two songs mixed, though once or twice the bird would utter three notes of the prelude of its ordinary song, and then stop, and start afresh on the second.

On May 31st I listened to another bird. This individual differed from the two preceding ones to a certain extent, as it mixed up the two songs occasionally. It was not in very full voice. During the time I stood listening it never sang song No. 2 properly. Several times it sang four syllables of this song, ending up with three notes of the prelude of No. 1, and once, without a break, it began with these four syllables, and ended up with No. 1 in full. But I failed to hear it sing more than four syllables of No. 2 at any time.

On June 7th I found a bird singing regularly, but not very fully. This one very seldom made use of song No. 2. Once it started on it and ran into the regular song (No. 1) without a break. I never heard it sing more than four syllables of song No. 2.

On June 19th I found a fifth bird in full song, and watched it carefully for 45 minutes, during which time it never moved far away, and never ceased singing. My notes on this occasion corroborated those I took in the first two cases. This bird differed slightly in one respect, as two or three times it sung the prelude to song No. 1 without the trill. In this individual the number of syllables in song No. 2 varied from 7–11—the average number being 8—and it mixed the two songs three times during the period that I had it under observation.

It would seem that when the Wood-Warbler is singing well, the number of syllables in its second song varies from 7–12. As regards the musical interval of this song, F sharp to E flat would represent a minor third; and I think that this interval in the song is approximately correct, though I cannot boast a musician's trained ear. When we consider that the double call of the Cuckoo constitutes an interval of a minor third, and that the ten-syllabled song of the Wood-Warbler (which is gradually falling in tone all through) represents an interval little, if at all, larger, it is easy to see that our diatonic scale is not well suited for gauging the musical intervals of the songs of birds.

ON THE MORE IMPORTANT ADDITIONS TO OUR KNOWLEDGE OF BRITISH BIRDS SINCE 1899.

BY

H. F. WITHERBY AND N. F. TICEHURST.

PART XV.

(*Continued from page* 150.)

LAPWING *Vanellus vulgaris* Bechst. S. page 555.

The wings of the two sexes have been shown by Mr. F. W. Frohawk to be different. Those of the male are rounder and broader than those of the female, a characteristic which may be distinguished in flight. The formulæ of the primaries are as follows :—

♂ 1st = 7th.   ♀ 1st = 4th.
2nd and 4th, equal.   2nd and 3rd, equal and longest.
3rd, longest.
7th, 8th, and 9th, 1¼ in. longer than in ♀.

" In the male the primaries are long and broad, giving a decidedly curved outline, while the secondaries, being considerably shorter, add greatly to the rounded appearance of the wing." Mr. Frohawk also points out that the bill of the female is longer and her crest shorter than in the male (F. W. Frohawk, *Ibis*, 1904, pp. 446-451, figs. 5-10).

AVOCET *Recurvirostra avocetta* L. S. page 561.

CORNWALL.—One was shot in the Cober Valley, Helston, on April 21st, 1900—" the only specimen recorded from Cornwall during the past twenty-seven years " (J. Clark, *Zool.*, 1907, p. 286).

NORFOLK AND KENT.—They still visit these counties with fair regularity every year in May or June.

ESSEX.—An immature female was shot at Leigh-on-the-Sea in November, 1908, and another was shot near the same place in August, 1901 (F. Cooper, *Field*, 1908, p. 888.)

NORTH WALES.—One seen and identified by Capt. Bailey on a marsh near Llanelltyd in 1901 (H. E. Forrest, *Vert. F. N. Wales*, p. 338).

BLACK-WINGED STILT *Himantopus candidus* Bonn.
S. page 563.

CHESHIRE.—An adult male was obtained on the Mersey at Latchford, but the date is unknown (Coward and Oldham, *B. of Cheshire*, p. 207).

YORKSHIRE.—A third specimen for the county was shot at Kilnsea in Holderness " many years ago " (T. Nelson, *B. of Yorks.*, p. 591).

NORFOLK.—Two were on the Broads on May 28th, 1905, and one on April 29th, 1906 (J. H. Gurney, *Zool.*, 1906, p. 127, and 1907, p. 127).

GREY PHALAROPE *Phalaropus fulicarius* (L.). S. page 565.

SCOTLAND.—One or two have occurred almost every year since 1899, and the following have been recorded from the *Outer Hebrides :*—Two on November 3rd, 1901, at Island Glass (*Ann. S.N.H.*, 1902, p. 193) ; and one at the Flannans on May 18th, and two on May 19th, 1906 (*t.c.*, 1907, p. 201). *Sule-Skerry :*—one on February 15th, 1903 (*t.c.*, 1904, p. 214).

IRELAND.—A male was shot on September 28th, 1899, near Logan, co. Armagh (A. W. Marsden, *Zool.*, 1899, p. 477).

RED-NECKED PHALAROPE *Phalaropus hyperboreus* (L.). S. page 567.

NORTH WALES.—One was obtained in Merioneth, and one was watched in Anglesea in 1902 (H. E. Forrest, *Vert. F. N. Wales*, p. 340).

IRELAND.—Breeding colony in the west discovered (*cf. Irish Nat.*, 1903, pp. 41 and 96 ; *Zool.*, 1903, p. 116 ; *B. B.*, Vol. I., p. 174).

WOODCOCK *Scolopax rusticula* L. S. page 569.

SCOTLAND.—Unusual numbers nesting in 1902, 1904, and 1908 (*cf.* J. A. Harvie-Brown, *Ann. S.N.H.*, 1904, pp. 191 and 245, and 1908, p. 142).

Weight.—Records of weights up to 17 oz., a few of 16 oz., and many of 15 to 15½ oz., in Shetland (R. C. Haldane, *Ann. S.N.H.*, 1906, p. 54).

GREAT SNIPE *Gallinago major* (J. F. Gm.) S. page 571.

SCOTLAND.—1901.—Sept. 25th, two Orkney ; Sept. 26th, one Shetland ; autumn, one Castle Douglas (*Ann. S.N.H.*, 1902, p. 54). [One Shetland Sept. 20th, 1904 (*t.c.*, 1905, p. 54)]. 1905—One Aberdeen Sept. 5th (*Zool.*, 1905, p. 466). One Orkney Sept. 12th (*Ann. S.N.H.*, 1906, p. 54). 1906— One Fair Isle Sept. 5th (*t.c.*, 1907, p. 79).

IRELAND.—Mr. R. J. Ussher gives thirteen records (*List of Irish B.*, p. 43.).

Weight.—Average of forty-three adult birds shot in August and September, 7 oz. 5 drs. Three were over 9 oz., and fifteen over 8 oz., the largest was 1 dram short of 11 oz. (N.F.T., *Field*, 13, v., 99).

PALLAS'S GRASSHOPPER - WARBLER (*LOCUS-TELLA CERTHIOLA*) IN IRELAND.

A NEW BRITISH BIRD.

BY

R. M. BARRINGTON, M.B.O.U.

AN immature example of Pallas's Grasshopper-Warbler was picked up dead at the Rockabill Lighthouse (five

Pallas's Grasshopper-Warbler, picked up dead at the Rockabill
Lighthouse (Co. Dublin), on September 28, 1908.

miles off the coast of co. Dublin) on September 28th, 1908, by the assistant light-keeper—Martin Kennedy. This is the first recorded occurrence of this bird in the British Isles, and so far as I can ascertain, it has only once before been obtained in Europe, viz., by the late Heinrich Gätke, in Heligoland, where a young bird was caught

at the lighthouse lantern on the night of August 12th—
13th, 1856. In 1858, Blasius, when on a visit to this
island, examined the specimen, and called it " the jewel "
of Gätke's collection (cf. H. Gätke, *Heligoland*, Eng.
Ed., pp. 310 and 312). The breeding range of this bird
appears to extend over Siberia, east of the Yenesei, to
the Pacific, and southwards to the Altai Mountains and
the Amur River, while it occurs in China on passage, and
winters in Burma, India, and the Malay Archipelago.
In habits it seems to be much the same as our Grasshopper-
Warbler, and in appearance it is somewhat similar. A
friend said it resembled a cross between a Hedge-Sparrow
and a Grasshopper-Warbler, but it is markedly larger than
the latter bird, and is of a reddish-brown on the upper
side, the feathers being striped with black, while the tail-
feathers are tipped with greyish-white. The bird was in
plump condition, and was no· "wind-driven," half-starved,
specimen.   Judging by lighthouse specimens it is probable
that many inconspicuous birds visit our shores more
frequently than other records would lead us to suppose.
In this case, however, the rarity of the species in Europe
scarcely suggests this possibility.

The specimen was exhibited on my behalf by Mr. W.
R. Ogilvie-Grant at the meeting of the British Ornithologists' Club held on October 21st last, and it was
shown by me at the scientific meeting of the Royal
Dublin Society on November 24th. Mr. Pycraft dissected
the body and it proved to be a male.

BARRED WARBLER IN LINCOLNSHIRE.

I FIND I have omitted to place on record the occurrence of the Barred Warbler (*Sylvia nisoria*) in Lincolnshire in 1905. On September 4th in that year I shot an immature specimen of this species in a hedge near the coast at North Cotes. It is a shy and wild bird, and takes wing more readily than any of the other warblers.

G. H. CATON HAIGH.

[This example is referred to in Vol. I., p. 56, of this Magazine, but as only the bare record was given by Mr. Gurney in the "Zoologist," from which the occurrence was taken, we are very glad to publish the details above.—EDS.]

GOLDCRESTS FROM EAST COAST LIGHTHOUSES.

DR. HARTERT regards the British-bred Goldcrest as sub-specifically distinct from the typical *Regulus regulus* of Continental Europe, and has described it under the name of *Regulus regulus anglorum* (*cf.* Vol. I., p. 218). This insular race he regards as resident (*l.c.*, p. 209). The North European form, he remarks, frequently crosses over to Great Britain in flocks in autumn and winter (*l.c.*, p. 218).

If the above views be correct, Goldcrests occurring at lighthouses on our east coast during the migration seasons ought to belong to the Continental form and be recognisable as such. To test this, I recently examined a number of specimens obtained at the Isle of May and Barnsness light-houses, at the mouth of the Firth of Forth, and could see no difference between them and examples from inland woods —in the north as well as in the south of Scotland—where the species breeds commonly, and is present all the year round. But to make sure I have submitted my specimens to Dr. Hartert for comparison with the series in the Tring Museum, and he writes me that he is unable to distinguish any of them from the British race ; " they are," he repeats, " *exactly* like British birds, their colour being darker than in Continental specimens." The specimens submitted included ten from the lighthouses as under :—

♂, Isle of May, September 17th, 1885 ; taken by myself at the lantern, with other migrants, about 11 p.m. A good many were seen in the course of the night.

♀, Isle of May, September 4th, 1908. (For this and sub-sequent specimens from the May, I am indebted to Mr. Ross, superintendent of the lighthouse.)

♂, Barnsness, night of October 1st, 1908 ; along with Larks, Starlings, etc. (For the Barnsness specimens I have to thank the lighthouse keepers and Mr. Pow.)

3 ♂ ♂ and one ♀, Isle of May, night of October 7th.

♀, Barnsness, night of October 7th ; several with other migrants.

♂ and ♀, Isle of May, night of October 31st, during a great rush of migrants, including besides Goldcrests, Redwings (very many), Fieldfares, Ring-Ousels, Owls, Woodcock, etc.

Thus it would seem either that many of our British Goldcrests do migrate, or that there are in some part of North Europe birds which in autumn plumage are indistinguishable from them. I have long regarded our British Goldcrests as in the main resident, and the flocks observed at our light stations in October as coming from Scandinavia or the adjacent parts of the continent ; and I still incline to this view. The subject, however, needs further investigation. An examination of specimens from stations in Orkney and Shetland, for instance, would be most interesting.

WILLIAM EVANS.

[In September, 1905, I obtained two examples of un-doubtedly migrating Goldcrests in Norfolk which clearly belong to the typical and not to the English race. There is much to be learnt regarding migration in conjunction with the study of local races.—H. F. W.]

YELLOW-BROWED WARBLER IN LINCOLNSHIRE.

On October 19th last I found a Yellow-browed Warbler (*Phylloscopus superciliosus*) dead in a hedge near the sea-bank at North Cotes. There was a great migration of birds in progress at the time. There were Song-Thrushes in thousands, hundreds of Robins and Goldcrests, and in less numbers Red-wings, Blackbirds, Ring-Ousels, Grey Crows, Chaffinches, Greenfinches, and Twites, with a few Bramblings, Wheatears, Rock-Pipits, Woodcocks, Merlins, and Black Redstarts.

This is the second appearance of the Yellow-browed Warbler in the county. G. H. CATON HAIGH.

THE EAST EUROPEAN CHIFFCHAFF IN THE ISLE OF WIGHT.

On April 15th, 1907, I received from the lighthouse at Niton, Isle of Wight, a single example of *Phylloscopus collybita abietina* (Nilss.)—the Eastern form of our Chiffchaff. According to Dr. Hartert, this form breeds in Scandinavia, East Prussia, Austria, and Hungary, southwards to Bosnia and Montenegro, and in Russia, south of 65° N. It winters in Greece, Asia Minor, and North-East and East Africa, but its migration route and western boundary are still uncertain. It may be distinguished from our native bird by its slightly larger size, paler coloration, and longer wing measurements, which are about 2.5 inches in the male, and 2.25 inches in the female. This is the first recorded example from this country, but it seems likely that solitary individuals may occur yearly in this country on migration, as it would be impossible to distinguish them from the common Chiffchaff unless they were obtained.

J. L. BONHOTE.

THE NORTHERN RACE OF THE WILLOW-WREN IN GREAT BRITAIN.

DURING the last two years, whilst examining birds and wings sent from the lighthouses and lightships on the south coast, I was struck by the fact that there frequently occurred a Willow-Wren which, though like our breeding Willow-Wrens superficially, was easily distinguishable from them. On going into the matter more carefully I found that these birds in spring differed from ours in the following characters :—

1. The colour of the dorsal parts has a greyish instead of a yellowish green tint, thus giving the bird a paler appearance.

2. Underparts almost entirely without the yellow which is seen on our birds in spring plumage, and much paler.

3. The superciliary stripe usually quite white, and not yellowish.

Further, I found that these birds do not begin to arrive in the south of England before the end of April, and that the majority pass through during the first two weeks of May— at a time when our own birds are busy breeding.

The race to which these birds evidently belong has been recognised by Dr. Hartert, and I think quite rightly, under the name of *Phylloscopus trochilus eversmanni* (Bonaparte) [in no way to be confounded with Eversman's Warbler], and the distribution which he gives (*Die Vög. pal. Faun.*, p. 509) is :—the breeding range begins in north Russia, east of Timan Hills, and extends south to the eastern parts of Perm and Orenburg ; eastward it is the breeding form of the Ob and Yenesei, and extends to the mouths of the Lena and Kolyma; passing through Roumania and Egypt on migration, it winters in South Africa. I have examined about a dozen examples obtained in Hampshire and Sussex, and half-a-dozen from the Shetlands, all obtained on the spring migration. Thus it seems certain that this form of the Willow-Wren occurs regularly on migration through England and Scotland, and since I have examples from Finmarken and have seen others from north Norway obtained in the breeding season, it seems that the breeding range must be extended further westward than Dr. Hartert states.

That this subspecies occurs also on the return migration in autumn is probable, but I know no certain way of differentiating them from our own birds in autumn plumage. It is not surprising that it should have hitherto been over-looked in Great Britain, since it arrives when the leaf is out and when our birds are nesting, and consequently at a time when few examples are obtainable for examination.

This is the species described by H. Seebohm as *Phylloscopus gaetkei* (*Ibis*, 1877, p. 92).

Six specimens were shown by me at the October meeting of the British Ornithologists' Club, all obtained from Hampshire. C. B. TICEHURST.

NESTING HABITS OF THE MARSH-WARBLER.

I AM interested in Mr. Bunyard's notes on the Marsh-Warbler in the November issue. I have had considerable experience with this species during the last three years in Gloucestershire. —and as regards the nest, its situation and construction, my observations confirm those of Mr. Bunyard. The nests I have seen have been in willow-herb, wormwood, figwort, meadow-sweet, and nettles, and the clutch generally consists of five eggs, occasionally only four. The 18th to 24th June I have found to be the best average date for fresh eggs.

I have, however, failed to notice the " extreme shyness " commented on by Mr. Bunyard. The sitting birds almost invariably allowed of close observation, and when building or feeding young were quite unusually careless. The song is freely uttered throughout the day, although certainly more so in the late afternoon, and is extraordinarily rich and melodious. The singing bird generally perches near the top of a low bush, frequently on the branch of a willow-tree, and seldom amongst the undergrowth, like the Reed-Warbler.

The eggs are certainly larger than Continental specimens, and present two distinct types, viz. : (1) the usual and well-known one ; (2) that in which the markings are uniformly brown. Neither type can possibly be mistaken for eggs of any other British species. NORMAN GILROY.

AQUATIC WARBLER IN SUSSEX.

On October 7th last I had the good fortune to shoot an Aquatic Warbler (*Acrocephalus aquaticus*) on the Eastbourne Crumbles. I have for the last fifteen years examined every Sedge-Warbler I have seen in the hopes of finding an Aquatic.

Aquatic Warbler, Eastbourne, Sussex, October 7, 1908.
(Drawn by E. C. Arnold.)

This bird put its head out of a single tamarisk bush on the shingle, and I at once felt sure it was a rarity, the eyestripe being most pronounced. The sketch which I made of it shows the wedge-shaped character of the tail, which seems to me a striking feature of the species. I judge the bird to be immature, and the legs were of a very light flesh-colour. The wind at the time was south-east, and the weather fine and hot.

E. C. ARNOLD.

BLUE-HEADED WAGTAIL IN NORFOLK.

ON September 23rd, 1908, one of the wildfowlers of Cley, Norfolk, shot, at that place, an adult male Blue-headed Wagtail (*Motacilla flava flava*). The bird was examined by Mr. Witherby, and exhibited by him on my behalf at the meeting of the British Ornithologists' Club, held on November 18th last.

F. I. RICHARDS.

AUTUMN AND WINTER SINGING OF BUNTINGS.

WITH regard to Mr. Oldham's note on the singing of a Cirl Bunting in October, it was recorded some years ago in the " Zoologist " by Professor Salter, that this species sings from time to time throughout the winter in Wales. My brother and I heard one at Reigate on October 31st this year, singing in the morning and afternoon at the same place. The following day I heard another. Between October 10th and 15th on very warm days I heard several Yellowhammers, one of them in full song, near Tunbridge Wells, and my brother heard a Corn-Bunting singing in Romney Marsh on October 13th. We had not previously heard either of these two species singing after the moult, and I think in the case of the Yellowhammer at any rate, it is abnormal. We have not heard any Reed-Buntings singing, however. As far as we have observed, all the Corn-Buntings leave this part of the country for the winter, so that there is no chance of hearing them ; Yellowhammers are also a good deal less common, and begin to sing as soon as they return in February. Between November 12th and 20th, while staying in Hayling Island, Hants. I have heard Cirl and Corn-Buntings singing a good deal, even on cool and sunless days. H. G. ALEXANDER.

WITH reference to a note on a Cirl Bunting singing in October (*antea*, p. 204), I find by my notes kept over some years at Clevedon, in Somerset, that I have records of this bird's song in *every* month in the year, my earliest date being January 8th, and my latest December 18th. H. MEYRICK.

IN the Mendip district of Somerset the Cirl Bunting sings intermittently throughout the winter. One has been singing here at Winscombe on more days than it has been silent during the last four weeks (I write on November 17th). I see that I noted it as singing on November 16th last year, and in January, 1907, I heard two birds in full song near Glastonbury. In my experience the Cirl Bunting does not sing in winter unless the weather is both still and mild—as an instance, the bird here was silent during the week of colder weather earlier in the month. The Corn-Bunting, on the contrary, may be tempted into song on a very cold frosty morning, provided there is bright sunlight.

C. I. EVANS.

LITTLE BUNTINGS IN IRELAND AND NORFOLK.

ON October 2nd, 1908, a female specimen of the Little Bunting (*Emberiza pusilla*) was picked up at the Rockabill Lighthouse (off the coast of co. Dublin) and forwarded to me. The bird was exhibited on my behalf by Mr. W. R. Ogilvie-Grant at the meeting of the British Ornithologists' Club held on October 21st last. The Little Bunting has not previously been recorded from Ireland.

R. M. BARRINGTON.

MR. H. N. PASHLEY, the well-known taxidermist of Cley-next-the Sea, Norfolk, has sent me an adult female example of the Little Bunting (*Emberiza pusilla*), which was brought to him on October 19th by a local gunner, who had shot the bird that day. Five examples of this bird have been previously recorded as occurring in England, thirteen in Scotland, and one in Ireland (*cf. antea*, Vol. I., pp. 249, 291, 383, 385, and above). This appears to be the first record for Norfolk.

Nine years ago, when Howard Saunders published the second edition of his "Manual," only one example of this species was known to have occurred in this country. Mr. Eagle Clarke, I may remind my readers, found on Fair Isle, the Little Bunting in some numbers amongst flocks of Twites, and it thus may very easily escape notice.

I had the pleasure of exhibiting the bird at the meeting of the British Ornithologists' Club, held on November 18th last.

H. F. WITHERBY.

THE GREAT SPOTTED WOODPECKER AS A BREEDING BIRD IN SCOTLAND.

WE have already referred to the interesting spread of the Great Spotted Woodpecker in Scotland (*cf.* Vol. I., p. 280). Mr. J. A. Harvie-Brown* now provides a valuable paper on

* "Ann. S.N.H.," 1908, pp. 209—216 (with map).

the subject, while Mr. W. Evans* discusses the very interesting question as to whether the birds, which are now thoroughly established in the south-eastern half of Scotland, owe their origin to England or Scandinavia. At one time nesting in the faunal area of Moray, the Great Spotted Woodpecker became extinct as a breeding species in Scotland between 1841 and 1851. Since that date there have been a number of autumn-winter irruptions, chiefly on the east coast, of presumably Scandinavian birds. In 1887 the first brood since the extinction of the bird as a nester was found, and this at Duns Castle woods, in Berwickshire. From that date it gradually extended, and its breeding range now embraces practically the whole of the south-eastern half of Scotland, although it seems strangely absent from Fife and Kinross. All this is well traced by Mr. Harvie-Brown, who illustrates his paper with an excellent map. It may be noted that a breeding record for Aberdeen (cf. antea, Vol. I., p. 281) is omitted.

Dr. Hartert has lately† shown us that the English Great Spotted Woodpecker (D. major anglicus) differs from the Scandinavian bird chiefly by its smaller and more slender bill and shorter wing, and Mr. Evans finds that three of these Scottish breeding birds belong to the English race. This fact points to the conclusion that Scotland is being repopulated from England, and not by the Scandinavian visitants, and it is hoped that more specimens may be examined to prove the contention conclusively.

The study of geographical races has only just begun in this country, but many of us have long been confident that a thorough appreciation of geographical forms would teach us very much (and notably in connection with migration problems) which is unknown, and even unsuspected, concerning the avifauna of the British Isles. Mr. Evans' observations are, therefore, very welcome.

H. F. WITHERBY.

COURTING PERFORMANCE OF THE CUCKOO.

IN his interesting notes on the Common Cuckoo in India, in your issue of November, Major Magrath calls attention to " a semi-upright attitude " assumed when uttering the call-note. The following note in my diary may be of interest :—
" May 12th, 1905.—Two Cuckoos alighted in one of the trees beside the lawn where I was sitting, a third alighting a little way off. Two of the birds I judged males from their behaviour. The one nearest me became very excited, uttering

* t.c., 216—218.   † BRITISH BIRDS, Vol. I., p. 221.

the ordinary as well as the three-syllable call-note all the time. His movements reminded me of the domestic male Pigeon paying court to his female. He kept raising his body to an upright position, spreading out his feathers, especially those of the tail, and spinning round on his perch, exactly as does the Pigeon. During this exhibition the female remained silent. Thereafter all three birds flew away."

T. THORNTON MACKEITH.

LITTLE OWL IN WARWICKSHIRE.

IN connection with the spread of the Little Owl (*Athene noctua*) which formed the subject of an article in a recent number of this Magazine (*vide B. B.*, Vol. I., p. 335), it is interesting to note that these birds have now reached Warwickshire. Messrs. Spicer & Son, taxidermists, Birmingham, now have in their possession a specimen which was killed at Sutton Coldfield quite recently, though I am unfortunately unable to give the exact date.

A. G. LEIGH.

A REMARKABLE VARIETY OF THE RED-LEGGED PARTRIDGE IN ESSEX.

THROUGH the generosity of Mr. Ruggles Brise, the British Museum of Natural History has just acquired a very remarkable variety of the Red-legged Partridge (*Caccabis rufa*) killed at Braintree on October 20th.

This bird, a male, has the crown, sides of the head and throat, dull black. The upper part normal. The neck, breast and flanks, however, are of a uniform rich dark brown, but show faint traces of the characteristic barring of the flanks when held in certain lights. On the breast is a white patch, recalling the horseshoe of the English Partridge. No similar variety has, we believe, ever been recorded, though white specimens have several times been met with. A bird " with a white breast-band," according to Yarrell's " British Birds," was obtained in the Haute Garonne in November, 1872, and similar varieties, it is interesting to note, " were captured at the same season in the years 1873 and 1874."

W. P. PYCRAFT.

GREY PHALAROPE IN CO. WEXFORD.

ON November 11th I was duck shooting with a friend on the south side of Wexford Harbour and saw a bird which I identified as a Grey Phalarope. We were by the side of a " pill " (an inlet from the sea containing brackish water, as it receives the drainage from the marshes) when a small bird flew past and alighted upon the water about thirty yards away. It then slightly lowered one wing, inclined its head to that side, bobbed, or ducked, two or three times, and turned partly around. I thought it was wounded, and so did our retriever, who dashed in to get it, and disturbed it before it completed the turning movement. The bird would fly about thirty yards and alight upon the water, swimming easily and lightly, and made six or seven flights, always within gunshot, during nearly ten minutes. It twice more started to turn and bob, causing the dog to rush at it, and once allowed the dog to get within a yard before rising. The bird was in a foot to fifteen inches of water. I was much struck with the compact, neat, and graceful appearance of the bird, while its tameness was in marked contrast to the wildness of the various other kinds of Plovers we saw. It was blue-grey above, and very pure white upon head and breast. Unfortunately I had not got my Goerz glasses with me, but I noted its two most conspicuous markings—a black patch on the nape of the neck extending partly forward with spots or a faint line, and, when flying, two rows, or one broad row, of light feathers across the secondaries (I could not notice whether the primaries were marked so) giving the appearance almost as if this portion of the wing was cut out.

R. C. BANKS.

[In his " List of Irish Birds " Mr. R. J. Ussher refers to this species as an " irregular visitor in autumn and early winter, chiefly in October and in bad weather."—H.F.W.]

BUFF-BREASTED SANDPIPER IN LINCOLNSHIRE.

THOUGH somewhat late to do so, it may be as well to place on record the occurrence of the Buff-breasted Sandpiper (*Tringites rufescens*) in Lincolnshire.

I shot an example of this American species on the foreshore at North Cotes on September 20th, 1906. The bird singularly resembled a Reeve both in appearance and flight, and but for its small size I should have paid little attention to it.

It was by no means shy, and allowed me to approach it within thirty yards on the open saltings. This is, I believe, the first appearance of this species in the county.

G. H. CATON HAIGH.

SABINE'S GULL IN LINCOLNSHIRE.

ON September 28th last I shot an example of Sabine's Gull (*Xema sabinii*) off Gramthorpe Haven. It was a young bird, in the plumage in which this species usually occurs in this country. It was sitting alone on a sandbank, though there were large flocks of other Gulls in the immediate neighbourhood. Its note was singularly like that of the Arctic Tern.

During the latter part of September and the beginning of October considerable numbers of Skuas, Gannets, Divers, and Shearwaters passed along the Lincolnshire coast.

G. H. CATON HAIGH.

LATE NESTS OF THE GREAT CRESTED AND LITTLE GREBES.

WITH reference to Mr. A. G. Leigh's note on the late nesting of Grebes (*antea*, p. 171), on searching my note-book I find the following entries : " Sept. 7th, 07, Roddlesworth Reservoir (Brinscall, Lancs.), a nest of the Little Grebe (*Podicipes fluviatilis*) containing four eggs ; the bird left her eggs uncovered, but I was unable to ascertain how much they were incubated because of several feet of deep water intervening." " Sept. 8th, 07, in a pond close by my house I discovered another nest of the Little Grebe with two eggs ; these eggs were not incubated and no more were laid ; they hatched safely." Even allowing that these were second, or possibly third nests, they were, I think, remarkably late for a moorland district.

W. MACKAY WOOD.

[In reference to Mr. Leigh's expression of doubt as to whether the Great Crested Grebe is double-brooded, the Rev. F. C. R. Jourdain writes that he has tolerably conclusive evidence of one case where two broods were reared by one pair of Great Crested Grebes. In 1907 a pair had large young on June 13th at Osmaston. On October 6th Mr. J. Henderson reported a pair on the same pond accompanied by four young in down, which looked not much larger than Dabchicks. Mr. Jourdain adds that the eggs of this bird have been taken from April to September, which is strong, though not conclusive, evidence—since late nests may be the result of the destruction of previous eggs or broods—of their being double-brooded. Mr. Jourdain gives the following references :—Early dates—April 13th, 1888, 1 egg (C. R. Gawen, *Zool.*, 1889, p. 19) ; April 26th, 1881, 2 eggs (J. H. Gurney, *t.c.*, 1881). Late dates—July 22nd, 1898, nests with 3, 4, and 5 eggs in North Ireland, and other nests with eggs on September 1st (C. B. Horsburgh, *Field*, October 29th, 1898). More recent records of late nesting were on September 18th, 1904, when Mr C. Oldham saw downy young ones (*Zool.*, 1905, p. 37), and this year, on October 10th, when Mr. O. V. Aplin saw an old bird in " practically full summer plumage," with two half-grown young (*t.c.*, 1908, p. 407). Howard Saunders does not appear to regard this bird as double-brooded, and perhaps it is only so when certain favourable conditions prevail.—H.F.W.]

SOOTY SHEARWATERS IN SUSSEX, KENT, AND YORKSHIRE.

DURING September and October three specimens of the Sooty Shearwater (*Puffinus griseus*) were shot on the coast of Sussex and Kent, and, as I was enabled to examine each of them in the flesh, I venture to place the occurrences on record. The particulars are as follows : (1) a ♂. East Bay, Dungeness, September 26th, 1908 ; (2) a ♀, Cliff End, Pett, near Winchelsea, October 14th, 1908 ; and (3) a ♂, off Bexhill, October 21st, 1908.

W. RUSKIN BUTTERFIELD.

I have a note from Mr. W. J. Clarke, of Scarborough, of one of these birds obtained some miles off the coast of Yorkshire on October 6th last.—H. F. WITHERBY.

* * *

LESSER WHITETHROAT NESTING IN FORFAR.—*A correction.* —In reference to the supposed nesting of this bird in Forfar in June, 1907 (*antea*, Vol. I., p. 126) Mr. T. L. Dewar has submitted one of the eggs to Mr. Eagle Clarke, who pronounces it to be that of a Common Whitethroat (*cf. Ann. S.N.H.,* 1908, p. 254).

PROBABLE NESTING OF WHITE WAGTAIL IN SCOTLAND.— Two were seen in the middle of July at Killilan, North-west Highlands, and one had food in its bill (P. Anderson, *Ann. S.N.H.,* 1908, p. 253).

PROBABLE NESTING OF BLUE-HEADED WAGTAIL NEAR ABERDEEN.—Between May 19th—July 31st a pair of *Motacilla flava flava* was often observed frequenting the banks of a burn on the links near Aberdeen. On July 8th the female was carrying food, and the behaviour of the birds always seemed to show that they had a nest, although this was not found. An accurate description of the birds is given (L. N. G. Ramsay, *Ann. S.N.H.,* 1908, p. 253).

HAWFINCHES IN SCOTLAND.—An old bird and a young one were seen at Lauder, Berwickshire, in August, 1908 (W. M'Conachie). A female was accidentally captured on April 9th, 1908, at Grove Gardens, Galloway, in which county the bird has been recorded from time to time for many years (R. Service, *Ann. S.N.H.,* 1908, pp. 252 and 253).

Rose-coloured Starling in Caithness.—A specimen of *Pastor roseus* was shot at Dunbeath on July 11th, 1907 (*Ann. S.N.H.*, 1908, p. 195).

Nightjars Breeding in Captivity.—An extraordinary case of a pair of Nightjars breeding in captivity is recorded by Mrs. Heinroth, wife of Dr. O. Heinroth, of the Berlin Zoological Gardens. In November, 1906, a male bird was obtained, and was kept with great care through the winter. In the following spring a mate was procured, and pairing took place at the end of May. The male made a nesting place by scraping in a peccary-skin rug in the dining room. An egg was laid on June 2nd and another on the 4th. The hen bird did most of the sitting; but the male occasionally relieved her. On June 18th the first egg was chipped, and hatched on the morning of the 20th, while the second egg hatched on the afternoon of the same day. The young fed at first by taking the parents' beaks as far as the nostrils into their own. On June 24th the old birds again paired, and on July 3rd and 5th eggs were laid in the same spot upon the rug. Incubation in this case lasted eighteen days as against sixteen and a quarter in the first case, the longer time being accounted for by the fact that the old bird allowed the eggs to cool several times. The tameness of these Nightjars is described as extraordinary—the six birds flying about the room, taking no notice of strangers, and being quite ready to settle on the shoulder or take food from the hand (*Die Gefiedeste Welt*, xxxvii., 29-31, 33-4, and *Field*, 17, x., 08, p. 717).

Red-footed Falcon in Norfolk.—The Rev. Julian G. Tuck records the occurrence of a female *Falco vespertinus*, which was shot near Sandringham about the middle of June last (*Zool.*, 1908, p. 394).

Scottish Heronries.—Mr. H. Boyd Watt gives a list of 230 Scottish breeding places of the Heron, but of these he marks forty-five as now not occupied, while many others appear to be tenanted by only a pair or two. (*Ann. S.N.H.*, 1908, pp. 218–223).

Purple Heron in Caithness.—A young male *Ardea purpurea* is reported on the Thrumster Estate on September 16th, 1907 (*Ann. S.N.H.*, 1908, p. 199).

Glossy Ibises in Northumberland and Cornwall.—A party of five *Plegadis falcinellus* arrived near Alnmouth on August 30th. Four (two of which are said to be in immature plumage) were secured during the following ten days (E. L. Gill, *Zool.*, 1908, p. 394). One was "lately" shot near Land's End (H. Welch, *Field*, 24, x., 08, p. 721).

MALLARD HATCHING IN OCTOBER.—A Wild Duck hatched out thirteen young in the middle of October at Thames Ditton (R. Porter, *Field*, 24, x., 08, p. 721).

PROBABLE NESTING OF THE GADWALL IN SCOTLAND.—Two pairs of *Anas strepera* were under observation in a certain loch in the east of Scotland this year from the middle of May until the end of June, and they were doubtless nesting there (W. Evans, *Ann. S.N.H.*, 1908, p. 254).

INCREASE OF SHOVELERS IN TIREE. — Reported to be increasing yearly as a breeding species (P. Anderson, *Ann. S.N.H.*, 1908, p. 252).

MARKED TEAL.—A hand-reared Teal marked at Netherby, Cumberland, this year, was shot on Lough Derg on September 28th (R. Graham, *Field*, 24, x., 08, p. 745).

GARGANEY IN SHETLAND.—A male *Querquedula circia* is reported from Baltasound on April 14th, 1907 (*Ann. S.N.H.*, 1908, p. 200).

PALLAS'S SAND-GROUSE IN YORKSHIRE.—Two records of a few birds each have been reported (*antea*, pp. 98 and 134) of *Syrrhaptes paradoxus* in Yorkshire during the recent irruption of this bird. Mr. W. H. St. Quintin now records (*Naturalist*, 1908, p. 420) that a flock of 30 to 40 was noticed early in June near Knapton. A considerable number remained at any rate until the beginning of October. The flock appears never to have broken up into pairs, although it certainly decreased, and there is no evidence that the birds ever attempted to breed.

PRATINCOLE AT THE FLANNAN ISLANDS.—An adult female *Glareola pratincola* was obtained on July 13th, 1908, at this out-of-the-way spot. It is the third example of the species obtained in Scotland (W. Eagle Clarke, *Ann. S.N.H.*, 1908, p. 256).

THE POSITION OF THE EAR IN THE WOODCOCK.—In the October issue of the "Ibis" Mr. W. P. Pycraft contributes a short paper on the position of the ear in the Woodcock, in the course of which he controverts the contention of Professor D'Arcy Thomson, that the peculiar conditions which prevail in the matter of the position of the ear in the *Scolopacidæ* are due to the shifting of the beak in relation to the base of the skull. Mr. Pycraft now shows that the matter is not thus to be explained ; but, on the contrary, is due to the shortening of the base of the skull, which has had the effect of drawing the hinder part of the skull, and with it the aperture of the ear, downwards and forwards, and this point is demonstrated by means of a series of diagrams.

MARKED WOODCOCK.—Mr. John Hamilton has for four seasons marked young Woodcock at Baron's Court, co. Tyrone, with a nickel ring engraved " B.C.," with the year in figures. The results, as far as known, are as follows, but, unfortunately, the dates of the captures are not given :—

		Number accounted for.					
Breeding season.	Number marked.	In first season.	Place.	In second season.	Place.	In third season.	Not traced.
1905	15	1	Home	1	Home	Nil	13
1906	68	1	Home	5 1 1	Home Cornwall Harrow }	—	60
1907	65	—	—	—	—	...	65
1908	63	1	{ Near Inverness }	—	—	—	62
Total	211						200

(*Field*, 17, x., 08, p. 717, and 24, x., 08, p. 745).

SABINE'S GULL IN THE INNER HEBRIDES.—A specimen of *Xema sabinii* is recorded from Skerryvore on November 30th, 1907 (*Ann. S.N.H.*, 1908, p. 205).

GREAT CRESTED GREBE IN SHETLAND.—An example of *Podicipes cristatus* was seen at Spiggie on January 11th, 1907 (*Ann. S.N.H.*, 1908, p. 207).

REVIEWS

Report on the Immigrations of Summer Residents in the Spring of 1907 : *also Notes on the Migratory Movements during the Autumn of* 1906. By the Committee appointed by the British Ornithologists' Club. (Forming Vol. XXII., Bull. B.O.C. Edited by W. R. Ogilvie-Grant). 31 Maps. Witherby & Co. 6s.

THIS, the third Annual Report of the B.O.C. Migration Committee, although drawn up in the same form as previous reports, is rather more ambitious in that it includes some notes on autumn movements. These are too incomplete, however, to have much value, but we welcome the promise of a more elaborate record of autumn movements in the next report. In our notice (*antea*, Vol. I., p. 30) of the second " Report," we questioned the accuracy of the table which shows the areas of the arrival of the various species, and curiously enough in the present " Report " this table (p. 11) is not free from blemish, the White Wagtail being entered as arriving *solely* in the western half of the south coast, whereas in the detailed summary on page 107, as well as in the map, it is shown to have been reported first in Kent, and similarly the table does not tally with the summaries and maps in the cases of the House-Martin and Common Sandpiper. It would be as well, perhaps, to omit this table in future, or it may become permanently misleading, since even when it is corrected it is obvious by a comparison of the three " Reports " that the points of *non*-arrival·seem due in a great measure to want of observation. Each successive " Report," indeed, makes one realise more and more how little even the best observer is able accurately to record of the movements of migrants, and only an average of the results taken over a long period, as the Committee have from the first insisted, can lead to any reliable conclusions.

Some interesting points recorded in this volume may here be summarized. March, 1907, was brilliantly fine, but the whole of April was wintry ; the effect being that stragglers arrived at early dates, but the main body of birds was delayed, with the result that the " waves " of immigrants were less marked, and the period of migration was extended. The *Blackcap* was noted by many observers to be less numerous than usual in 1907. *Chiffchaffs* were seen at Penzance throughout the winter. The *Cuckoo* was reported on March 26th (Gloucester), 29th (Hereford), 30th–31st

(Wilts.), 31st (Dorset, Hants., Gloucester). The *Land-Rail* was neither heard nor seen in 1907 by observers in Hants., Sussex, Middlesex, Essex, Bucks., Herts., or Suffolk, and only once in Kent, twice in Berks. and Lincoln, and thrice in Norfolk. It seemed practically confined during the year under notice to the western counties. A specimen of the Continental Robin (*Erithacus rubecula rubecula*), which is common on migration on the *east* coast, was taken on April 7th at St. Catherine's Point, Isle of Wight. On page 180 the curious statement is made that the *Dartford Warbler* is rarely seen in winter in Hampshire !

In conclusion, we can unreservedly recommend the present and the two previous " Reports " to every student of migration, and we may add that the B.O.C. Committee and its many helpers all over the country by no means labour in vain.

H.F.W.

A List of Irish Birds. By R. J. Ussher, M.R.I.A., M.B.O.U. Dublin : A. Thom & Co. 4d.

THIS is a very useful up-to-date " abbreviated text book " on Irish birds. Mr. Ussher has placed within square brackets those American *land* birds which have been recorded from Ireland, and, on the whole, this is perhaps wise, although in a case such as the Yellow-billed Cuckoo, which has now occurred so many times in England, although seldom in Ireland, the rule might, perhaps, have been relaxed. Amongst other birds placed within square brackets we may mention the *Noddy Tern*, which has long been accorded a regular place upon the British list on the basis of two examples recorded by Thompson as having been obtained between the Tuskar Lighthouse and the Bay of Dublin about 1830. The birds were brought into port *skinned*, and we think that Mr. Ussher is perfectly right in not admitting them, more especially as the taxidermist who was responsible for the record was proved to have been unreliable in the case of two Belted Kingfishers supposed to have been shot a few years later. Of positive information additional to that given in the author's well-known " Birds of Ireland," there is very little, but we note the following, which do not appear to have been recorded elsewhere :—Five occurrences (against three in Saunders' *Manual*) of the *Red-breasted Flycatcher* are noted, but no dates are given ; an example of the *Serin Finch*, the second for Ireland, was taken on January 31st, 1907 ; a third specimen of the *Lapland Bunting* was taken alive at Kilbarrack on December 12th, 1907 ; the *Jay* is extending its range, and has spread into Kildare and Meath ; a pair of *Pochards*, with their young, were identified by Mr. R. Patterson in June, 1907, in Monaghan.

H.F.W.

BRITISH BIRDS

BIRDS

AN·ILLUSTRATED·MAGAZINE
DEVOTED·TO·THE·BIRDS·ON
THE·BRITISH·LIST

JANUARY 1,
1909.

Vol. II.
No. 8.

MONTHLY·ONE·SHILLING·NET
326·HIGH·HOLBORN·LONDON
WITHERBY·&·C°

SEP 8 1909

BRITISH·BIRDS

EDITED BY H. F. WITHERBY, F.Z.S., M.B.O.U.

ASSISTED BY W. P. PYCRAFT, A.L.S., M.B.O.U.

A TAME SNIPE AND ITS HABITS.

BY

HUGH WORMALD.

HAVING been asked to write a few notes on my tame
Snipe for BRITISH BIRDS, I cannot do better than relate
his history from the beginning. He was hatched in my
incubator on May 11th, 1908, incubation having lasted

twenty days, at a temperature of 102° Fah. He
remained in the incubator for twenty-four hours or so,
drying off, before he had his first meal.   There is no
prettier young bird than a Snipe in down, the colour of
which is a rich reddish-brown, speckled with black, and
here and there tipped with white.   Unfortunately a
pen and ink drawing (Fig. 1) cannot do justice to the bird
at this stage.   The combination of colours renders the
chick extremely difficult to find in its natural surroundings,
even when one knows to within a foot or so where it is
hiding, and I may mention that a spaniel is a very

FIG. 1.— A day old.
(*Drawn by* H. Wormald.)

useful aid in searching for both eggs and young of
Plover, Redshank, and Snipe.

For the first two days of his existence my young Snipe
ran backwards instead of forwards.   I believe this is
the case in a wild state.   The young do not pick up food
for themselves, like most young waders, but the parents
feed them from the bill.   I had for some time believed
this to be the case, and was glad to have my opinion
verified a short time ago by Mr. Richard Kearton, who
informed me that he had watched a male Snipe feeding
his offspring in this way.   In consequence of this habit,
I had to feed my young Snipe entirely by hand for the

first fortnight, the food then consisting of small worms, of which he devoured an enormous quantity.   As soon, however, as he had learnt to feed himself he took to maggots, and any small animaculæ that he could find while probing at the edge of a pond, or in mud which I dug up and gave to him in a pan.

The first signs of feathers appeared on the shoulders on May 17th.   The feathering was very rapid, the feathers of the tail and the back of the neck being the last to appear.   Fig. 2 shows the bird at this stage of development.   By the beginning of July he was quite grown up

FIG. 2.—As he appeared at the end of May.
(*Drawn by* H. Wormald.)

and fully feathered.   During the last week in September he commenced his first moult by losing his tail-feathers, the two outer ones being the last to fall.   The moult was completed about a month later.   On October 18th my brother winged a Common Snipe, which I took home alive, and this bird I take to be also a bird of the year, owing to the fact that it (I do not know the sex) was in exactly the same state of moult as my hand-reared bird.   Adults begin to moult during the end of July, and I have constantly seen them during the first week in August with their wing-feathers in full moult, but immature birds, as is commonly the case with waders, do not moult their primaries at all in their first autumn.

"John" (as my tame Snipe is christened) is exceedingly sluggish, and I believe that all Snipe are naturally so when undisturbed. He lives in a cage in the smoking-room, and sits every evening on a board in front of the fire. On being taken out of his cage and placed

FIG. 3.—Preening his Feathers.
(*Photographed by* P. H. Bahr.)

on the board his usual procedure is to give himself a shake (this he always does after being handled), and then eat two or three worms, after which he retires as near the fire as he can get, and "suns" himself for some little time. He then has another worm or two, preens his feathers (Fig. 3), and rests, either standing on one leg or squatting down on the board. Occasionally he varies this procedure by taking a bath, and very rarely he will hover round the room. His attitudes while sunning himself are very extravagant. He leans right over to one side and spreads his tail out into a fan, the *outside* tail-feathers nearest the fire *only* being extended beyond the rest. This is curious, for while bleating *both* outer tail-feathers are extended far beyond

Fig. 4.—Giving his Feathers a shake.
(*Photographed by* P. H. Bahr.)

the rest. He also raises the wing nearest the fire to get all the heat possible under the feathers. He continues in this attitude for a few minutes, then gives his feathers a shake (Fig. 4), turns round, and " suns " the other side.

The bill of the Snipe is known to be extraordinarily

flexible, and this is well shown as the bird yawns, when the last inch or so of the upper mandible is raised upwards. This movement is thought to be effected by the endotympanic muscle first described in 1748 by Hérissant* who, however, did not realise its function. Later the movement of the bill was described and figured in Bronn's "Thier Reich" (Taf IV., fig. 1). Mr. Pycraft† has observed the same thing in the Dunlin, and Dr. R. W. Shufeld,‡ in Wilson's Snipe and the American Woodcock. It would appear that in all the *Trochili* and *Scolopacidæ* the anterior part *only* of the upper mandible is movable.§ Mr. W. H. Workman‖ has written a paper on this subject, and has proved the endotympanic to be especially well developed in this species, and suggests that it acts by pulling the quadrate and maxillary bones forward, thus tilting the premaxillary upwards, which then gives at its most flexible portion, situated one inch from the tip of the bill. The use of this movement is obvious, in that it enables the bird when probing to grasp a worm underground, without even opening its bill, so that the tongue can draw the prey into the mouth. The flexibility of " John's " bill can also be noticed when he is trying to take a worm off a· hard flat surface, for then the tip of the upper mandible bends downwards.

His food, now that maggots are not procurable, consists entirely of worms, though I am endeavouring to teach him to eat raw liver, for worms will be very difficult to obtain during prolonged frost. He feeds entirely by " feel," being unable to see a worm right under him, but if one is placed two or three inches in front of him, he catches sight of it at once and walks up to it, then feels about with his bill until he touches it, when it is instantly swallowed. This shows the sensibility of the bill. He can also instantly distinguish between raw liver and a worm as soon as they come in contact with his

* "Histoire de l'Academie des Sciences," 1748, pp. 345-386.
† "Ibis," 1893, p. 361.   ‡ "Ibis," 1893, p. 563.
§ Gadow, "Dict. Birds," p. 877.   ‖ "Ibis," 1907, p. 614.

·bill, but this is not so surprising when one realises that the last inch or so of the bill is a mass of nerves.   Fig. 5 shows him toying with a worm held in front of him.   He feeds at intervals throughout the whole day and night, and eats a large quantity of grit and small pebbles, which

FIG. 5.—Toying with a Worm.
(*Photographed by* P. H. Bahr.)

can be heard grinding in his gizzard quite distinctly at a distance of several feet, especially immediately after feeding ; the gizzard grinds twelve times to the minute. The digestion is wonderfully rapid, so much so that I do not think a worm stays in the bird for more than ten minutes.   His hearing is very acute, and I have seen him listen like a Thrush, then drive his bill into the

turf and bring out a worm, which is sucked down with
no apparent exertion, and the bird does not throw back
his head as one constantly sees depicted, but rather
stretches out his neck, the bill pointing downwards.
Fig. 6 shows him in the act of swallowing a worm. If
the worm is too large to be swallowed whole, it is hammered
and pinched until broken up, when the pieces are
swallowed separately. He will eat any kind of worm
except brandlings, and is very fond of the grubs of

FIG. 6.—In the act of Swallowing a Worm.
(*Photographed by* P. H. Bahr.)

daddy long-legs. While feeding he keeps up a perpetual
twitter.

" John " is not so large as a wild Snipe, nor is his bill
so long as it should be, and I put both these defects down
to his being hand-reared. He is exceedingly tame,
and will let me do anything with him (*cf*. Fig. 7). He
will even " display " to me, walking round and round my
hand, uttering the spring note, with his tail spread out in

a fan, gently poking my hand with his bill. Then he squats down flat on the ground with his neck stretched out, which makes me wonder whether " he " is not really a female. Mr. Bahr's photographs are the best proof that could be given of his tameness to anyone who has not seen him, for they were taken almost on the window sill, the window being wide open, with the camera held three feet from the bird, which did not even flinch at the click of the shutter. Fig. 8

FIG. 7.—A Proof of his Tameness.
(*Photographed by* P. H. Bahr.)

depicts him looking out of the open window at a passing Rook.

When at rest he almost invariably stands on one leg, hopping about, and even feeding in this attitude, a habit common with most waders. Constantly he will play by himself, commencing by standing bolt upright and then squatting down flat, with his tail raised and spread out into a fan (the two outer tail-feathers not extended beyond the others). Then he will suddenly take two or three jumps to either side with wings closed. After going through this performance perhaps half-a-dozen times he strolls leisurely off.

As far as I can judge his eyesight is about equal during day and night.   I have been asked how weather affects him, but this I am unable to answer, because he lives,

FIG. 8.—Looking at a passing Rook.
(*Photographed by* P. H. Bahr.)

as before stated, in my smoking-room, out of the reach of weather.

It should be stated that " John " is in perfect health and plumage. So many so-called tame birds are really ill, which is the cause of their tameness, but I think the illustrations show that my Snipe's tameness is not caused through ill-health.

SOME EARLY BRITISH ORNITHOLOGISTS AND THEIR WORKS.

BY

W. H. MULLENS, m.a., ll.m., m.b.o.u.

VI.—THOMAS PENNANT (1726—1798).

ALTHOUGH the fame of Thomas Pennant both as a naturalist and as an author, has suffered somewhat by the lapse of time, he nevertheless must ever hold a somewhat prominent position amongst the British ornithologists of the past. This he would, perhaps, be entitled to by reason of his being the author of the first important history of British birds, which was illustrated with coloured plates* (*i.e.*, *The British Zoology*, London, 1766, one vol., folio). But this point, interesting as it is, is quite overshadowed by the fact that it was owing to Pennant's undoubted position as the leading British zoologist of his time that Gilbert White was led to address to him, in the shape of letters, those notes and observations which afterwards formed part of the immortal " Natural History of Selborne." The numerous zoological works of Pennant had, moreover, a very marked effect on the production of ornithological literature in Great Britain. The period which had elapsed from the death of the celebrated John Ray in 1705, till the publication of Pennant's " British Zoology " in 1766 is among the leanest in the history of British ornithology, but the publication of Pennant's works seems to have given an impetus to the production of such literature, and though many of the books that followed his " British Zoology," in quick succession, such as John Berkenhout's " Outlines of the Natural History of Great Britain " (London, 1769, three vols., 8vo) ; William Hayes' " Natural History of

---

* The first book treating of British birds, illustrated with coloured plates, would appear to be " A Natural History of English Song Birds," by Eleazar Albin, London, 1737, 1 vol., 8vo.

British Birds " (London, 1775, one vol., imp. folio) ; John Walcott's " Synopsis of British Birds " (London, 1789, two vols., 4to); William Lewin's "Birds of Great Britain " (London, 1789, seven vols., imp. 4to); Thomas Lord's " Entire New System of Ornithology, or Œcumenical History of British Birds " (London, 1791; 1 vol., folio); Bolton's " Harmonia Ruralis " (London, 1794, two vols., folio); and Edward Donovan's " Natural History of British Birds " (London, 1794, ten vols., 8vo); were little more than compilations, and of no particular interest save to the collector and bibliographer. Exception must be made in favour of such valuable works as John Legg's " Discourse on the Emigration of British Birds " (one vol., 8vo), anonymously published at Salisbury in 1780, and afterwards erroneously attributed to George Edwards ; Tunstall's " Ornithologia Britannica,"* which also appeared anonymously in 1771 (London, one vol., folio); the well-known "General Synopsis of Birds," by John Latham (London, 1781), which contained in the second volume of its supplement " A List of the Birds of Great Britain," and the still more famous " History of British Birds," by Thomas Bewick, the first volume of which appeared in 1797.

Thomas Pennant, the son of a country gentleman, was born at Downing, in Flintshire, in the year 1726, and was educated at Queen's College, Oxford. Our principal source of information for the particulars of Pennant's life is his own work :—

" The / Literary Life / of the late / Thomas Pennant, Esq. / By Himself. / [Latin quotation] London : / Sold by Benjamin & John White, Fleet-Street, / and Robert Faulder, New Bond-Street. / MDCCXCIII.

1 vol., 4to, pp. 144 & IV. Plates.

From this quaint and somewhat self-laudatory work we learn that Pennant having received as a present from a kinsman, when twelve years old, a copy of the

---

* A similar but much rarer work by Charles Fothergill was published at York in 1799.

" Ornithology of Francis Willughby," early developed
a " taste for that study, and incidentally a love for that
of natural history in general, which I have since pursued
with my constitutional ardor." Pennant began the first
of his many " Tours," his accounts of which from their
topographical interest are more read at the present day
than his other writings, from Oxford in 1747. His first
literary work, an extract from a letter written to his uncle,
James Mytton, concerning an earthquake at Downing
in 1750, appears in the 10th volume of the " Abridgement
of the Philosophical Transactions," and thenceforward
his active pen knew no rest until the time of his death,
when he was engaged on an ambitious work entitled
" Outlines of the Globe," of which he had projected some
fifteen quarto volumes, only four of which would seem
to have been published. It is here only possible to deal
with a few of the zoological books of this prolific author,
but it may afford some idea of the vast output of his
writings if we mention that the number of plates engraved
for his several works totals no less than eight hundred
and two (cf. Literary Life, p. 38). In 1755 Pennant
commenced a correspondence with the great Linnæus,
and in 1757, as he tells us, received " the first and greatest
of my Literary honors," being elected " at the instance
of Linnæus himself," a member of the Royal Society of
Upsal. In 1761 Pennant began to publish his " British
Zoology," which, when completed in 1776, contained
one hundred and thirty-two coloured plates, engraved
by Peter Mazel, and coloured by Peter Pallou, " an
excellent artist, but too fond of giving gaudy colours to
his subjects." This work which, as Pennant himself
observes, would have been more useful in quarto size,
he produced chiefly at his own expense, devoting the
proceeds to the " benefit of the Welch Charity-School on
Clerkenwell Green " (cf. adv. to the second edition of
The British Zoology, 1768). The publication of the
first edition of the " British Zoology " had been delayed
by a journey, which Pennant made to the continent in

1765. In the course of his travels he visited Buffon
(1707-1787) at Paris, and informs us that " the celebrated
naturalist was satisfied with my proficiency in natural
history, and publickly acknowledged his favourable
sentiments of my studies in the fifteenth volume of his
' Histoire Naturelle.' Unfortunately long before I had
any thoughts of enjoying the honour of his acquaintance
I had in my ' British Zoology ' made a comparison between
the free-thinking philosopher and our great and religious
countryman, Mr. Ray, much to the advantage of the
latter . . . . . but such was his irritability, that in the
first volume of his ' Histoire Naturelle des Oiseaux,' he
fell on me most unmercifully, but happily often without
reason." From France, Pennant passed on to Germany
and Holland, and at The Hague met Pallas (1741–1811),
the famous traveller, " a momentous affair, for it gave
rise to my ' Synopsis of Quadrupeds,' * and the second
edition, under the name of the ' History of Quadrupeds,' †
a work received by the naturalists of different parts of
Europe in a manner uncommonly favourable."

To return to the " British Zoology," the full title is
as follows :—

" The / British Zoology / Class I. Quadrupeds. /
II. Birds. / Published under the Inspection of the /
Cymmorodorion Society, / Instituted for the / Promoting
Useful Charities and the knowledge of / Nature among
the Descendants of the / Ancient Britons. / Illustrated
with / one hundred and seven Copper Plates. / London : /
Printed by J. & J. March, on the Tower Hill, for the
Society : / and sold for the Benefit of the British Charity
School on / Clerkenwell-Green. MDCCLXVI.

1 vol., imp. folio. Collation : pp. 14, un. + pp. 162 +
pp. 4, Index and list of " Encouragers to this Undertaking," + CXXXII. Plates. (A fifth part containing
twenty-five plates was added to the one hundred and
seven enumerated in the above title, thus making one

---

* Chester, 1771, 1 vol., 8vo.　　† London, 1781, 2 vols., 4to.

hundred and thirty-two in all, viz., eleven of quadrupeds
and one hundred and twenty-one of birds.)

In 1768 appeared the second edition of the above. This
was published in two volumes by Benjamin White (brother
of Gilbert White, the naturalist), who paid Pennant £100
for the right of publication.

In 1770 an octavo volume was published of ninety-six
pages, "including a list of European Birds extra
Britannic," and CIII. Plates. This must rank as the
third edition of the "British Zoology"—it was incor-
porated in the fourth edition, published in 1776, four
volumes 4to and 8vo. This edition was printed at
Warrington for Benjamin White, and is sometimes found
with the plates coloured. A fifth edition, also in four
volumes, 4to and 8vo, appeared in 1812.

It may here be mentioned that the folio edition of the
"British Zoology" had been translated into German and
Latin by "M. de Murre, of Nurenbergh," and published
in the same size as the original, but the colouring of the
plates is an improvement on that in the English edition.

The summer of the year following the publication of
the "British Zoology," viz., August, 1767, saw the commencement of the celebrated correspondence between
Gilbert White and Thomas Pennant ; White's share of
which (Pennant's is lost) was afterwards published in
his "Natural History of Selborne." This correspondence
continued down to November, 1780, and consisted in all
of forty-four letters, the first actually addressed to Pennant
by White being numbered ten in the series, the preceding
nine being added for the sake of uniformity when White
published his book in 1789. The correspondence was
commenced by White, who was prompted to address
his observations to Pennant both on account of the latter's
leading position as a naturalist, and also because "of
your repeated mention of me in some late letters to my
brother" (*i.e.*, Benjamin White, Pennant's publisher).
There does not seem to have been any great friendship
between White and Pennant—Gilbert White appears to

have been hurt at Pennant's making full use of the
material contained in White's letters for his second and
subsequent editions of the " British Zoology," without
due acknowledgment ;* and Pennant makes no mention
of the Selborne naturalist in his " Literary Life."
" Little did he anticipate," says Professor Bell, ...
" that his correspondent would be commemorated with
ever-increasing admiration and esteem, while his own
more pretentious book is only regarded of value because,
at the time of its publication, it filled a gap in British
Natural Science, and contained some matter of importance, the best of which was really not his own."

It was, however, probably to Pennant that White owed
his first introduction to Daines Barrington, his other
correspondent ; and to whom the remaining sixty-six
letters of the " Natural History of Selborne " were
addressed. Writing to Pennant in 1768, White says,
" I have received from your friend Mr. Barrington one
of the naturalist's journals, which I shall endeavour to
fill up in the course of the year."

In 1766 Pennant made the acquaintance of another
very eminent man, Sir Joseph Banks (1743–1820), the
zoologist, and companion of Cook in his circumnavigation
of the globe.   The commencement of Pennant's friendship
with Sir Joseph Banks was signalised by a gift from the
latter of a copy of Turner's " Avium Historia," a book
which even at that time was described as scarce.   From
Sir Joseph Banks, Pennant received much kindness and
help, notably in the case of his " Arctic Zoology,"
published in 1785 (three volumes and supplement, 4to),
which, although mainly a compilation, proved to be
by far the most valuable of Pennant's zoological works,
and which was translated into German, French, and
Swedish.   Of Pennant's contributions to natural history
there is but little to be said ; they derived their great

* But such acknowledgment was rare at that time, and Pennant
does refer to the help he received from White, p. xiii., preface, and
p. 498, appendix to the 1768 edition of the " British Zoology."

THOMAS PENNANT.

1726—1798.

(From the Engraving by J. Romney, after the Painting by T. Gainsborough)

popularity partly from their very brief and formal
descriptions, and partly from the lack of standard works
available both at that time, and for many years to come.
The charm of Gilbert White had yet to be discovered,
and though the woodcuts of Thomas Bewick proved a
great incentive to the study of ornithology, it was not
until the genius of George Montagu produced in 1802
the " Ornithological Dictionary " that the work which
had been begun by Willughby and Ray, was properly
continued. The very productiveness of Pennant's work
no doubt also detracted from its utility—as he himself
tells us, " I am often astonished at the multiplicity of
my publications, especially when I reflect on the various
duties it has fallen to my lot to discharge, as a father of a
family, landlord of a small but very numerous tenantry,
and a not inactive magistrate." * Towards the close of
Pennant's active life he was confined to his ancestral
seat at Downing by an accident which broke the patella
of his knee, but he continued to work with unabated
energy at the revision of his " Outlines of the Globe,"
but his health was rapidly failing, and he passed away
on December 16th, 1798, at the advanced age of
seventy-two.

* Besides the Zoological works already mentioned, Pennant wrote
"Indian Zoology," 1769-1790; "Genera of Birds," Edinburgh, 1773,
and London, 1781; "Indexes to the Ornithologie of the Comte de
Buffon," 1786, while the observations on natural history contained
in the various Tours, notably in " The Tour to Scotland," 3 vols.,
1776, and that "in Wales," 3 vols., 1810, are of considerable interest,
and this principally from the fact that they were jotted down without
any attempt at scientific treatment.

ON THE MORE IMPORTANT ADDITIONS TO OUR KNOWLEDGE OF BRITISH BIRDS SINCE 1899.

BY

H. F. WITHERBY and N. F. TICEHURST.

PART XVI.

(Continued from page 229.)

COMMON SNIPE *Gallinago cœlestis* (Frenzel). S. page 573.

Weight.—Ninety shot in Shetland averaged 5.78 oz. Have been killed up to 7¾ oz. (*Ann. S.N.H.*, 1906, p. 53, and 1905, p. 55).

"*Sabine's Snipe.*"—Some examples regarded as mutations or discontinuous variations and not melanoid varieties (W. P. Pycraft, *Ibis*, 1905, p. 289).

BROAD-BILLED SANDPIPER *Limicola platyrhyncha* (Temm.). S. page 577.

KENT.—An immature female was procured near Littlestone-on-Sea, on August 31st, 1901 (L. A. Curtis Edwards, *Zool.*, 1901, p. 390).

SUSSEX.—An immature female was shot at Rye on August 29th, 1904 (M. J. Nicoll, *Bull. B.O.C.*, XV., p. 12).

AMERICAN PECTORAL SANDPIPER *Tringa maculata* Vieill. S. page 579.

SUFFOLK.—One was shot at Aldeburgh on September 13th, 1900 (E. C. Arnold, *Zool.*, 1900, p. 521). [A "Pectoral Sandpiper" was reported in the "Field" to have been shot at Southwold on September 2nd, 1904 (J. H. Gurney, *t.c.*, 1905, p. 96).]

CORNWALL.—Two have been obtained on the mainland of the county, the last at Porthgwarra on April 30th, 1906 (J. Clark, *t.c.*, 1907, p. 286).

SCILLY ISLES.—Ten are recorded in place of four mentioned in the "Manual." The last was shot by Captain Dorrien-Smith in September, 1891 (J. Clark and F. R. Rodd, *t.c.*, 1906, p. 339).

IRELAND.—A young bird in full autumn plumage was shot at Belmullet, co. Mayo, early in October, 1900 (H. Saunders, *Bull.B.O.C.*, XI., p. 34), and another was shot near the same place in September, 1900 (R. J. Ussher, *List of Irish Birds*, p. 44).

CHANNEL ISLANDS.—*Jersey.*—One in Mr. Romeril's collection was shot from a party of four about thirty years ago (A. Mackay, *Zool.*, 1904, p. 379).

AMERICAN STINT *Tringa minutilla* Vieill. S. page 587.

CORNWALL.—One " was killed by a fisherman near Mousehole in September, 1890, and was bought in the flesh by W. E. Baily, of Paull, in whose collection the writer saw it in February, 1902, incorrectly labelled ' *Tringa minuta* ' " (J. Clark, *Zool.*, 1907, p. 286).

CURLEW-SANDPIPER *Tringa subarquata* (Güld.).
S. page 591.

NESTING.—Found nesting numerously in June, 1901, by the late Dr. H. Walter in the Taimyr Peninsula (H. E. Dresser. *Ibis*, 1904, p. 231).

KNOT *Tringa canutus* L. S. page 595.

NESTING.—Found nesting in June, 1901, by the late Dr. H. Walter in the Taimyr Peninsula. The eggs vary greatly in form, size, and coloration; the nests—depressions lined with a few dry grass-bents and white tangle—were placed in grassy places on the Tundra; the incubating male (or female) did not leave the nest until almost trodden on, when it puffed out its feathers until it appeared almost double its normal size; the male was most careful of the young, but the female appeared as an uninterested spectator (H. E. Dresser, *Ibis*, 1904, p. 232). A clutch of eggs was taken in Hrisey, in the north of Iceland, on June 17th, 1898, and the bird belonging to it is stated to have been watched at a few yards' distance by a competent observer—E. Möller, a collector in Iceland, now dead (Otto Ottosson, *t.c.*, 1905, p. 105).

DISTRIBUTION.—*T. canutus* is an irregular visitor to India on migration as well as *T. crassirostris* (F. Finn, *t.c.*, p. 351). Dr. V. Bianchi has informed us that *T. canutus* is common on the Yenesei and Lena Rivers.

RUFF *Machetes pugnax* (L.). S. page 599.

DURHAM.—Nested in 1901, 2 and 3 near the mouth of the Tees, and not on the Yorkshire side as stated *antea* Vol. I., p. 68 (T. H. Nelson, *Ibis*, 1906, p. 735 and *in litt.*).

HEBRIDES.—Six records for the Outer Hebrides are detailed (J. A. Harvie-Brown, *Ann. S.N.H.*, 1903, p. 13). *South Harris.* —A male, autumn, 1906 (*Field*, 29, IX., 06, p. 580). *Coll.*— One about September 16th, 1905 (*t.c.*, 1906, p. 201).

IRELAND.—For a detailed account of the various occurrences *cf.* C. J. Patten, " Irish Nat.," 1900, p. 187.

BUFF-BREASTED SANDPIPER *Tringites rufescens* (Vieill.).
S. page 601.

NORFOLK.—An immature male was shot near Wells, on September 7th, 1899 (E. C. Arnold, *Zool.*, 1899, p. 475).

BARTRAM'S SANDPIPER *Bartramia longicauda* (Bechst.).
S. page 603.

CORNWALL.—One was found hanging in a poulterer's shop at Falmouth in October, 1903, by Dr. Owen (J. Clark, *Zool.*, 1907, p. 286).

SPOTTED SANDPIPER *Totanus macularius* (L.).
S. page 605.*

KENT.—A pair were shot on May 5th, 1904, in a ditch between Lydd and Brookland, in Romney Marsh (J. L. Bonhote, *Bull. B.O.C.*, XIV., p. 84).

WOOD-SANDPIPER *Totanus glareola* (J. F. Gm.).
S. page 607.

IRELAND.—One was obtained on August 26th, 1899, by Mr. J. F. Knox, on the Black Strand, Trancore, co. Waterford (E. Williams, *Irish Nat.*, 1899, p. 231). One was shot on August 19th, 1901, near Sutton, co. Dublin (W. J. Williams, *t.c.*, 1901, p. 205).

ORKNEY.—One was shot on Eday on September 1st, 1902 (C. S. Buxton, *Zool.*, 1902, p. 391).

GREEN SANDPIPER *Totanus ochropus* (L.). S. page 609.

SCOTLAND.—*South Uist.*—One was obtained in the autumn of 1901, and was the first recorded for the Outer Hebrides (J. MacRury, *Ann. S.N.H.*, 1902, p. 55). *Fair Isle* (Shetlands.) —One or two (the first for the Shetlands or Orkneys) were seen in early September, 1905 and 1906 (W. E. Clarke, *t.c.*, 1906, p. 76, and 1907, p. 79).

IRELAND.—A solitary bird was shot at Foxford, co. Mayo, on June 30th, 1903 (G. Knox, *Irish Nat.*, 1903, p. 248), and another at Malahide, co. Dublin, on April 28th, 1906. The species is chiefly known in Ireland as a casual autumn and winter visitor (R. J. Ussher, *List of Irish Birds*, p. 46).

SOLITARY SANDPIPER *Totanus solitarius* (Wilson).
S. page 611.

SUSSEX.—One was shot at Rye Harbour on August 7th, 1904 (C. B. Ticehurst, *Bull. B.O.C.*, XV., p. 12).

Nesting.—The eggs were first taken in 1903, and again in 1904 in Northern Alberta by Mr. Evan Thompson. They are described as being like those of the Green Sandpiper, but considerably smaller, and like that bird, this species lays in the old nests of other birds. One set of eggs was found on June 16th, 1903, in the old nest of an American Robin, some fifteen feet from the ground ; another on June 9th, 1904, in a Bronzed Grackle's nest, in a similar position, and another on June 24th, 1904, in the old nest of a Cedar-Waxwing (cf. F. C. R. Jourdain, *Ibis*, 1905, p. 158, and 1907, p. 517, Pl. XI., Figs. 1, 4).

SPOTTED REDSHANK *Totanus fuscus* (L.).
S. page 617.

Scotland.—*East Renfrewshire.*—One was seen in October, 1898, and a pair in September, 1899 (*Ann. S.N.H.*, 1899, p. 51, 1900, p. 51). *Dumfriesshire.*—One was shot on February 13th, 1899, on the Solway (*t.c.*, 1899, p. 112). One was *seen* in October, 1903, at Carsethorn (*t.c.*, 1904, p. 216).

North Wales.—Very rare, only occurred three or four times—the last on the Dovey Estuary in September, 1899 (H. E. Forrest, *Vert. F. N. Wales*, p. 362).

Nesting.—Nests found by the late S. A. Davies and Mr. J. Stares on the River Muonio (Lapland) were always in the marshes (*Ibis*, 1905, p. 84).

RED-BREASTED SNIPE *Macrorhamphus griseus* (J. F. Gm.).
S. page 621.

Yorkshire.—One shot in September, 1864, on Norland Moor has been examined by Messrs. Eagle Clarke and Nelson (*B. of Yorks.*, p. 638).

Hampshire.—Two, said to have been got in the county, are in Mr. Hart's collection, one being dated September, 1872, and the other October, 1902 (J. E. Kelsall and P. W. Munn, *B. of Hants.*, p. 320).

BLACK-TAILED GODWIT *Limosa belgica* (J. F. Gm.).
S. page 625.

Scotland.—*Outer Hebrides.*—Two have apparently been obtained (cf. J. A. Harvie-Brown, *Ann. S.N.H.*, 1903, p. 14). *Lanarkshire.*—Three were identified near Lenzie on May 4th, 1907 (J. Paterson, *t.c.*, 1907, p. 184).

COMMON CURLEW *Numenius arquata* (L.). S. page 627.

Surrey.—A nest was found and two eggs taken on Chobham Common in 1897 (H. Saunders, *Bull. B.O.C.*, XI., p. 34).

(*To be continued.*)

THE GREENLAND WHEATEAR *SAXICOLA ŒNANTHE LEUCORRHOA* (GMELIN).

BY

C. B. TICEHURST, M.A., M.R.C.S., M.B.O.U.

It¯is curious that so little attention should have been paid to this bird of late years, and that its migrations through Great Britain should be so little known. Gould, in his " Birds of Great Britain," seems to have noted the occurrence of this large Wheatear, but it was not until 1879 that Lord Clifton (now the Earl of Darnley) pointed out that this race did not arrive on the Kent and Sussex shores till May and, besides being larger, differed from the small race in having a deeper reddish buff throat and breast ; further, he did not know of its occurrence west of Sussex (*Ibis*, 1879, pp. 256–7).

As far as I have been able to ascertain no one, since Lord Clifton wrote on the subject, has described its range in Great Britain. I have examined 460 Wheatears or Wheatears' wings, obtained in various parts of Great Britain, and in many other parts of the world, and I think that it can be said with certainty that the Greenland Wheatear *passes through the whole of Great Britain* on migration, for I have seen specimens of it from Yorkshire, Suffolk, Norfolk, Kent, Sussex, Hants, Middlesex, Cornwall, Scilly Isles, Channel Isles, Pembrokeshire, co. Wexford, and Shetland, whilst Mr. Barrington (*Migration of Birds at Irish Lights*) records Wheatears with large wings from cos. Cork, Donegal, Antrim, Dublin, and Wicklow, which evidently belong to this race.

It usually arrives in the south of England during the last week in April, and the first week in May, and continues passing through till the end of that month ; a few early ones may sometimes be seen migrating with the small race in the second and third weeks of April, and the earliest record I have is April 15th. The return journey takes place usually during the latter half of September, though a few examples are recorded during the last week in August, and the first part of September ; they continue to pass .during October, and the latest date of which I have a record is October 31st.

Of the distribution of this Wheatear outside Great Britain more is known, for Herr Stejneger, in reviewing the whole subject (*Proc. U.S. Nat. Mus.*, Vol. XXIII., No. 1220) states that it migrates *via* France, Great Britain, Shetlands, and Faroe Islands, to Greenland and the opposite portions of North America, as well as to Iceland, where it is the breeding species, whilst the western part of North America is inhabited by the *small* (typical) race, which reaches these parts *via* the Asiatic continent.

Exactly where the Greenland Wheatear passes the winter is not yet completely known. Hitherto it has been supposed to have been more or less confined to the western part of North Africa, Senegal (where probably the original type specimen was obtained), and Gambia districts, but I have seen undoubted specimens from Khartoum, Nubia, and Fashoda ; so that, although the majority may winter in West Africa, some at least spread as far east as the Nile Valley. It apparently passes through the Azores on migration.

The first Wheatears arrive in Greenland, according to Herr Winge (*Groenland's Fuglefauna*) about the end of the first week of May ; in early years it may be seen in the first few days of May, in late years not till the third week. The return migration lasts from mid-August to mid-September, and few are seen by the end of that month ; it has frequently been met with flying over the open sea south-west of Iceland. Whether this race breeds in the Faroe Islands or not must, I think, at present remain doubtful.

Taking into consideration the difference in coloration and size, migration, and breeding area, I have not the slightest hesitation in agreeing with Lord Clifton and Herr Stejneger as to the distinctness of the Greenland race. The following diagrams give the results of my measurements of 450 Wheatears' wings :—

Length of Wings of Males in millimètres.

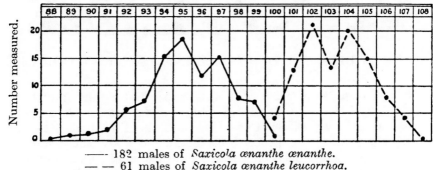

—— 182 males of *Saxicola œnanthe œnanthe.*
— — 61 males of *Saxicola œnanthe leucorrhoa.*

Length of Wings of Females in millimètres.

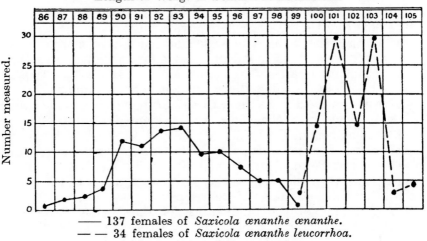

—— 137 females of *Saxicola œnanthe œnanthe.*
— — 34 females of *Saxicola œnanthe leucorrhoa.*

It will be seen that there is some slight overlapping in measurements, but I find that this does not amount to more than 2 per cent. of individuals—which agrees exactly with Herr Stejneger's results. Herr Winge states that no Wheatears from Greenland which he has examined measure less than 100 mm. in the wing, with which statement I can quite agree (the two or three females of 99 mm. which I measured probably being rather worn specimens) ; moreover, I have not been able to find any Wheatears shot outside the range of the Greenland Wheatear which do not conform in wing measurement to the small race.

All these birds were measured by myself, and only those with which full data were recorded have been utilised.

THE JUBILEE OF THE BRITISH ORNITHOLOGISTS' UNION.

THE meeting to celebrate the Jubilee of the foundation of the B.O.U. in 1858 was held at 3, Hanover Square, on December 9th, 1908. Dr. F. Du Cane Godman, the President, was in the chair, and the proceedings commenced by the reading of a number of congratulatory messages from other Ornithologists' Unions. Dr. Godman then gave a short address, which showed how intimate had been the relation between the progress of ornithology and the progress of the B.O.U. Dr. P. L. Sclater gave a history of the Union, its journal, the " Ibis," of which he has for so many years been editor, and its founders, chief amongst whom was the much-lamented Alfred Newton. Mr. A. H. Evans spoke very briefly of the life and work of some of the founders.

Mr. Henry Upcher, as the earliest (surviving) elected member (1864), then took the chair, and presented on the behalf of the members of the Union, a gold medal to each of the four (surviving) founders, viz., Dr. F. Du Cane Godman, Dr. P. L. Sclater, Mr. Percy Godman, and Mr. W. H. Hudleston. The medal bears on the obverse the well-known figure of the Ibis, on the reverse the name of the recipient.

A facsimile of the original list of the twenty founders, written by Newton and corrected by Dr. Sclater in 1859, was handed round. Amongst the names famous in ornithology, besides those already mentioned, may be noted Lieut.-Col. H. M. Drummond (first President), T. C. Eyton, J. H. Gurney (Senr.), Hon. T. Lyttleton Powys (afterwards Lord Lilford), Osbert Salvin, Rev. (afterwards Canon) H. B. Tristram, and John Wolley.

In the evening a largely attended commemorative dinner was held.

A special volume of the " Ibis " commemorating the Jubilee and containing a history of the Union, with lives of the founders and principal members, together with portraits, will be published shortly.

THE BRITISH ORNITHOLOGISTS' UNION AND RARE BREEDING BIRDS.

AT the annual meeting of the British Ornithologists' Union, held in May last, H. F. Witherby proposed a new rule, the

effect of which · was to exclude from the Union any member
who took or connived at the taking of any bird or egg of
certain species which were extremely rare as breeding birds in ·
the British Isles. The proposer explained that his rule was
founded on purely scientific grounds, his opinion being that
it was unscientific, and, therefore, directly contrary to the
interests of the premier Ornithologists' Union of the world,
to exterminate or risk the extermination of any bird in any
particular portion of its breeding area, and so alter its natural
geographical distribution. The details of the rule were much
criticized, and it was generally thought to be too drastic in
character, although the majority at a largely attended meeting
were without doubt in favour of the " spirit " of the proposed
rule. It was decided to refer the matter to the Committee
for consideration.

At a Special General Meeting of the Union held on December 10th Dr. F. Du Cane Godman, the President, being in
the chair, the Commitee communicated their report, and
submitted a new rule for the consideration of the members.
The proposal to adopt the new rule was seconded by H. F.
Witherby, who withdrew his proposed rule. Amendments moved
by the Hon. Walter Rothschild and Dr. J. Wiglesworth were
carried, and amongst others who took part in an exhaustive
discussion were the following:—Messrs. R. M. Barrington,
W. Bickerton, P. F. Bunyard, W. Eagle Clarke, Dr. F. D.
Drewitt, Messrs. J. Gerrard, N. Gilroy, A. F. Griffith, Dr. E.
Hartert, Sir T. Digby Pigott, Mr. A. Trevor-Battye, Lt.-Col.
R. G. Wardlaw-Ramsay, and the Honorary Secretary, Mr. J. L.
Bonhote. The rule as amended was then put to the meeting
and was carried *unanimously*. The new rule will require confirmation at the next annual meeting of the Union. As finally
amended it reads as follows :—

" If, in the opinion of the Committee, any member shall have
acted in a manner injurious to the interests of or good name of
the Union, or shall have personally assisted in, or connived at,
the capture or destruction of any bird, nest or eggs in the British
Isles, by purchase or otherwise, likely, in the opinion of the
Committee, to lead to the extermination or serious diminution
of that species as a British bird, the Secretary shall be
directed to send a registered letter to the member, stating
the facts brought before the Committee, and asking for an
explanation of the same, but without mentioning the source
from which such information was obtained. After allowing
a reasonable time (not less than a clear fortnight after the
receipt of the Secretary's letter) for reply, or for appearing in
person before the Committee if he so desire, the Committee,
provided that not less than four are agreed, shall have
power to remove that gentleman's name from the List of

Members without assigning any reason.　Such member may, if he so desire, stand for re-election by ballot at the next Annual Meeting, and in the event of his re-election, no fee for re-admission shall be required."

The action of the British Ornithologists' Union in condemning in such unhesitating fashion the practice of collecting the birds and eggs of *rare British breeding species* will be received with the most intense satisfaction by all who have the science of Ornithology at heart.—EDS.

IRISH BIRDS.

IN noticing Mr. Ussher's " List of Irish Birds " in the last number of this Magazine, I much regret to have done an injustice to Irish ornithologists—quite unintentionally—by stating that little information had been added since the publication of Messrs. Ussher and Warren's " Birds of Ireland." I fully intended to add, " which had not already been referred to in these pages." Mr. Ussher has very kindly supplied me with particulars of the information additional to that in the " Birds of Ireland " contained in his " List," and I am glad to be able to draw attention to the following records which have not already been mentioned in BRITISH BIRDS :—

WATER-PIPIT.—A specimen shot by the late Canon Tristram on Rockabill, co. Dublin, in June, 1861, has hitherto been unrecorded. This, the first and only Irish specimen, is now in the Dublin Museum.

HONEY-BUZZARD.—One was shot in King's County, on September 28th, 1903.

AMERICAN BITTERN.—Has now occurred fifteen times, as against eleven given in the " Birds of Ireland."

SPOONBILL.—Has occurred in thirty-five instances, while only thirty-three were mentioned in the " Birds of Ireland."

CRANE.—In the " Catacombs " cave at Edenvale, co. Clare, several bones of Crane have been discovered.

There is also additional information with regard to some of the Terns and Shearwaters and other birds which will be noticed in the articles on " Additions." Mr. Ussher also points out that in the case of the *Rose-coloured Starling* he made a slip in stating that only about twenty had been recorded—the number should have been twenty-eight.　　　　　　H.F.W.

RARE BIRDS IN IRELAND.

BLACK REDSTART (*Ruticilla titys*).

One was shot near Mountrath, Queen's co., on November 4th, 1908.

NOTES. 277

HONEY-BUZZARD (*Pernis apivorus*).

An immature male was shot near Ardee, co. Louth, on October 13th, 1908, and, being only winged, was forwarded by its captor to the Dublin Zoological Gardens where, however, it died within a week of its arrival.

OSPREY (*Pandion haliaëtus*).

On November 1st, 1908, a bird in immature plumage flew on board a fishing-boat coming to Wexford, and was captured, but died soon after reaching the shore. I examined the bird and found it thin, although the plumage was in good order.

BUFFON'S SKUA (*Stercorarius parasiticus*).

A bird in first year's plumage was shot in a wood bordering Lough Neagh, co. Antrim, on November 18th, 1908.

GREAT NORTHERN DIVER IN SUMMER PLUMAGE IN OCTOBER.

A very large specimen of *Colymbus glacialis* in full summer plumage was shot on the River Moy, co. Sligo, on October 31st, 1908. The bird showed no trace whatever of winter plumage, and was in fact in better plumage than birds I have examined in the month of May.　　　W. J. WILLIAMS.

. ALBINISTIC VARIETY OF THE REDWING.

EARLY in November we received from Filey, Yorkshire, an albinistic variety of the Redwing. This bird was pale cream-coloured all over, the bases of the feathers being, however, grey. Its beak was yellowish, and the legs and feet were very pale brown. We sent it to Mr. Eagle Clarke, who identified it as *Turdus iliacus*, and we have presented the bird to the Royal Scottish Museum.

EVELYN V. BAXTER AND LEONORA JEFFREY RINTOUL.

THE NORTHERN MARSH-TITMOUSE IN ENGLAND.
A NEW BRITISH BIRD.

AN undoubted example of the Northern Marsh-Tit (*Parus borealis* De Selys) was shot at Tetbury, Gloucestershire, in March, 1907, by Mr. J. H. Paddock, who presented it to the British Museum. I had the pleasure of exhibiting this bird at the meeting of the British Ornithologists' Club held on November 18th, 1908 (*cf. Bull. B.O.C.*, XXIII., p. 34). In January, 1908, I observed a small lot of four or five Marsh-Tits, undoubtedly of this species, at Welwyn, Herts. My attention was first attracted by their Linnet-like song, composed of a number of broken ascending notes, entirely different to the call of the common Marsh-Tit. I watched the birds at very close range, and had no doubt in my own mind that they were Scandinavian Marsh-Tits, the white sides of the face and the pale upper-parts being very conspicuous. I was, however, unable to procure a specimen; and, although I wrote to Mr. H. F. Witherby, describing my experience, I did not venture to place my observations on record until receiving this undoubted example of *P. borealis* from Mr. Paddock. It is difficult to account for the appearance of this North-west European Titmouse in Great Britain, for, so far as is at present known, it is not a migratory species. It must now, however, be added to the list of our accidental visitors.

W. R. Ogilvie-Grant.

THE FIRST BRITISH EXAMPLE OF THE RED-THROATED PIPIT.

The first recorded "British" example of the Red-throated Pipit (*Anthus cervinus*) is said to have been obtained near Brighton on March 13th, 1884.* This example went into the "Monk" Collection, and finally passed into the Booth Museum at Brighton.

A few months ago I had the opportunity of examining the specimen in question, and I have no hesitation in saying that it is not a Red-throated Pipit at all, but merely a brightly-coloured example of the Meadow-Pipit (*Anthus pratensis*). During the spring (March and April) large flocks of Meadow-Pipits arrive on the coast of Sussex, and all the males of these immigrants are very brightly coloured—in some the coloration of the throat and upper breast is almost as red as in some examples of *Anthus cervinus*—and it is undoubtedly owing to this fact that the bird in question has been wrongly identified.

If we exclude the Red-throated Pipit which was formerly in the collection of the late Mr. Bond, labelled "Unst, May 4th, 1854" (Saunders' *Manual*, p. 135), the first British example is either the bird obtained by Mr. Prentis at Rainham, Kent, in April, 1880, or the undoubted example of *A. cervinus* shot near St. Leonards, Sussex, on November 13th, 1895 (*cf. Zool.*, 1896, p. 101).

The Red-throated Pipit may be readily identified at all stages of plumage—except, perhaps, that of the nestling—by the clear black marking to the centre of the feathers of the rump and upper tail-coverts. The dark streaks on the longest pair of under tail-coverts are not a reliable feature, as these markings are frequently absent in *Anthus cervinus* and often present in *Anthus pratensis*.

While on the subject of Pipits, I should like to point out

* Borrer, "Birds of Sussex," p. 101; and Saunders, "Manual," p. 135, 2nd ed.

that all the examples of red-breasted Rock-Pipits in Case 16 in the Booth Museum are "Scandinavian" Rock-Pipits (*Anthus rupestris* Nilss.) ; none of them are Water-Pipits (*Anthus spipoletta*), as has been formerly suggested.

M. J. NICOLL.

During the past summer I paid a visit to the Rochester Museum, which contains the admirable and excellently-cared for collection of the late Mr. Walter Prentis, of Rainham, and at Mr. Nicoll's request carefully examined the bird to which he refers in the above letter. With much regret I came to the same conclusion with regard to it, as he has done with regard to the Sussex specimen : it is undoubtedly nothing more than an unusually bright Meadow-Pipit (*A. pratensis*). The breast is pinkish-yellow, *not* red, and the rump and upper tail-coverts are *entirely devoid* of the large black centres to the feathers, which are such a characteristic feature of *A. cervinus*.

N. F. TICEHURST.

RICHARD'S PIPIT IN NORFOLK.

THIS bird is a not infrequent autumn visitor to Norfolk, but it is worthy of record that a female was obtained at Cley on October 31st, and another on November 18th last, as I am informed by Mr. H. N. Pashley. H. F. WITHERBY.

SOME SUSSEX RAVENS.

I AM indebted to Mr. Walter Hewett, who was then game-keeper to the lessees of Heathfield Park, for the following interesting account of the nesting-places of the Raven on that picturesque estate during the seventies of last century. There were two nesting sites used alternately by a pair of Ravens in the park itself ; the one in a clump of old Scotch firs on the Tower plain, the other in the Gravel Pit clump, also ancient Scotch firs. This pair of Ravens were so destructive to lambs and ewes during the lambing season—at times destroy-ing the mother, during parturition—that deadly war was waged against them. The old Ravens were so wary that it was difficult to shoot them, but when the young were nearly ready to fly the nest was riddled with bullets, and the brood destroyed annually. In 1876 the Ravens deserted Heathfield Park and built their nest a mile or so away, in a group of Scotch firs, called the Mare and Foal, a very prominent object in the landscape, situated on the ridge that runs from Pun-nett's Town, overlooking Cade Street. In April, 1876, Hewett took up his position in Slaughter Lane, on the south side of the Mare and Foal clump, sending a companion to the nest to disturb the birds. The male Raven fell to Hewett's gun, but, being only winged, recovered, and lived for many years in the Devonshire Park, Eastbourne. The next day, taking advantage of a fog, Hewett shot the female from her nest. This closes the history of the Heathfield Park Ravens, doubtless descendants of those that feasted on the bodies of Cade and his followers who perished near the same spot—"Leaving thy trunk for crows to feed upon."—*Henry VI.,Act IV.*

H. W. FEILDEN.

LITTLE OWL IN NORTH-WEST OXFORDSHIRE.

A PAIR of Little Owls established themselves at Kingham, Chipping Norton, last spring, and continued with us all the summer, attracting much attention by their loud cries, uttered repeatedly while hunting after sunset and during the early part of the night. I may mention that I have had great difficulty in finding a good description of this cry in English ornithological works ; but in Fatio's " Oiseaux de la Suisse " I have at last found an excellent one. Professor Fatio is gifted with a very keen ear for the utterances of birds, and has had the experience of a long life among them. He writes (Vol. I., p. 194) : " Son cri, souvent répété, et qui passe volontiers pour un mauvais présage, peut etre traduit de diverses manières, selon les circonstances et les appréciations ; c'est souvent : *kuitt* ou *kuwitt*, parfois *kuück* ou *kouuk*, ou *keuw-keuw* ou encore *poupou-poupou.*" The second of these descriptions agrees almost exactly with the way in which I attempted to syllable the cry myself last summer.

No doubt the birds bred here, but we thought it advisable to refrain from making an elaborate search. As we are on the borders of Gloucestershire, I think their appearance here marks the farthest point to the west that the birds have as yet reached. W. WARDE FOWLER.

[In western counties the Little Owl has previously been reported from Goring and Henley, in Oxfordshire (*B. B.*, Vol. I., p. 338), Fairford, Willey, and Shrewsbury, in Shropshire (Vol. I., pp. 388 and 339), Avebury, in Wiltshire (Vol. II., p. 100), and Sutton Coldfield, in Warwickshire (Vol. II., p. 240).—H.F.W.]

SCAUP-DUCKS IN NOTTINGHAMSHIRE IN THE SPRING AND SUMMER OF 1908.

ON March 21st, when fishing in the large lake in Thoresby Park (this piece of water is over ninety acres, and is situated in the middle of a 2000 acre deer park) I saw one male and three female Scaups. I had my binoculars, and got pretty near to them in the boat. On May 2nd Mr. H. E. Forrest and I saw three, and on August 14th the Rev. B. D. Aplin and I saw two females. Of course I cannot say if they nested, but I may mention that the lake is full of pike, and very few ducks rear many young ones. On all three dates I also saw a pair of Pochards, and one solitary male Goosander. Even the female Scaup, when once known, cannot be mistaken. It is much coarser about the head and bill than the Tufted, and shows the white, or pale yellow, face very distinctly.   J. WHITAKER.

AMPUTATION OF LAPWING'S TOES BY MEANS OF WOOL.

A FRIEND of mine shot a Lapwing (in good condition) on September 28th in Wigtownshire, N.B., which, when we picked it up, was found to have the following condition of its feet :—
Right foot—Amputation of inner two digits at the metatarso-phalangeal joints. *Left foot*—Amputation of internal digit at metatarso-phalangeal joints, and a tight constriction, caused by sheep's wool, round the tarsus, just distal to the

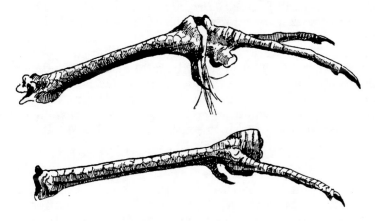

hallux, the wool cutting deeply into the tarsus, especially on the outer side, and causing an everted edge to the furrow, similar to that observed on the remaining proximal end of the right foot. I am indebted to Mr. P. H. Bahr for kindly drawing the condition for me.   HENRY B. ELTON.

[A series of legs of the Lapwing affected in the same way as described above was shown by Dr. C. B. Ticehurst at the meeting of the British Ornithologists' Club held on October 19th, 1904 (*cf. Bull. B.O.C.*, XV., p. 12).   These specimens were from birds shot on Romney Marsh, and of eight Lapwings shot, four were thus affected. A similar case is recorded of a bird shot in co. Armagh (*cf.* A. R. Nichols, *Irish Nat.*, 1905, p. 32).—H. F. W.]

BLACK-NECKED GREBES IN NORTH LANCASHIRE.

THE occurrence of the Black-necked or Eared Grebe (*Podiceps nigricollis*) in north Lancashire is, I venture to think, worthy of note, especially so of three specimens. The first, an adult in full summer plumage, I mentioned in the "Zoologist" of September, 1904, as having been captured alive on a pond at Middleton, near Morecambe, on July 28th, 1904. The second was shot on October 24th of the same year on the tidal part of the Lune below Lancaster, at Snatchems, and was an adult in full winter plumage, whilst the third, an immature bird, was shot in the same place as the last specimen in February, 1907. All the three specimens occurred within a couple of miles of one another. H. W. ROBINSON.

LEACH'S FORK-TAILED PETRELS IN CUMBERLAND AND LANCASHIRE.

DURING the week between November 18th and 25th the coasts of north Lancashire and Cumberland were visited by numbers of Fork-tailed Petrels (*Procellaria leucorrhoa*). The week was a very stormy one, and the birds were probably blown inshore by heavy winds, some being found some distance inland. They were specially numerous off the slag-tip at Carnforth, near Lancaster, on the 23rd and 24th, and occurred all the way up the coast, and inland as far north as Carlisle, and into the West Riding of Yorkshire. From the dates of their capture it would seem that they were travelling down the coastline from north to south. H. W. ROBINSON.

BULWER'S PETREL IN SUSSEX.

A MALE example of Bulwer's Petrel (*Bulweria bulweri*) was picked up much exhausted, but still alive, at Cliff End, near Winchelsea, Sussex, on September 4th, 1908, after strong south-westerly gales. The bird was taken to Mr. Bristow, of St. Leonards, for preservation, and was there seen in the flesh by Mr. W. R. Butterfield. It was eventually purchased by Mr. C. J. Carroll, by whose courtesy we have been allowed to photograph the stuffed bird. It was exhibited on Mr. Carroll's behalf by Mr. W. R. Ogilvie-Grant at the meeting of the British Ornithologists' Club held on November 18th, 1908. There have only been three previous occurrences of Bulwer's Petrel in the British Isles, and of these two have also been in Sussex. A Yorkshire example is quoted in Howard Saunders' "Manual" (2nd ed., p. 749), and on February 3rd, 1903, one was found dead at Beachy Head, while on February 4th, 1904, another was picked up dead at St. Leonards (*Bull. B.O.C:*, XIII., p. 51,

and XIV., p. 49). This Petrel breeds on the islands of the
Madeira and Canary groups, and is also found in the middle

Bulwer's Petrel, picked up near Winchelsea, Sussex, on
September 4th, 1908.

of the Pacific in the Hawaiian group, as well as in the Bonin
and Volcano Islands far to the south-east of Japan.   H.F.W.

* * *

WRYNECKS IN NORTH LANCASHIRE.—An example of *Iynx
torquilla* was captured alive on September 3rd, and anoth r
on October 2nd, 1908, near Lancaster. At one time ᴜ ne
Wryneck seems to have nested in Lancashire, but it now rarely
visits the county (H. W. Robinson, *Zool.*, 1908, p. 428).

HONEY-BUZZARD IN ENGLAND.—The Rev. F. L. Blathwayt
records that two were shot near Grantham and one near
Lincoln between September 24th and October 5th, 1908
(*Zool.*, 1908, p. 428). On September 26th an immature bird
was shot near Oldham (F. Stubbs, *Nat.*, 1908, p. 456). One
was shot near Beccles and another near Great Yarmouth (? in
September), (B. Dye, *Zool,*, 1908, p. 468).

LINCOLNSHIRE AND SOMERSETSHIRE HERONRIES.—The Rev.
F. L. Blathwayt ·gives some interesting particulars of these
(*Zool.*, 1908, p. 450).

NESTING OF THE SCAUP-DUCK IN SCOTLAND.—*Correction.*—
The nest found by Captain (now Colonel) R. Sandeman and
Mr. Heatley Noble was in Sutherlandshire, as mentioned on
page 85 (*supra*), and not in Speyside, as stated by Mr. P. H.
Bahr on page 211.

REVIEWS

Bird-Hunting through Wild Europe. By R. B. Lodge. Illustrated. R. Culley. 7s. 6d. net.

In this book Mr. R. B. Lodge, who will be well-known to most of our readers as a successful bird-photographer, describes his recent experiences in Spain and the Balkans.

In Albania and Montenegro especially the author had to undertake much travelling in a decidedly difficult country before he was able to find the birds he was particularly in search of, consequently, his time was limited for making observations on the habits of the birds he met. Nevertheless there is a good deal in the book which will interest the student of British birds, because Mr. Lodge was fortunate enough to see and photograph, in their breeding haunts, many species which very rarely wander to this country. Herons (he was successful in photographing the rare *Ardea alba* at its nest) and birds of prey seem to have engaged Mr. Lodge's chief attention, and his list of successful photographs of these birds is remarkable. We may here mention that in making the statement on p. 210 that the Common Bittern's nest has not been photographed, Mr. Lodge has overlooked Mr. Wade's successful photographs already published in this Magazine (Vol. I., p. 329).

The simply-told narrative of the author's experiences and adventures is well worth reading, while the photographs, as we should expect, are both numerous and excellent. The book is nicely got up and well printed, but a tint of an extremely inartistic yellow has been printed under some of the best pictures, and forming a wide frame gives a most displeasing effect, and altogether spoils these really beautiful photographs.—H.F.W.

BIRDS

AN·ILLUSTRATED·MAGAZINE
DEVOTED·TO·THE·BIRDS·ON
THE·BRITISH·LIST

FEBRUARY 1,
1909.

Vol. II.
No. 9.

MONTHLY·ONE·SHILLING·NET
326·HIGH·HOLBORN·LONDON
WITHERBY & Co.

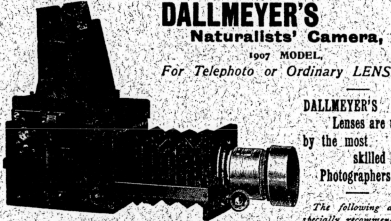

BRITISH·BIRDS

EDITED BY H. F. WITHERBY, F.Z.S., M.B.O.U.
ASSISTED BY W. P. PYCRAFT, A.L.S., M.B.O.U.

FIELD NOTES ON THE "POWDER-DOWN" OF THE HERON.

BY
J. M. DEWAR.

THE Heron preens its plumage comparatively seldom, and in this respect it differs from the majority of British birds, which may be seen to do so at least once every day. A summary is here given of observations which were made in the autumn of 1907, after I had been on the look-out during some half-a-dozen years for a Heron in the act of attending to its plumage.

September 20th—a calm, sunny day. Two Herons perched on a log-fence, and a third near them on grass at the edge of a meadow. They faced the south and were preening their feathers at 10 a.m., when first I saw them. From the shelter of a wood about fifty yards away I could see the bird on the grass, but not the other two without risk of exposure. Ruffling out its plumage, the Heron separated the right wing from the body and insinuated the bill under the feathers in, as nearly as I could judge, the position of the right breast powder-patch, where it rubbed the bill slowly up and down, applying the sides, the upper and the lower surfaces. It withdrew the bill and preened the breast in the ordinary way, leisurely drawing the bill among the feathers, biting at their bases, and brushing them on both surfaces. With intervals of rest and watching for signs of danger, it preened its plumage, and had frequent recourse to the areas of the powder-downs, where the movements of the bill were always of the nature of a gentle to and fro rubbing, directed to the whole surface of the bill. Occasionally after preening it rubbed the bill by applying the adjacent surfaces of two toes, and drawing them slowly downwards over the bill. Before beginning to preen again it rubbed the bill in one of the powder-downs. The neck, breast, and ventral feathers received the most attention, and the Heron spread the wings one at a time and drew the bill downwards between each pair of remiges. Then I noticed for the first time that the bill was coloured pale blue, and had a dull appearance. On looking up cautiously at the Herons on the fence I saw that their bills were blue also, while the bill of a Heron which was watching for food in sedge behind the three had a yellowish colour. When the preening was finished the plumage was fluffed out very much, and the bird stood erect with its neck fully outstretched. The feathers hung loosely in frills round the neck, across the breast, and encircling each leg. The wings were allowed to fall downwards and outwards from the shoulders, while the tips remained crossed over the tail. Having completed its arrangements it indulged in an unmistakable yawn. It stood thus for fully half-an-hour, in the warm sunshine, without a movement except the occasional turning of its head, as it surveyed the neighbourhood. About noon a pony which, as I thought, had been watching the Herons for some time ran across the field and drove them away. Before they went I noticed that their bills were still bluish in colour.

October 1st—mild, sunny weather. Six Herons in the sedge. Two came into the meadow and for a while preened desultorily. The bill of each bird was yellowish in colour. One bird stood in the usual attitude, without shaking out its plumage, and at intervals smoothed feathers here and there without having recourse to the powder-downs. Its bill remained yellow. The second bird began by shaking out its plumage to a marked degree of fulness. Then it pressed its bill into the region of the right breast powder-down and rubbed the bill up and down gently. After preening a few feathers on the breast it devoted its attention to the right wing, where it preened the coverts, especially the lower, and drew its bill over and among the remiges of the half-opened wing. The left wing was preened less carefully. It was sunny then, and at intervals the Heron held out its wings horizontally, as Cormorants do. The wings trembled visibly at these times, and the bird soon let them fall, as if tired. When the bill was lifted from the powder-down the lower mandible alone was of a bluish colour and, in the case of the under coverts which were turned towards me, the bill was introduced below each feather at the outer border and drawn to the tip, so that the under-surface of the bill came into contact with the under-surface of each feather. This bird shook its bill vigorously sidewise at times as if to get rid of something.

October 4th—a calm, sunny day. A Heron which had been watching for food in the estuary stepped out of the water and walked slowly some way over the sands. There it shook itself so as to fluff out the plumage. At that time the bill had a shining appearance, and was of a straw-yellow colour. The Heron pressed its bill into the region of the breast powder-downs and rubbed it gently up and down. When the bill was withdrawn the lower mandible was seen to have a dull bluish-white appearance against the dark background of muddy sand. It preened the feathers of the foreneck and breast, drawing out each long feather by a slow movement of the bill from base to apex and arranging the feathers parallel to one another. It smoothed the anterior margins of the wings, and went gradually over the whole of the lower plumage, the shoulders and the wing coverts. During this lengthy operation it had frequent recourse to the breast powder-downs and to the inguinal areas latterly. The bill gradually became yellow and glistening during the preening of the feathers, and after being in the region of a powder-down a bluish-white colour appeared by contrast. At the end of the general preening the Heron pressed the bill into the region of the right breast powder-patch, and rubbed the bill slowly up and down about twenty times, the greatest number of times I had seen yet at a single application. When the bill emerged its dull, blue-white appearance was plain. Thus dusted, the Heron applied its fore-neck and under mandible to the outer surface of the right wing, beginning at the shoulder and drawing the neck and bill gently over the coverts towards the remiges four or five times. When this was done the bill was bright yellow in colour and glistening in appearance. After some further arranging of the feathers on the anterior margin of the right wing, the Heron drew in its plumage and walked back to the channel.

October 25th—mild and sunny. Two Herons alighted in a pool on the shore and, in a little while, began to preen. Their actions were similar to those already described. Several times one of them reached over to the area of the right femoral powder-patch and rubbed its bill there. Once or twice they dipped their bills in the sea-water before preening the feathers. They spent about half-an-hour at this occupation and then went inland. Their bills were yellowish in colour when they came. During their stay the bill of the bird which had been applied to the powder-patches so much became light blue in colour, and remained so, while the other's bill was yellowish in colour, streaked at times with blue.

November 18th—a calm sunny day. A Heron perched on the same fence, facing the south and with a fairly warm sun shining on it. For about half-an-hour it preened in a leisurely manner. At first the bill was yellow, and yellow it remained. The Heron attended to the plumage on the breast, the legs, the shoulders, and especially the neck. It worked with the point of the bill at the bases of the feathers, and then drew it among them to their tips. At no time was the flat of the bill used, nor did "rubbing" occur, and I did not see the bill once in the areas of the powder-downs.

I had in mind two sources of error in these observations : first, in determining the colour of the bill, and, secondly, in estimating the relative importance of the powder-downs to the general toilet. The colours remained constant when the light was reflected from the bill at various angles, and at equal and different distances from two or more birds. Lately I had an opportunity of examining a Heron not long dead, and found that the bill was coated readily with the powder, and while the blue colour was not displayed prominently with the bird in the hand, it became much more distinct when looked at from any distance within reasonable limits. "Rubbing" may be the ordinary way of treating the powder-down patches. On this point more observation is necessary, but considering the sequence of events as I have outlined them, we may suppose that the powder is carried to other parts of the body by means of the bill. On the dead bird, after making some of the wing feathers ragged, I found it easier to mat the rami with the aid of the powder than without such help, and the powder has the further merit of rendering the plumage highly waterproof, which is no small advantage to a wading bird, whose plumes, in the absence of powder, easily become draggled with wet.

SOME EARLY BRITISH ORNITHOLOGISTS AND THEIR WORKS.

BY

W. H. MULLENS, M.A., LL.M., M.B.O.U.

VII.—JOHN RAY (1627—1705) AND FRANCIS WILLUGHBY (1635—1672).

(PLATE VI.)

THE names of John Ray and Francis Willughby, the founders of scientific ornithology in this country, must ever be held in equal honour and esteem. Of very different origin—Willughby being a country gentleman of means, descended from a long line of illustrious ancestors, and Ray the poor son of a village blacksmith— a common devotion to the study of natural history made them close friends and zealous fellow workers. "Together they studied, together they travelled, and together they collected." To separate their joint work or to credit one with a greater share in devising the scientific classification of the subjects they studied, is as impossible as it is invidious. The misfortune of Willughby's premature death, and the fact that his posthumous works were edited by his friend, and that the latter became not only an eminent ornithologist, but also world-famous as a botanist, have undoubtedly tended to obscure Willughby's claim to an equal recognition. Had he, however, been spared to accomplish his allotted share of their joint labours he would undoubtedly have achieved as great a reputation as his famous friend. In the course of their investigations these two eminent men having become " dissatisfied with the status of natural history, agreed to attempt a systematic description of the whole organic world," in which their different parts were apportioned according to the following method, as is detailed by Dr. Derham from information he received when he visited John Ray at Black Notley in May, 1704 (*Memorials of Ray*, p. 33) : " For these two gentlemen, finding the 'History of Nature' very imperfect, had agreed between themselves, before their travels beyond sea, to reduce the several tribes of things to a method; and to give accurate descriptions of the several species, from a strict view of them. And forasmuch as Mr. Willughby's genius lay chiefly to animals, therefore he undertook the birds, beasts, fishes, and insects, as Mr. Ray did the vegetables. And how each of these two great men discharged his province, the world hath seen in their works; which show that Mr. Ray lived to bring his part to great perfection; and that Mr. Willughby carried his as far as the utmost application and diligence of a short life could enable him." The period in which Ray and Willughby flourished is justly described by Linnæus as the dawn of the golden age in natural history. Before their great work was undertaken, ornithology as a science could scarcely be said to exist. It is true that an Englishman, Edward Wootton (1492–1555), had in a folio work entitled :—

Edoardi Wotto- / ni Oxoniensis de / Differentis Ani- / malium Libri / Decem. / Ad Sereniss. Angliae Regem / Edoardum VI. / . . . . itemque singulae eorum partes recensentur, ... Lutetiae Parisiorum / apud Vascosanum. / M.D.LII. / Cum privilegio Regis.

made some attempt at a systematic arrangement of birds, but he did not profess to do more than give a compilation from the classical authors, while the standard authorities of the day, Gesner and Aldrovandus, were full of obscurity and mistakes.

In England itself the study of zoology had hitherto received but scant attention, hence " observing in this busie and inquisitive age the History of Animals alone to have been in a great measure neglected by English men (for that since Turner* and Mouffet† none that I know of have performed anything worthy of commendation). . . . . Our main design was to illustrate the History of

---

* William Turner (1500-1568), author of " Avium Historia."

† Thomas Mouffet (1553-1604), author of " Insectorum Theatrum."

Birds, which is (as we said before of Animals in general) in many particulars confused and obscure, by so accurately describing each kind, and observing their Characteristics and distinctive notes, that the Reader might be sure of our meaning, and upon comparing any Bird with our description not fail in discerning whether it be the described or no. Nor will it be difficult to find out any unknown Bird that shall be offered : for Comparing it with the Tables first the Characteristic notes of the genus's from the highest or first downward will easily guide him to the lowest genus ; among the species whereof, being not many, by comparing it also with the several descriptions the Bird may soon be found " (Preface to the *Ornithology*).

John Ray, the son of Roger and Elizabeth Ray, was born in the parish of Black Notley, in Essex, in the autumn of 1627, possibly on November 29th ; and was baptized on December 6th of that same year. The date of Ray's birth and baptism have proved a stumbling block to most of his biographers. This arises from the coincidence that on the same page of the parish register at Black Notley are recorded the baptisms of two John Rays, in the successive years of 1627 and 1628, as will be seen from the reproduction of these entries here given. They run as follows :—

(1627) John [son] of Roger and Elizabeth Ray December 6.

(1628) John son of Thomas and Dorothie Wray bapt. June 29.

The latter of these two entries has apparently been mistaken as referring to John Ray the naturalist. William Derham, in his " Select Remains and Life of Ray " * gives the date of Ray's birth as November 29th, 1628, and then in a footnote informs us that on " searching the parish registers " it was discovered that " he was baptized on the 29th of June, 1628 ; consequently the

* Included in the "Memorials of John Ray." London, 1846. 1 vol., 8vo.

ORNITHOLOGY

O F

FRANCIS WILLUGHBY

O F

Middleton in the County of *Warwick* Efq;

Fellow of the Royal Society.

In Three Books.

Wherein All the

HITHERTO KNOWN,

Being reduced into a M ET H O D futable to their Natures, are accurately defcribed.

The Defcriptions illuftrated by moft Elegant Figures, nearly refembling the live B I R D s, Engraven in LXXVIII Copper Plates.

Tranflated into Englifh, and enlarged with many Additions throughout the whole W.O R K.

To which are added,

Three Confiderable DISCOURSES,

I. Of the Art of F o w L I N G : With a Defcription of feveral N ɛ T s in two large Copper Plates.

II. Of the Ordering of S I N G I N G B I R D s.

III. Of F A L C O N R Y.

B Y

J O H N RAY, Fellow of the R O Y A L S O C I E T Y.

Pfalm 104. 24.

How manifold are thy works, O Lord? In wifdom haft thou made them all : The Earth is full of thy richtes.

LONDON:

Printed by *A.C.* for *John Martyn*, Printer to the *Royal Society*, at the *Bell* in St. *Pauls* Church-Yard, MDCLXXVIII.

above date, as the supposed one of his birth, is incorrect."
Acting on this ingenious hypothesis, Ray's subsequent
biographers have fixed his birth on November 29th,
1627, and his baptism on June 29th, 1628.*

Ray was the son of the village blacksmith, and the
house which now stands opposite the forge in Black
Notley is said to have been his birthplace. Although of
humble origin. he received an excellent education : first
at the Grammar School in the neighbouring town of
Braintree, and afterwards at St. Catherine's Hall (where
he only remained for a short time), and Trinity College,
Cambridge. At Trinity he obtained a fellowship in 1649,
and afterwards filled many important offices in his College.
Ray remained at Cambridge for several years. From
the University he commenced his earlier " Itineraries,"
journeys which he undertook for the sake of observation
and the collection of plants, and of which he kept an
account. The first of these he undertook alone in 1658, but
in many of the subsequent ones he was accompanied by
Francis Willughby, proceeding on different occasions
as far as Scotland and Cornwall. At Cambridge, Ray
published the first of his numerous works, a small 8vo
volume entitled " Catalogus Plantarum Circa Cantabrigiam nascentium." This appeared in 1660, and in
the same year Ray entered into holy orders. Two years
later his connection with his College came to an end.
Refusing to subscribe to the " Act of Uniformity " of
1662, he resigned his fellowship, and being now at greater
liberty he resolved to pursue his studies in Natural history
still more ardently, and for that purpose to extend his
travels beyond the confines of his own country.

Accordingly in the spring of 1663, Ray, with two of
his pupils,† and accompanied by Willughby, left England
for France, and after " passing through divers parts of

---

* *Vide* art. " Dict. Nat. Biogr." " Ray, John (1627-1705), naturalist,
was born at Black Notley, near Braintree, Essex, probably on
29th Nov., 1627. He was baptized on 29th June, 1628."

† Mr. Skippon (afterwards Sir Philip) and Mr. N. Bacon.

Europe " returned to this country in 1665, having parted
company from Willughby during the latter part of the
journey.   On his return to his native country, Ray
devoted his serious attention (as he wrote to Dr. Martin
Lister) to " gathering up into a catalogue all such plants
as I had found at any time growing wild in England
. . . . possibly one day they may see the light : at present
the world is glutted with Dr. Merrett's bungling ' Pinax.' *
I resolve never to put out anything which is not as perfect
as it is possible for me to make it."   These labours bore
fruit in after years, when Ray published his " Catalogus
Plantarum   Angliae,"   and   his   yet   more   famous
" Synopsis   Methodica   Stirpium   Britannicarum,"   the
second edition of which, published in 1696, set the seal
on his fame as a botanist.   In 1667 Ray was persuaded
to become a Fellow of the Royal Society, and in 1670
he changed the spelling of his name, which he had hitherto
written Wray, by dropping the initial " W," thus returning, as he informed his correspondent Martin Lister, in
a letter written August 22nd, 1670, to the style used by
his ancestors.   In 1672 Ray suffered a great blow by the
death of his intimate friend and companion, Francis
Willughby, who died in this year " to the infinite and
unspeakable loss and grief of myself, his friends, and all
good men."   To Ray's guardianship Willughby com-
mitted his two sons, and further left him an annuity of
sixty pounds, which formed his chief means of support
during the remainder of his life.   Faithful to his trust,
Ray now took up his residence at Middleton Hall, the
Warwickshire seat of his late benefactor, and in 1673
he was married to Margaret Oakley, in Middleton Church.
The year 1674 saw the publication of Ray's first contribution to ornithology, entitled :—

A / Collection of English Words / not generally used
. . . . / and catalogues of English Birds / and Fishes
. . . . / London . . . . 1674. 1 vol. 12mo.

* "Pinax Rerum," by Christopher Merrett.   London, 1666,
1 vol., 8vo.

JOHN RAY.

(From the Engraving by H. Meyer, after a Picture in the British Museum.)

Entry in Parish Register of John Ray's Baptism,
and that of another John Ray.

The catalogue of Birds, which was an imperfect one, was omitted in subsequent editions, and Ray now commenced to engage himself on a far more important work, the arrangement and publication of the notes and observations about birds which had been left by Willughby as his share of their undertaking in the study of natural history. This posthumous work of Willughby's Ray published in Latin in 1676, under the title " Ornithologia libri Tres," the English edition of the same appearing in 1678. This, the well-known " Ornithology of Francis Willughby," was edited by Ray " with large editions." A facsimile of the title-page is here given, the collation of the book being : 1 vol. folio ; pp. 12. un + pp. 441. + pp. 6, Index. 2 plates of fowling and LXXXVIII. of birds.* A catalogue of English birds appears on pp. 21–28, some 190 species in all being mentioned. How much original matter Ray added to Willughby's labours it is impossible exactly to determine, but it was evidently not inconsiderable, for not only did he, as he tells us in his preface, add the " descriptions and histories of those [birds] that were wanting," these being principally those recently discovered in the Indies and the New World, but he also added a good deal of information which he received from certain of his fellow-countrymen, notably from Sir Thomas Browne, of Norwich, who " frankly communicated the Drafts of several rare Birds, with some brief notes and descriptions of them," and also from Mr. Jessop and Sir Philip Skippon. His most important assistance, however, came from Mr. Ralph Johnson, of Brignal, in Yorkshire, who is described by Ray as " a person of singular skill in Zoology, especially the History of Birds," and who appears to have been not only an observer of nature far in advance of his time, but also to have in-

* The engravings which were executed at the expense of Mrs. Willughby, are poor, and Ray laments that although he employed good workmen the great distance he was from London necessitated all directions and descriptions passing by letter. and observes that " in many Sculps they have not satisfied me."

vented a " method of Birds " which was freely used by
Ray " in the divisions and characteristic notes of the
genera." In 1676 Ray left Middleton Hall, his two young
pupils having ceased to be under his tuition, and removed
to Falborne Hall, near Black Notley, the residence of
Mr. Edward Bullock, to whose son Ray probably acted
as tutor, and to whom he dedicated the " Stirpium
Europeanarum . . . . Sylloge " in 1694.

In 1678 Elizabeth Ray, the naturalist's mother, died,
and Ray then took up his abode at her house on
" Dewlands," in Black Notley, where, said he, " I intend,
God willing, to settle for the short pittance of time I have
yet to live in this world." Ray now settled down to un-
interrupted work, and in 1682 published his " Methodus
Plantarum nova," in which he proposed a

" new method of classifying plants, which when altered and
amended as it subsequently was by himself at a later period,
unquestionably formed the basis of that method which under
the name of the system of Jussieu is universally received at
the present day."

In addition to his own numerous labours Ray also
continued to deal with the mass of material left to him
by his friend, Francis Willughby, and in 1686 he published
the " Historia Piscium," 1 vol. folio, which " he had
extracted out of Mr. Willughby's papers, revised, supplied,
methodized and fitted for the press." This Ichthyology
was, by the assistance of Bishop Fell, printed at Oxford,
the Royal Society defraying the expense. The " History
of Fishes " was, as Ray laments in a letter to his friend,
Dr. Tancred Robinson, far from being as complete as it
should have been, most of the notes which he and
Willughby had made in the course of their travels having
been mislaid.*

It is here quite impossible to enumerate all the works
which came year after year from Ray's pen ; a list of
them will be found in the " Memorials of Ray " (p. 111).

---

* This refers to their joint notes ; of Willughby's notes Ray
writes " it is almost impossible to procure a sight of them."

Ray's final contribution to ornithology was the " Synopsis Methodica Avium," which was completed in 1694, but not published till after his death in 1713. " In this synopsis Mr. Ray added many species of birds and fishes which were omitted in Mr. Willughby's histories of them," and by way of a supplement added a short catalogue with figures of " Avium Maderaspatanarum," or " Indian Birds about Fort St. George," compiled by James Petiver (1663–1718), and of interest as being " the first attempt to catalogue the Birds of any part of the British possessions in India " (*cf.* Newton, *Dict. of Birds*, Introduction, p. 7).

Ray's health now began to fail, and in 1704 he passed away at Black Notley " in a house of his own building " called " Dewlands " (destroyed by fire in 1900). He lies buried in the churchyard of his native parish, where his sadly neglected grave requires prompt attention, if the several inscriptions that his monument bears are to remain decipherable.

Of Ray's influence on natural history it is impossible to say too much. His works on zoology, in the words of Cuvier, " may be considered as the foundation of modern zoology," and by Haller he was termed " the greatest botanist in the memory of man."

Of the short but busy life of Francis Willughby it is necessary to say but little. He was born in 1635, the only son of Sir Francis Willughby, Knt., of Middleton Hall, Warwickshire. At the age of seventeen he became a fellow Commoner of Trinity College, Cambridge, and there formed the acquaintance of John Ray, with whose labours in natural history his name will ever be associated. It has been generally asserted that Willughby was Ray's pupil at the University, but what little evidence exists on the matter is rather against this supposition. In 1655 Willughby took his degree as B.A., and proceeded M.A. in 1659. That he had early begun to assist Ray in his work is evident from the allusions in the latter's " Catalogus Plantarum Circa Cantabrigiam," which was published in 1660. In the preface to that work Ray writes as follows :—

" Jam quoniam honestum est fateri per quos profeceris, generossimi Juvenes, D. Franciscus Willughby et D. Petrus Courthorpe * Armigeri, natalium splendore ingenii sublimitate. Suavite morum, fide, virtute illustres, non rei duntaxat herbariae callentissimi, sed in omni literarum genere versatissimi, amici nostri, plurimum honorandi, non sunt a nobis silentio transmuttendi, ni ingrati & arrogantes esse velimus. Horum opera nos saepius usos & ab his non mediocriter adpitos fuisse in hoc opusculo, Concinnando, libere & ingenue profitemur."

In 1663 Willughby, who had already accompanied Ray in some of his expeditions in Great Britain, went with him on his journey to the continent, but they parted company the next year at Montpelier, and Willughby continued his journey through Spain alone. It should here be mentioned that Willughby's name appeared as one of the original Fellows of the Royal Society on its incorporation in 1663–4. In 1665, on the death of his father, Willughby succeeded to the estates of Wollaton and Middleton, and in 1668 married Emma, daughter of Sir Thomas Bernard, by whom he had two sons and a daughter. His great devotion to work seems to have overtaxed his strength, and on December 22nd, 1670, Dr. Martin Lister, writing to Ray, says, " I am very glad Mr. Willughby is near well again. Methinks he is very valetudinary, and you have often alarm'd me with his Illnesses." In the beginning of June, 1672, " he fell into a pleurisie which terminated in that kind of fever called Cattarrhalis."

He died on July 3rd, 1672, " to the immense grief of his friends and all good men that knew him, and the great loss of the commonwealth in general." Thus was frustrated his project of a voyage to the New World, " that he might perfect the History of Animals."

* Mr. Peter Courthorpe, of Danny, in Sussex ; a friend and pupil of John Ray's, to him Ray dedicated the " Collection of English Words," published in 1674.

BIRD-LIFE IN A SPRING SNOWSTORM.

BY

THE REV. A. ELLISON, M.A., B.D., M.B.O.U.

THE series of snowstorms which visited nearly all parts
of the British Islands in the fourth week of April, 1908,
was probably unprecedented for so late a date in the
spring. Coming at the very height of the spring migration, and at a time when nearly all our resident birds
were breeding, the result must have been, for the time
at least, extremely disastrous ; and it furnished a good
illustration both of the calamities to which wild creatures
may be exposed, and also of Nature's wonderful recuper-
ative powers.

In mid-Hertfordshire the season, though cold and
changeable, was not on the whole unfavourable to bird-
life up to April 18th. The spring, however, was distinctly
backward. Chiffchaffs had appeared in their usual
numbers by April 1st. Willow-Warblers and Swallows
were first observed on the 15th, but only one or two ;
the majority had not come; while the great body of
April migrants still held back, waiting for kinder conditions. But resident birds were, perhaps, a week late on
the average in breeding, not more. An early Robin had
young well advanced on April 15th. Several others had
hatched out by the 22nd. I knew of a good number
of Lapwings' nests with incubated eggs between those
two dates ; one had been hatched off on the 22nd, while
every hedge or plantation was full of nests of Thrush or
Blackbird with eggs or callow young.

Heavy snow showers had occurred, sufficient to whiten
the country, on the 19th and 20th, with sharp frosts
at night ; but not enough to cause any serious danger
to bird-life. However, early in the afternoon of the
23rd a cold rain gave place to snow, which increased to
a blizzard, and lasted without cessation for sixteen hours.
On the morning of the 24th the country was covered
with snow to an average of eight inches in depth, while

in many lanes the northerly gale had heaped up drifts of three or four feet. Temperature had fallen to 28°, the result being that the snow, which at first had partially thawed, had frozen and clung to trees and bushes to such an extent as to break down large branches, or to bend shrubs prostrate to the ground. Even at noon it still froze in the shade, and roofs and stacks, where the sun did not reach, were decorated with long fringes of icicles.

What the effect of such an occurrence was upon bird-life it was not easy fully to ascertain. But there can be little doubt that young birds which had recently left their nests for the most part perished. I saw no trace, after the snow, of broods of young Thrushes which had left the nest just before. A great many nestlings also perished, but no inconsiderable number survived, where the nests were in sheltered situations. But it must have been extremely difficult for the parents to feed them during the two days when the ground was snow-covered.

However, the manner in which many nests escaped was truly marvellous, and indicated a wonderful devotion and intelligence on the part of their owners. Lapwings' nests, which, of all others, were as one would think, most exposed to the fury of the elements, passed through the ordeal uninjured. One, found by me on April 16th, with four eggs far advanced in incubation, had been hatched off on the 22nd, the day before the snowstorm. On the 27th I visited the spot, and, although I could not find the young, I felt convinced that they had survived from the clamorous cries of the parent birds whenever I approached the neighbourhood of the nest.

A nest, with four eggs, in a neighbouring field, discovered on the 17th, was quite intact, the bird still sitting and the eggs warm. This clutch afterwards hatched off safely, and yet the nest must have been surrounded with snow six inches deep. The case of this nest would suggest that in the other instance, where the young had left the nest, they might have been kept together and covered by the parents, and so protected from the inclement conditions. But the question how they could have been fed is a difficult one.

Early in April I had found a Robin's nest on a wayside bank. It was situated, not in a crevice, but on the flat ground at the top of the bank, under a thick tuft of old grass. It was at the extremity of a kind of tunnel, roofed over by some pieces of stick and the thick tangled grass.

The Robin's nest was at the top of the bank on the right hand of the photograph.

On the 21st it contained six eggs, and I placed in it two more, from a deserted nest not far off, so that it contained the large number of eight eggs. I had never yet seen the bird on the nest, and the eggs were cold and remained so, so that I thought the bird had deserted in consequence of my intermeddling in her domestic matters. On the 23rd came the great snow, and the following morning the nest was buried under drifts from one to two feet deep. The snow had drifted through the hedge, and formed wreaths over the spot. Yet when I looked at the nest on the 29th, after the snow had disappeared, the bird was sitting, as if nothing had happened, and the eggs were slightly incubated. In due time all the eight eggs hatched out, and the young were safely reared. It is quite impossible that the bird can have sat on this nest during the snow. Indeed, the place was so thoroughly drifted over that I could not identify the exact situation of the nest. There was nothing but a wilderness of compact snow wreaths, and if the bird had been sitting she must have 'been a prisoner for nearly three days. The eggs, remaining fresh, were protected by the thick roof formed by the grass tuft and sticks, from the super-incumbent snow; while the inherent warmth of the ground underneath kept them from being chilled or frozen. So soon as the snow was gone they were ready for incubation, and the bird returned to them.

The rush of migrants which came as soon as the weather changed was most remarkable. On the 25th the Blackcap was singing in my garden, although the country was still snow-covered, and there were heavy snowstorms again that evening. On the 26th the call of the Wryneck was heard from the tall elms. But on the 28th the temperature at last rose to 58°, and nearly all the summer birds appeared at once. Cuckoos were calling loudly, and the country seemed alive with Willow-Warblers in full song. Many Swallows were about, and the Sand-Martin colonies were in full force. Tree-Pipits were singing, and at 11 p.m. the songs of Nightingales could be heard in all directions. Encouraged by a summer temperature on the first two days of May, the Lesser Whitethroat, Swift, and Spotted Flycatcher were all up to time, on the 2nd, 3rd, and 6th of the month respectively. And at the end of the first week of May, the sycamore, beech, and other trees, which had been still bare at the close of April, were thick with leaves, while the general aspect of the country showed little trace of the wintry ordeal through which it had so lately passed.

ON THE MORE IMPORTANT ADDITIONS TO OUR KNOWLEDGE OF BRITISH BIRDS SINCE 1899.

BY

H. F. WITHERBY AND M. F. TICEHURST.

PART XVII.

(Continued from page 270.)

BLACK TERN *Hydrochelidon nigra* (L.).   S. page 633.

SCILLY ISLES.—Seen every now and then on the pools of Tresco in immature plumage in autumn, and sometimes in August.   Seven were seen on St. Mary's on April 10th, 1903. and four at Tresco on April 26th, 1905 (J. Clark and F. R. Rodd, *Zool.*, 1906, p. 342).

CORNWALL.—A flock of twenty-five to thirty seen first on April 19th, 1901, frequented Marazion Marsh, near Penzance, for some days (A. W. H. Harvey, *t.c.*, 1901, p. 188).   Until the last few years, rarely recorded in spring, but since 1900 it has been observed every year in April (J. Clark and F. R. Rodd, *t.c.*, 1906, p. 342).

HAMPSHIRE.—Two immature birds were shot near Ringwood in August, 1905 (G. B. Corbin, *t.c.*, 1905, p. 394).

OXFORDSHIRE.—Mr. O. V. Aplin considers it may be an annual visitor to the Thames in this county (*t.c.*, 1903, p. 453). One was seen on June 26th, 1903, near Bampton (O. V. Aplin, *t.c.*, 1905, p. 449).   One was seen at Oxford on June 11th, 1904 (*id.*, *t.c.*, 1906, p. 447).

DERBY.—One was killed at Etwell in the late summer of 1900 (F. C. R. Jourdain, *t.c.*, 1902, p. 455), and another at Aston Hall on August 27th, 1908 (*id. in litt.*).

CHESHIRE.—Three in breeding plumage were seen at Budworth Mere on June 4th, 1900, and an immature bird was seen there on September 6th, 1903, and others on August 19th and 26th, 1905 (F. S. Graves, *t.c.*, 1901, p. 188; C. Oldham, *t.c.*, 1903, p. 393, 1905, p. 393).

BRECONSHIRE.—Two were shot on Llangorse Lake in 1889 (E. Cambridge Phillips, *B. of Brecon*, p. 134).

NORTH WALES.—Somewhat rare; met with chiefly on the estuaries (H. E. Forrest, *Vert. F. N. Wales*, p. 369).

ISLE OF MAN.—An immature specimen was shot on October 15th, 1903, on Langness (P. G. Ralfe, *Zool.*, 1903, p. 461).

YORKSHIRE.—Not uncommon in spring and autumn (T. H. Nelson, *B. of Yorks*, p. 648).

CUMBERLAND.—One was seen at Ravenglass on May 6th, 1907 (H. W. Robinson, *Field*, 22, VI., 07).

On the eastern side of England, south of Yorkshire, the Black Tern would appear by the records to be an annual bird of passage, and we have not thought it worth while to give the records.

SCOTLAND.—The following have been recorded in Scotland during the period under notice :—1899, August 7th, Forth, one ; 1901, end of September, Tay, two ; 1902, June 2nd, Tweed, several near Hawick ; 1904, September 7th, Midlothian, one at Gladhouse Reservoir ; November 26th, East Lothian, one at Gullane ; 1908, May 30th, Tweed, one.

WHITE-WINGED BLACK TERN *Hydrochelidon leucoptera* (Schinz). S. page 635.

NORFOLK.—Eight were seen on Breydon by Mr. Jary on April 22nd, 1901, and a single bird on May 15th (J. H. Gurney, *Zool.*, 1902, p. 88).

KENT.— Five were shot out of a small flock at Dungeness on May 29th, 1904 (N.F.T.).

WHISKERED TERN *Hydrochelidon hybrida* (Pallas). S. page 637.

KENT and SUSSEX.—An adult male was shot at Rye Harbour on August 9th, 1905, and is now in the Booth Museum. Four or five others were shot about the same time near Lydd and Pevensey (N.F.T.).

CASPIAN TERN *Sterna caspia* Pall. S. page 641.

NORFOLK.—One was watched by Messrs. Patterson, Eldred, and Jary at Breydon on July 21st and 22nd, 1901 (J. H. Gurney, *Zool.*, 1902, p. 91). One was seen on Breydon by Mr. G. Jary on July 24th, 1902 (*id., t.c.*, 1903, p. 132, and A. Patterson, *t.c.*, 1902, p. 391).

KENT must for the present be struck out of the list of counties in which this bird has occurred, as Mr. J. H. Gurney's record (*t.c*, 1887, p. 458) from Thompson's " Notebook of a Naturalist" (p. 265) was based on a misreading of the name *cantiaca* for *caspia*.

NOTTS.—One at Caythorpe on May 17th, 1863 (J. Whitaker, *B. of Notts*, p. 279) does not seem to have been noted in the " Manual."

SANDWICH TERN *Sterna cantiaca* J. F. Gm. S. page 643.

GUERNSEY.—" Fairly plentiful here [Guernsey], and I know places where it breeds " (Gordon Dalgleish, *Zool.*, 1903, p. 277).

SCILLY ISLES.—In 1903 at least one pair hatched a brood,

but the bird is no longer a regular breeder at Scilly (J. Clark and F. R. Rodd, *t.c.*, 1906, p. 343).

LANCS.—Mr. T. Hepburn could not find any on Walney Island in June, 1901 (*t.c.*, 1902, p. 377).

CUMBERLAND.—The Ravenglass colony is steadily increasing in numbers (C. Oldham, *t.c.*, 1908, p. 165).

NORFOLK.—" Probably bred in Norfolk in 1893 " (*Vict. Hist. Norfolk*).

IRELAND.—A colony (twenty pairs were seen) was found on Lough Erne (*co. Fermanagh*) in 1900 (R. Warren, *Irish Nat.*, 1900 (p. 220). A few nests were found by Mr. H. S. Gladstone on an island in Lough Conn (*co. Mayo*) in 1903, and in May, 1906, a considerable colony (thirty-seven nests found) was discovered by Mr. Warren on another island in the same lough (R. Warren, *Zool.*, 1906, p. 277). In 1906 two small colonies were found in *co. Down* (R. Patterson, *I. Nat.*, 1906, p. 192).

ROSEATE TERN *Sterna dougalli* Mont. S. page 645.

NORTHUMBERLAND.—*Farne Islands.*—In an article on the history and status of this species by Rev. F. L. Blathwayt (*Zool.*, 1902, p. 52), it is stated that two pairs bred up to 1897. In 1898 five or six pairs were seen, but in 1899 only two pairs were reported ; while in 1900 it was thought that only one pair inhabited the Islands.

NORFOLK.—One seen at Blakeney Point and at Wells throughout May, June and July, 1902, and is thought to have paired with a Common Tern and nested. Its identity seems to have been fairly established (J. H. Gurney, *t.c.*, 1903, p. 131). One seen at Blakeney on May 29th, 1903 (*id., t.c.*, 1904, p. 208).

NORTH WALES.—The colony on the Skerries has been considerably reduced, but steps have now been taken to preserve the birds. There is another colony, the locality of which is not divulged (H. E. Forrest, *Vert. F. N. Wales*, p. 371), and the numbers here are well maintained (F. C. R. Jourdain, *in litt.*).

IRELAND.—As a breeding species it *seems* to have ceased to exist. One " was shot on the coast of Connaught on August 3rd, 1904 " (R. J. Ussher, *List of I. Birds*, p. 48).

COMMON TERN *Sterna fluviatilis* Naum. S. page 647.

SHETLAND.—Nesting in some numbers in 1901 for the first time (W. E. Clarke, *Ann. S.N.H*., 1902, p. 121).

FAIR ISLE (Shetlands).—Numbers migrating on September 11th and 12th, 1906 (*id., t.c.*, 1907, p. 79).

BARRA (Outer Hebrides).—Seen in summer in 1900 and 1903 (*t.c.*, 1901, p. 143; 1903, p. 15; 1904, p. 216). There is apparently little previous evidence of its occurrence in the islands strictly included in the Outer Hebrides.

[NORFOLK.—One is said to have been picked up dead in January, 1906, on the edge of Thetford Warren (J. H. Gurney, *Zool.*, 1907, p. 122).]

LITTLE TERN *Sterna minuta* L. S. page 651.

SCILLY ISLANDS.—Though it breeds in Cornwall, it appears to be only a casual visitor to the Scillies (J. Clark, *Zool.*, 1906, p. 343).

ISLE OF MAN.—A small colony was found in 1898 (P. Ralfe, *t.c.*, 1899, p. 32).

SCOTLAND.—From Tay to Dee, common and *increasing* (J. A. Harvie-Brown, *Fauna of Tay Basin*, p. 334).

SHETLAND.—Six adults at Grutness Voe on September 20th, 1900. Not previously recorded from Shetland (W. E. Clarke and T. G. Laidlaw, *Ann. S.N.H.*, 1901, p. 11).

OUTER HEBRIDES.—*Barra.*—Five pairs nesting on a small island in 1901 (W. L. MacGillivray, *t.c.*, 1901, p. 237) : also noted there in 1902-3, but since then has not been seen (N. B. Kinnear, *t.c.*, 1907, p. 81 ; *cf.* also *antea*, Vol. I., pp. 193 and 232).

SOOTY TERN *Sterna fuliginosa* J. F. Gm. S. page 653.

SUFFOLK.—An adult in good plumage, which had apparently died from exhaustion, was found on the heathland between Thetford and Brandon, at the end of March or beginning of April, 1900, by Messrs. J. Nunn and G. Mortimer. It was stuffed and remained wrongly identified until 1903, when Mr. W. G. Clarke saw it and identified it as a Sooty Tern, which was confirmed by Mr. T. Southwell (W. G. Clarke, *Zool.*, 1903, p. 393).

LANCS.—One was picked up alive, but in an exhausted condition, in the early morning, on October 9th, 1901, in a street in Hulme, near Manchester (H. Saunders, *Bull. B.O.C.*, XII., p. 26 ; see also C. Oldham, *Zool.*, 1902, p. 355).

These are the fourth and fifth examples recorded in this country of this species.

NODDY TERN *Anous stolidus* (L.). S. page 655.

Mr. R. J. Ussher's reasons, as expressed in his lately published " List of Irish Birds," for excluding this bird from the British avifauna, have already been given (see *antea*, p. 248). For further details with regard to the specimen said to have been shot on the Dee marshes see Coward and Oldham, " Birds of Cheshire," p. 229.

(To be continued.)

NOTES

WOOD-PIGEON DIPHTHERIA.

WE would remind our readers that the schedules relating to this enquiry, which were attached to the November number of BRITISH BIRDS, should be filled in and posted to us by March 1st. We may here reiterate the hope expressed on page 199, that our readers will make such a response to the request for information that the enquiry may be made really useful. We cannot remind observers too often that negative evidence which affects the distribution either of the birds or the disease is of equal importance to positive evidence.—EDS.

UNUSUAL BIRDS IN HERTFORDSHIRE.

LONG-TAILED DUCK (*Harelda glacialis*).

An immature male Long-tailed Duck was shot on the reservoirs at Tring on November 20th, 1908. We have previously had one visit from this species, an adult ♂ shot November 12th, 1906.

PALLAS'S SAND-GROUSE (*Syrrhaptes paradoxus*).

On December 1st I saw a flock of seven or eight Pallas's Sand-Grouse near Tring (Parish of Buckland). We were shooting pheasants, and the Sand-Grouse rose out of a turnip field.

SHAG (*Phalacrocorax graculus*).

On October 22nd, 1908, a Shag, the first recorded for Hertfordshire, was shot on Tring reservoirs.

BITTERN (*Botaurus stellaris*).

A Bittern stayed on the reservoir for ten days, but left last week (January 14th, 1909). It allowed a keeper to stand watching it within three yards for some time.

BLACK-NECKED GREBE (*Podicipes nigricollis*).

A female and a young male were shot on Tring reservoirs on November 21st and 24th, 1908. A female was shot on November 24th, 1903.
L. W. ROTHSCHILD.

SONG-THRUSH'S NEST IN DECEMBER.

ON December 17th, 1908, a Song-Thrush's nest containing two newly-laid eggs was found at Forton, near Lancaster.
H. W. ROBINSON.

EVERSMANN'S WARBLER (*PHYLLOSCOPUS BOREALIS*) AT FAIR ISLE.

A NEW BRITISH BIRD.

ON September 28th, 1908, while Mr. W. Eagle Clarke was pursuing his investigations (frequently referred to in these pages) of the migration of birds at Fair Isle, he put up out of a patch of potatoes a dark-coloured Willow-Warbler, which he at once suspected belonged to a species he had never seen before. He was fortunate enough to secure the bird, which proved to be an undoubted example of Eversmann's Warbler (*Phylloscopus borealis*), a species which has not previously been detected as occurring in the British Isles. This species has only once before occurred in Western Europe, viz., at Heligoland on October 6th, 1854. It summers in Finmark, Northern Russia and Siberia, and winters in Burma, Malaya, China, etc., and, as Mr. Clarke remarks, it would be interesting to know where the European contingent passes the winter, "for it is difficult to believe that there are no winter retreats for this species nearer than the eastern sections of Southern Asia" (*cf. Ann. S.N.H.*, 1909, pp. 1 and 2).

H. F. W.

LITTLE OWL IN HAMPSHIRE.

HAVING seen the article on the spreading of the Little Owl in BRITISH BIRDS, I thought it worth while to send you a record of one being shot near Petersfield, Hants, on December 26th, 1908. While driving one of the hangers it dashed out from amongst some thick ivy bushes, and was shot by a friend who mistook it for a Woodcock or large Snipe.

A. W. MARRIAGE.

MONTAGU'S HARRIER IN IRELAND.

A MONTAGU'S HARRIER (*Circus œruginosus*) in immature plumage was shot about September 10th, 1901, at Castle Flemyng, Queen's co. My informant is the owner of the specimen, on whose estate it was shot. It has not hitherto been specially recorded in print, but it makes the eleventh occurrence of this bird in Ireland. to which I allude in my "List of Irish Birds" (p. 27). All the other examples have been taken in or near the co. Wicklow (*cf. antea*, Vol. I., p. 318).

R. J. USSHER.

ICELAND FALCON IN SCOTLAND.

A FEMALE (? adult) Iceland Falcon (*Falco islandicus*) was received by me in the flesh on December 19th, 1908. It was killed some ten days earlier on or near Callanish Light, Flannan Isles, by Stornoway, Lewis, N.B.

FRED. SMALLEY.

FOOD OF THE RED-BREASTED MERGANSER.

On November 27th, 1908, on dissecting a Red-breasted Merganser drake (*Mergus serrator*) I found in the crop a small round crab a little larger than a shilling, and in the gizzard two more crabs of the same size, one whole and the other slightly digested. Besides these there were two or three claws, one being that of a much larger crab, and the ground-up remains of a number of crab-shells. There was also some flesh in all stages of digestion, most of which was that of crab, but the more digested pieces were difficult to determine, although I think they were crab, as there was no trace of fish-bones whatever. The absence of fish remains was all the more interesting as the bird was shot off a shore swarming with coal-fish fry.

H. W. ROBINSON.

SMEW IN MONTGOMERYSHIRE.

On January 1st I received for identification from Churchstoke a young male Smew, which had been shot there the previous day. It is the first ever recorded in the county of Montgomery. A similar bird was shot near Shrewsbury just a week earlier. As the Smew has occurred over a dozen times in Shropshire, it must almost certainly have visited the neighbouring county of Montgomery, but has hitherto apparently escaped notice.

H. E. FORREST.

RED VARIETY (*P. MONTANA*) OF THE COMMON PARTRIDGE.

WITH reference to Mr. W. P. Pycraft's description of the remarkable variety of Red-legged Partridge (*Caccabis rufa*) killed this season in Essex (*cf. antea*, p. 240), it is perhaps interesting to record the two somewhat similarly marked varieties of the Common Partridge (*Perdix cinerea*) in the possession of Lord Forester. These birds were killed a number of years ago on the Willey Park Estate, near Broseley, Salop, and it is believed that they were both shot from the same covey. The two birds are very much alike, and a description of one will perhaps suffice. The lower neck, breast and flanks—indeed all the underparts save the centre of the belly—are of a uniform chestnut-brown; the back and wing-coverts are also abnormal, these parts being profusely, and more or less evenly, blotched with dark brown. The rest of the plumage is of normal coloration, but, owing to long exposure to the light, both specimens are sadly faded.

C. INGRAM.

[Mr. Ingram kindly forwarded us an accurately coloured sketch of the bird in question, and this we have submitted to Mr. J. H. Gurney, who has for some years taken a great interest in this variety of the Partridge, which persistently recurs in Norfolk. Mr. Gurney writes as follows :—

" On comparing Mr. Ingram's coloured sketch with a good example of the *Perdix montana* of Brisson (*Orn.*, I., p. 224), killed a few years ago in Norfolk, it appears certainly to represent an immature but faded example of that variety. This singular race or breed seems to be best described as an erythrism, or abnormal replacement of the natural colours by red."]

THE AVERAGE WEIGHT OF SNIPE.

THE average weight of ninety Snipe shot in Shetland, as given on p. 267, Vol. II., seems to suggest that they must have been specially selected specimens, or at least secured during a specially favourable season. It would have been interesting had you given the date and time of year when the ninety 5·78 oz. Snipe were secured in Shetland. Most careful and accurate statistics, taken by a friend of mine in Orkney during September, October, and November, 1908, work out at 4·15 oz. for 1679 Common, and 2·24 oz. for 328 Jack Snipe.

MAURICE C. H. BIRD.

[The weights given by Mr. Haldane were quoted because they were above the usually accepted average. Saunders (*Manual*, p. 574) gives the average weight of the Common Snipe as 4 ozs. : Mr. Harting (*Handbook*, p. 200) 4 to 4½ ozs. In the note quoted Mr. Haldane suggests that the heavy weight of the Shetland Snipe and Woodcock might be due to the fact that food is always plentiful, and that the weather is usually open. In reply to the Rev. Bird's questions, Mr. Haldane kindly writes that he has re-read his note in the " Annals," and agrees with all he there stated. As corroboration he gives the following weights, taken at random from his diary :—" Dec. 22, 1904, 4 Snipe, 6½, 6½, 6, 7 ozs. One day in December three Snipe, 6, 7, 7 ozs. Dec. 27, 1901, eleven Woodcock weighed an average of 13·6 ozs., the lightest being 12 and the heaviest 16½ ozs." Mr. Haldane adds :—" About the end of November and December they seem to attain their greatest weight. I never pick birds, but have them weighed as they come in. I know nothing about Orkney. The two groups of islands are quite unlike."—EDS.]

POMATORHINE SKUA IN LANCASHIRE.

AN adult female Pomatorhine Skua (*Stercorarius pomatorhinus*) was killed with a stone near Cockersand Light, Lancashire, on November 28th, 1908.

I may also mention that an adult male was shot near Graemsay Lighthouse, Orkney, on November 4th, and an immature male at Great Yarmouth, Norfolk, on November 17th.   FRED. SMALLEY.

TWO NORFOLK LEVANTINE SHEARWATERS.

MR. H. N. PASHLEY, of Cley, Norfolk, has kindly sent me word of two Levantine Shearwaters which were shot by George Long (a local wildfowler) on September 22nd, 1891, on the bar at Blakeney. One of these birds is in the collection of Mr. E. M. Connop, of Wroxham, who permits me to record it, and the other is in the collection of Mr. Percy Evershed, of Norwich, and both have been examined by Mr. T. Southwell. Mr. Pashley states that both birds were seen by Howard Saunders and Lt.-Col. H. W. Feilden, and were pronounced by the former to be true Levantines. It may have been due to a slip that they were not referred to in the second edition of the "Manual," but in any case their history and identification seem perfectly satisfactory.

H. F. WITHERBY.

BLACK REDSTART IN THE OUTER HEBRIDES AND IN FIFE.—On November 6th, 1908, the Duchess of Bedford saw a specimen of *Ruticilla titys*, a scarce visitor to Scotland, on South Uist (*Ann. S.N.H.*, 1909, p. 4). On October 22nd, 1908, a fine male was seen at Balcomie, Fife, by the Misses Rintoul and Baxter (*t.c.*, p. 49).

GARDEN-WARBLER AT SULE-SKERRY (N.W. of Orkney).—On September 22nd, 1908, a specimen of *Sylvia hortensis* was taken at the Sule-Skerry Lighthouse (W. Eagle Clarke, *Ann. S.N.H.*, 1909, p. 48).

RED-BREASTED FLYCATCHER IN BARRA, AND AT THE BUTT OF LEWIS.—On November 3rd, 1908, while at Barra, the Duchess of Bedford saw a small brown bird which, coming well into view, was seen to have the basal half of the tail white with the exception of the centre feathers, which were dark. It thus became clear that the bird was either a female or young male *Muscicapa parva* (*Ann. S.N.H.*, 1909, p. 3). Although of late years Mr. Eagle Clarke has recorded several of these birds from Fair Isle, only two other instances (one at the Monarch Lighthouse in 1893, and the other at the Bell Rock on October 25th, 1907) of its occurrence in Scotland were known. Mr. Robert Clyne, who obtained the bird at the Bell Rock, now writes that he is certain he saw a bird of the same species on November 1st, 1908, on the cliff edge at the Butt of Lewis, where he is now stationed (*t.c.*, p. 48).

HAWFINCH IN SUMMER IN EAST LOTHIAN.—The Rev. H. N. Bonar records that an immature male *Coccothraustes vulgaris* was found dead at Tyneholm, Pencaitland, on July 3rd, 1908 (*Ann. S.N.H.*, 1909, p. 48). For Scottish breeding records, see *antea*, Vol. I., p. 151.

LITTLE BUNTING AT SULE-SKERRY (N.W. of Orkney).— On September 22nd, 1908, a specimen of *Emberiza pusilla* was taken at the Sule-Skerry Lighthouse (W. Eagle Clarke, *Ann. S.N.H.*, 1909, p. 48).

WHITE-TAILED EAGLE IN HEREFORD.—Mr. H. E. Forrest writes that a male " four-year-old White-tailed Eagle was shot near Hereford on December 31st, 1908, and is being preserved for the Hereford Museum."

HONEY-BUZZARDS IN ENGLAND.—Mr. F. Smalley kindly informs us that the specimen of *Pernis apivorus* mentioned in our last number (p. 283) as having been shot near Beccles, Norfolk, was obtained on September 23rd, and was in dark chocolate-coloured plumage. Mr. H. E. Forrest writes that specimens were shot at Ashbourne (Derbyshire) on September 10th, near Cardigan on September 24th, and near Tamworth on September 30th. One was shot near Carlisle on October 23rd, 1908 (L. E. Hope, *Nat.*, 1909, p. 30).

THE STONE-CURLEW IN YORKSHIRE.—In an interesting article (*Nat.*, 1909, pp. 11–16) Mr. E. W. Wade deplores the fact that the Stone-Curlew, once so plentiful in Yorkshire, now barely exists in two districts only—one in the North Riding and the other on the wolds. Cultivation of what was once " waste " land or warren, is responsible for the banishment of the bird, and although there is great danger of their extinction in the county owing to their present very small numbers, there should be good hope for them on account of their well-known persistence in returning to ancient breeding haunts, however changed. Moreover, the cultivation of the wolds " appears to have reached its highest point." Mr. Wade confirms Mr. Meade-Waldo's observation of the incubation period, viz., 26–27 days (*cf. antea*, Vol. I., p. 92).

BLACK-NECKED GREBE ON THE SOLWAY.—A female specimen of *Podicipes nigricollis* in winter dress, but showing traces of nuptial dress on the neck and cheeks, was shot at Bowness on the Solway on December 3rd, 1908. The species rarely occurs on the Solway (L. E. Hope, *Nat.*, 1909, p. 30).

REVIEWS

The Food of Some British Birds. By Robert Newstead, M.Sc., A.L.S., &c. Supplement to the Journal of the Board of Agriculture. December, 1908. Price 4d.

WE are glad to find that the Board of Agriculture is at last beginning to realize the importance of the study of Economic Ornithology, for the Report they have just issued does credit alike to the author and the authorities under whose auspices it is published.

Before proceeding to indicate the scope of Mr. Newstead's careful and valuable work, we may say that if there is one thing more than another which it serves to demonstrate, it is this—that his conclusions are of local value only; and Mr. Newstead, probably more than anyone else, would be the first to insist on this. But his work, we trust, will be taken as a model to be followed in every county throughout these islands; then, and not till then, shall we be in a position to draw reliable data from the facts collected, whereon to base legislation, or to adopt measures for the control of any given species in any particular area. The conclusions which Mr. Newstead has drawn from his study of the food of birds in the county of Chester, for example, will not apply with equal truth in, say, a fruit-growing county.

Mr. Newstead's paper contains the results only of some 871 *post-mortem* examinations of birds representing 128 species, some of which are but rare visitants, such as the Hoopoe, Waxwing, Bittern, and Crane; while in the case of many common species he has examined but a single stomach.

In the first eighteen pages of this Report he gives a general summary of his work, which has extended over the last twenty years, concluding this section with a few brief generalizations as to the relative value of our commoner British birds, in relation to the farmer and gardener, and though we agree in the main with his summary, we feel that in some cases his condemnation of certain species is premature and based on insufficient evidence.

The Blackbird, Bullfinch, Sparrow-Hawk, and Raven he brands as " doubtfully of any utility," while the Carrion Crow, House-Sparrow, and Wood-Pigeon are " species which are wholly destructive and useless." At any rate, of the last-named it may be said that it is good to eat, and, therefore, not useless.

Of many fish-eating birds, such as the Kingfisher, Auks, and so on, he gives no particulars as to the kind of fish eaten, apparently because their remains were too fragmentary to make identification possible. But this is not necessarily so, for most, if not all, of these fish could have been identified by means of the " otoliths," or ear-bones.

Of some other birds his findings are curiously interesting. Thus an analysis of the stomachs of four Common Snipe yielded seeds, grass, fragments of beetles, and small land shells ! while five Jack Snipe gave similar results. The only Woodcock examined contained an earwig, a beetle, and a little sand !

Of the Black-headed Gull we are glad to note he remarks :— " Fortunately the birds were, and are still, strictly protected in this area (Chester)." And this because, among other things, it devours enormous numbers of crane-flies, and their larvæ—" leather jackets." During the plague of these insects, which devastated the Dee Marshes in 1901, these Gulls gathered in hundreds to the feast, and gorged themselves so completely that the pellets, or castings, thrown up were left scattered over the land, " looking like little bundles of dead grass " !

Mr. Newstead is to be congratulated on his work, which, so far as it goes, is excellent ; but what is wanted is an exhaustive analysis of a larger number of the commoner species of our native birds continued through every month of the year, including the nesting season, for our knowledge of the food of nesting birds is peculiarly meagre. And this work, to be convincing, must be carried on by experts, and with scrupulous accuracy and attention to details. Results obtained at haphazard, and from a single example of any given species, are practically useless.—W.P.P.

BRITISH BIRDS

AN·ILLUSTRATED·MAGAZINE DEVOTED·TO·THE·BIRDS·ON THE·BRITISH·LIST

MARCH 1,
1909.

Vol. II.
No. 10.

MONTHLY·ONE·SHILLING·NET
326·HIGH·HOLBORN·LONDON
WITHERBY & C°

LITTLE TERN ON THE NEST, SPURN, YORKSHIRE.

(Photographed by Oxley Grabham.)

BRITISH·BIRDS

EDITED BY H. F. WITHERBY, F.Z.S., M.B.O.U.

ASSISTED BY W. P. PYCRAFT, A.L.S., M.B.O.U.

THE COLONY OF LITTLE TERNS AT SPURN POINT, YORKSHIRE.

BY

OXLEY GRABHAM, M.A., M.B.O.U.

(PLATE VII.)

I DO not propose to enter here into the peculiarities of the Little Tern (*Sterna minuta*) in general, but merely to give a few facts relative to the last Yorkshire breeding colony, which at present is, I am glad to say, in a flourishing condition. Spurn Point, at the mouth of the

Humber, is an isolated spit of land, bordered on one side
by the North Sea and on the other by the vast mud-
flats of the Humber. It has been a happy hunting
ground of mine for many years, in autumn and winter,
wild-fowling along the river and coast and in the
marshes, and spending nights in the lighthouse to view
the enormous flocks of birds that pass on migration, and
in the summer watching and photographing the Little
Terns and other birds that breed there.

The Little Terns have bred there as long as living

FIG. 1.—The Eggs in a Slight Scoop on Fine Sand.
(*Photographed by* Oxley Crabham.)

memory goes back, and doubtless for a great many years
before ; but a decade or so ago the birds were in danger
of extinction owing to the raids made upon them by
collectors, and also owing to the thoughtlessness of
excursionists who used to pick up the eggs and throw
them at one another for fun ! ! We did not mind anyone
taking a clutch of eggs for scientific purposes, but this
sort of thing was too much, and so a few of us, with Mr. W.
H. St. Quintin, of Scampston, at our head (than whom no
one living has done more to preserve the birds of our

county that needed protection), subscribed together and put on a watcher. We also received great assistance from Mr. Consett Hopper, Mr. J. W. Webster, the lighthouse keepers, and others who live at the Point. Things had got so bad that the colony had dwindled to about a dozen pairs, and these few were so harried and disturbed that they hardly ever came near their eggs during the daytime, and had to trust to the heat of the sun and the sand, only settling down when night fell.

The birds arrive at their breeding grounds almost

Fig. 2.—Newly-Hatched Young.
(*Photographed by* Oxley Grabham.)

to a day at the end of April. In the cold spring of 1907 they did not appear till May 1st, the latest date that Robinson, our watcher, has ever known, and most of them leave at the end of August. They sit on their eggs for about seventeen days, and the young can toddle away as soon as they are out of the shell, which the old birds remove at once. I have noticed two types of chicks: one much yellower than the other. High tides often do much damage to the eggs, which are placed too near the

ordinary high-water mark. But Robinson, if he scents
danger, moves the eggs a considerable distance inshore,
and the birds easily find them. If the first clutch be
destroyed the Little Terns always lay again, and occasionally even when they have hatched off one clutch they
will lay again. Sometimes just at hatching time we
have had two or three days of very cold rough weather,
and then I have seen the poor little chicks, just out of
the shell, huddling together under the lee of a big stone,
an old boot, piece of wood, or any flotsam and jetsam

FIG. 3.—Little Tern calling to her Mate.
(*Photographed by* Oxley Grabham.)

washed ashore by the sea that will afford them protection
and at such times a few always succumb to exposure.
But, as a rule, there is very little mortality amongst
either the old or young birds, if the weather is propitious.
They have few natural enemies here, and their eggs are
very fertile—one seldom comes across a bad one. The
young are fed largely on very small plaice about the size
of a penny, sand-eels, sprats, etc. During 1908 between
fifty and sixty pairs of birds bred here, and in spite of
some very cold weather, just at hatching time, a good

percentage of young arrived at maturity. There is
nothing peculiar to this particular colony in the nests of
eggs. The usual clutch is two, occasionally three, and
rarely I have found four. Owing to the drifting sand-
storms to which this coast is exposed, the eggs frequently
get covered up to the depth of several inches, but the old
birds almost invariably scratch them out again, and
make all right.

In connection with our Spurn colony one further item
of interest may be mentioned. The late John Cordeaux,

Fig. 4.—Little Tern on the Nest.
(*Photographed by* Oxley Grabham.)

who took a very great interest in the birds of the Humber,
told me that a good many years ago he sent some eggs
of the Little Tern—as this species is wanting in those
otherwise Tern-favoured islets—to the Farne Islands.
They were put in the nests of the Common and Arctic
Terns, but although they hatched out all right, and
eventually went away with their foster-parents, they
never returned to the scenes of their youth ; and so the
attempts to introduce this pretty little species into the
Farnes resulted in failure.

ON A PLAN OF MAPPING MIGRATORY BIRDS IN THEIR NESTING AREAS.

BY

C. J. AND H. G. ALEXANDER.

AFTER some years' observation of the birds in the neighbourhood of Tunbridge Wells, we came to the conclusion that in many species each pair inhabits a definite area, into which other pairs do not intrude. In the spring of 1907, therefore, we decided to mark in the positions of these pairs on 6 in. Ordnance Survey Maps. In that summer we mapped a considerable area round Tunbridge Wells, while in the summer of 1908 we increased this area, and also began mapping at Wye (near Ashford).

In placing a pair of birds on the map we generally relied on the singing of the male, though in many cases we saw the female as well, and occasionally found the nest. In the migratory species of the *Turdidæ* most individuals sing persistently from their arrival (provided the weather is suitable) to the time of pairing, less during the time of nest-building, again more while the females are sitting, and less after the young are hatched. A Chiffchaff at Wye apparently did not sing at all after it began to build, and from this extreme all gradations occur up to individuals which sing nearly as much all through the rest of the song-period as just after their arrival.

When a bird of any migratory species appears at a place on one day and is gone again on the next, it is safe to assume that it is only on its way to its own breeding-ground. Of such individuals we only see a very few, both of species which breed in our districts, and of species such as the Ring-Ouzel, Greenland Wheatear, Redstart, and Common Sandpiper, which do not; the only occasion on which we have seen any number was at Wye on April 23rd, 1908 (the day before the snow), when seven passing Willow-Wrens were observed where only eighteen local ones had arrived.

The method of arrival which we have observed in these two districts agrees with what has been made clear in the " B.O.C. Migration Reports," namely, that each migratory " wave " drops a few individuals of a species in a district. Thus the filling of these two districts with any one species takes several weeks. In 1908, for instance, the first male Chiffchaff belonging to the Tunbridge Wells district was seen on March 30th, while another of the Chiffchaffs of the district did not arrive until about April 30th ; some Willow-Wrens at Wye did not arrive until after May 4th ; two Tree-Pipits arrived near Tunbridge Wells on April 11th, while some did not come until early May ; and one Blackcap reached the same district on April 10th, others not until the beginning of May.

So far as we have been able to judge, certain males are habitually among the earliest, others among the later, arrivals.

The males inhabiting one small district (such as a wood, or stream valley) often appear to arrive together. In the case of such a district which contained several pairs of a particular species in 1907, but none in 1908, we conclude that the males were travelling together and were overtaken by some calamity. At Tunbridge Wells, the only five Willow-Wrens of a district known as Bishopsdown, and three Chiffchaffs close together near Langton, were absent in 1908, and at Wye three Sedge-Warblers, which had inhabited a part of the river for at least three seasons, likewise did not appear in 1908.

The females seem to be always a few days behind the males, and we have observed the curious fact that those males which arrive first are first joined by the females, so that an early pair may be building before the male of a late pair has arrived. This suggests to us that the same female returns to the haunt of its mate year after year.

The young seem to disappear earlier than their parents
in most species, the Red-backed Shrike being an exception.   In those species in which the sexes are alike,
and in which the males do not sing in the autumn, we
cannot tell which sex leaves first.   In the case of the
Chiffchaff the males are the last to leave ; a pair of
Whinchats left together in 1908 between September
30th and October 1st ; the last two Blackcaps seen in
1908 were females, but there is a possibility that these
were merely passing.

It seems that certain individuals habitually stay
latest, just as certain ones habitually come earliest.   We
have not been able to detect any correlation between
early arrival and late departure.

Our mapping also provides a census of certain species
in the two districts, and shows the variation in numbers
from year to year.   It will be seen in the subjoined table
that the numbers in 1908 were practically the same as
those in 1907, except in the Chiffchaff and Red-backed
Shrike, which show considerable losses.   In almost all
cases the losses appear slightly greater than the gains,
but this is probably due to the fact, that it is easier to be
sure that a bird which certainly was here last year, is
not here this year, than to be sure that one which is here
this year was not overlooked last year.   The minus
number in 1908 compared with the 1907 number gives
the proportional loss on migration of the adult males,
and hence their length of life ; but it would not be safe
to work this out on one season's difference.

What becomes of young birds we cannot pretend to
say ; the individuals shown in the plus number in 1908
are presumably young of the previous year.   We have
occasionally found two males of a species arrived at the
same place and singing at one another.   Eventually, as
with Robins in the autumn, one has disappeared, or else
has settled in an unoccupied place near by.

The observations of only two years are insufficient to
form a basis for definite conclusions in all cases, but in

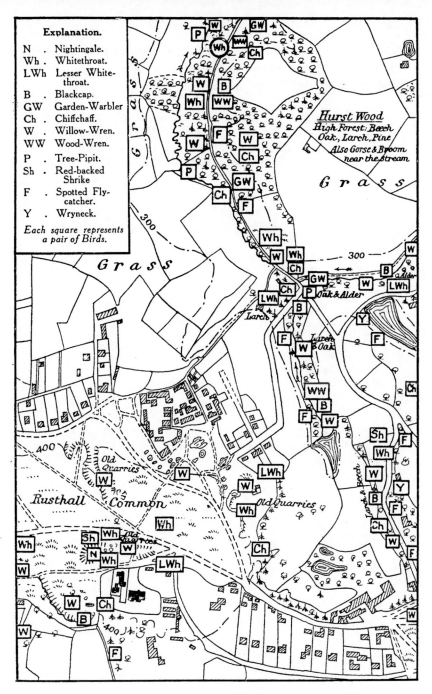

MAP SHEWING SOME OF THE MIGRANTS IN PART
OF THE BOROUGH OF TUNBRIDGE WELLS.

(Six inches equals one mile).

this article we have given some of the results, as well as an idea of what may be the outcome of more complete observations on these lines.

TABLE showing the difference in number of pairs of certain migrants at Tunbridge Wells between 1907 and 1908, and the total number of pairs of migrants mapped in a larger district at Tunbridge Wells and in a district at Wye in 1908.

	Number of pairs in 1907.	Difference in 1908.	Number of pairs in 8 sq. miles at Tunbridge Wells (1908).	Number of pairs in 4 sq. miles at Wye (1908).
Wheatear ..	0 .	0	0	4
Whinchat ..	0	0	3	0
Stonechat ..	3	+1	4	0
Nightingale ..	4	—1	6	25
Whitethroat ..	29	—3	101	30
Lesser Whitethroat	15	+2—4	24	15
Blackcap	17	+2—4	32	20
Garden-Warbler ..	13	—4	22	14
Chiffchaff	42	+1—10	56	10
Willow-Wren ..	123	+9—14	188	88
Wood-Wren ..	7	0	11	0
Grey Wagtail ..	0	+1	2	0
Tree-Pipit ..	38	+4—5	55	35
Red-backed Shrike	7	—3	5	1
Spotted Flycatcher	18	+1—3	27	1
Wryneck	10	+1—1	13	10
Corncrake ..	0	+1	1	1*

* Not present in 1907.

ON THE MORE IMPORTANT ADDITIONS TO OUR KNOWLEDGE OF BRITISH BIRDS SINCE 1899.

BY

H. F. WITHERBY and N. F. TICEHURST.

PART XVIII.

(Continued from page 308.)

SABINE'S GULL *Xema sabinii* (Sab.). S. page 657.

Immature birds appear to occur almost regularly in autumn on the Norfolk coast, while occurrences have been recorded of recent years from Cornwall, Somerset, Yorkshire, Derby, and Hants. An adult bird was shot near Rye, Sussex, on October 20th, 1891. (N. F. T.)

SCOTLAND.—*Skerryvore.*—One immature bird on February 10th, 1905, and one on November 30th, 1907 (*Ann. S.N.H.*, 1906, p. 202, and 1908, p. 205). *Argyllshire.*—An immature bird received for preservation on October 30th, 1903, from Easdale (C. H. Bisshop, *t.c.*, 1904, p. 57).

The breeding of this species on Spitzbergen has now been proved beyond doubt by the discovery in 1907 by Professor König's expedition of a nest with two eggs, from which the parent birds were shot (F. C. R. Jourdain *in litt.*).

WEDGE-TAILED GULL *Rhodostethia rosea* Macgill. S. page 659.

Nesting Habits.—In the delta of the Kolyma River, N.E. Siberia, it was found breeding numerously by Mr. S. A. Buturlin in 1905. Though snow was still deep, and the ice had only just broken up on the river, incubated eggs were found on June 13th. The birds nest in small colonies of ten to fifteen pairs. Early in July young in down were found. The eggs, young in down, and young in first plumage, are fully described (*Ibis*, 1906, pp. 131–139, 333–337, 610, and Pl. XX. (eggs), 661–666, 1907, Pl. XII. (young in down)).

LITTLE GULL *Larus minutus* Pall. S. page 663.

This species being of fairly regular occurrence on the east and south-east coasts of England, and especially so on the Norfolk coast and the east coast of Scotland in autumn and winter, we have not quoted the records.

SCILLY ISLES.—One was shot on St. Mary's in December, 1905 (J. Clark, *Zool.*, 1906, p. 343).

CORNWALL.—One was obtained at Swan Cove in November, 1904 (J. Clark, *t.c.*, 1907, p. 287).

CHESHIRE. — One (apparently adult) seen on December 26th, 1902, on the Manchester Ship Canal (T. A. Coward, *t.c.*, 1903, p. 172). One seen flying round the "Conway," on December 16th, 1903 (F. C. R. Jourdain, *t.c.*, 1904, p. 193).

NORTH WALES.—One in 1898 and two in 1901 are mentioned (H. E. Forrest, *Vert. F. N. Wales*, p. 379).

SHETLAND.—One at Nyra Sound, May 3rd, 1904 (T. E. Saxby, *t.c.*, 1904, p. 230).

[IRELAND.—One was seen by "G. W." on the coast of Connaught on several days between July 13th and August 25th, 1906, in company with a number of Black-headed Gulls (*Field*, 13, x., 06, p. 650).]

Has recently been found breeding in the Ringkjöbing Fjord, Denmark (*vide Field*, 17, XII., 04, and 23, XII., 05 ; and *Vid. Med. nat. For. Kbhvn.*, 1905. p. 245), and also at Rossitten in East Prussia (*J.f.O.*, 1903, p. 186), thus it seems to be extending its range westward.

MEDITERRANEAN BLACK-HEADED GULL *Larus melanocephalus* Natt. S. page 667.

YORKSHIRE.—An adult in winter plumage was obtained on the Yorkshire coast in November, 1895. No further details are permitted by the owner (T. H. Nelson, *Birds of Yorks.*, p. 675).

[CORNWALL.—Two examples, stated in a manuscript catalogue by Harry Shaw to have been killed near Falmouth in March, 1851, are now in the possession of Mr. Beville Stanier, of Peplow Hall, Salop (H. E. Forrest, *Zool.*, 1907, p. 33).]

YELLOW-LEGGED HERRING-GULL *Larus cachinnans* Pall. S. page 674.

[One was seen in Dover Harbour on April 18th, 1904, amongst some other Herring-Gulls, and came close enough for its orange-coloured legs to be noticed (N. C. Rothschild, *Bull. B.O.C.*, XIV., p. 91).]

GLAUCOUS GULL *Larus glaucus* O. Fab. S. page 679.

This species occurs so frequently in Scotland and on the coasts of England as not to require special mention.

IRELAND.—One seen on January 1st, 1901, and an immature bird obtained on February 14th, 1905, at Moyview, *co. Sligo*

(R. Warren, *Irish Nat.*, 1905, p. 71), and an immature bird was shot on Rathlin Island on February 19th, 1907 (W. C. Wright, *t.c.*, 1907, p. 224). One was found dead at Bartragh Island on December 8th, 1906 (R. Warren, *Zool.*, 1907, p. 73).

ORKNEY.—December 25th–26th, 1901, " over fifty, mostly adult birds ; never saw more than four at a time before " (*Ann. S.N.H.*, 1902, p. 197).

ICELAND GULL *Larus leucopterus* Faber. S. page 681.

Winter occurrences of this species are too frequent to require special notice.

Late Dates.—One was seen from April 30th to May 2nd, 1903, and another on May 17th, 1904, in Mull (D. Macdonald, *Ann. S.N.H.*, 1904, p. 247). One was seen on April 7th, 1902, at Londonderry (D. C. Campbell, *Irish Nat.*, 1902, p. 151). One was shot on April 26th, 1905, in the Moy Estuary (R. Warren, *t.c.*, 1905, p. 135).

IVORY GULL *Pagophila eburnea* (Phipps). S. page 685

YORKSHIRE.—One was seen at Flamborough on April 5th, 1904, and was ultimately obtained (T. H. Nelson, *Birds of Yorks.*, p. 694).

NORTHAMPTONSHIRE.—A bird in immature plumage was shot at Weston-by-Weedon on or about February 7th, 1901 (O. V. Aplin, *Ibis*, 1901, p. 517).

CORNWALL.—Two were seen, and one (adult male) was shot in the Hayle Estuary on January 24th, 1907 (J. Clark, *Zool.*, 1907, p. 287).

SCOTLAND.—In January, 1890, the first from the Outer Hebrides was obtained at *Stornoway* (J. A. Harvie-Brown, *Ann. S.N.H.*, 1903, p. 16). Early in February, 1901, one was obtained at Broadford, *Skye* (T. E. Buckley, *t.c.*, 1901, p. 116). One was identified on a close view at Largo Bay, *Fifeshire*, on September 14th, 1904 (L. J. Rintoul and E. V. Baxter, *t.c.*, 1905, p. 53). One was seen "lately " (spring, 1906) in *North Uist* (N. B. Kinnear, *t.c.*, 1907, p. 85).

IRELAND.—The third specimen for Ireland was shot at Belmullet on March 27th, 1905 (R. J. Ussher, *List of Irish Birds*, p. 50).

GREAT SKUA *Megalestris catarrhactes* (L.). S. page 687.

HANTS.—One was picked up dead at Lainston, in March, 1904 (*Birds of Hants*, p. 339).

KENT.—A female was shot on Dungeness on October 4th, 1900 (W. R. Butterfield, *Zool.*, 1900, p. 521).

NORFOLK.—Five were seen on the coast by Mr. Long on August 31st, 1899. Only once before been seen so early, October being the usual time (J. H. Gurney, *t.c.*, 1900, p. 109).

LINCOLNSHIRE.—A bird, probably of this species, was seen off Donna Nook, on September 21st, 1901 (G. H. Caton Haigh, *t.c.*, 1902, p. 132).

YORKS.—One was shot near Robin Hood's Bay on June 29th, 1904 (W. J. Clarke, *t.c.*, 1905, p. 74). One was obtained at Bridlington in the autumn of 1904 (*Birds of Yorks.*, p. 696).

SCOTLAND.—*Fair Isle.*—The natives assured Mr. W. Eagle Clarke that they had it from their fathers that the " Bonxie " long ago bred on the island (*Ann. S.N.H.*, 1906, p. 78). *Shetland.*—" Has increased in numbers . . . . there being at least eighty-four birds on this island " [Unst] (T. E. Saxby, *Zool.*, 1901, p. 391). Twenty-one nests at Hermaness in 1901 (*Ann. S.N.H.*, 1902, p. 197). At least thirty-four nests with eggs in June, 1905, one new colony started (*t.c.*, 1905, p. 182). Forty-two nests in 1907 (Duchess of Bedford, *t.c.*, 1908, p. 4). A pair breeding on Burrafirth Voe had their eggs taken by the Rev. Sorby in 1904, and another pair breeding on Hascasay were robbed by Major Stirling in 1907. These records seem to show that the bird has now a tendency to form new colonies, as is the case with the Fulmar. *Outer Hebrides.*—One was shot in North Harris on January 8th, 1894, the first for the Outer Hebrides (J. A. Harvie-Brown, *t.c.*, 1903, p. 17). *Ayr.*—One was seen on October 22nd, 1907, near Lendalfoot, the first for the Clyde area (*t.c.*, 1908, p. 206).

ISLE OF MAN.—One was caught at Douglas in the late autumn of 1903 (P. Ralfe, *Zool.*, 1904, p. 33).

IRELAND.—One was seen in Holyhead Harbour on July 20th, 1903 (C. J. Patten, *t.c.*, 1904, p. 75). Mr. G. P. Farran has on six occasions at various seasons and at from 30-70 miles off the isles of Kerry observed these Skuas (R. J. Ussher, *List of Irish Birds*, p. 50 ; *Irish Nat.*, 1907, p. 184).

POMATORHINE SKUA *Stercorarius pomatorhinus* (Temm.).
S. page 689.

This species has been recorded from all the east coast counties except Essex of recent years.

IRELAND.—A bird in entirely brown plumage with twisted tail-feathers was shot on May 6th, 1902, on Inniskeal Island, *co. Donegal* (D. C. Campbell, *Irish Nat.*, 1902, p. 187). One was picked up dead at Lough Kiltooris, *co. Donegal*, on May

29th, 1902 (J. Steele-Elliott, *Zool.*, 1906, p. 154). One was shot on June 6th, 1906, at Loop Head, *co. Clare* (R. M. Barrington, *Irish Nat.*, 1906, p. 193). The following were seen in 1906 by Mr. G. P. Farran, of the Fisheries Board, while 20–30 miles off the south-west coast :—One on October 16th off Drogheda ; four on November 6th off Tearaght, co. Kerry ; also seen in May (R. J. Ussher, *t.c.*, 1907, pp. 163 and 184).

LONG-TAILED SKUA *Stercorarius parasiticus* (L.). S. page 693.

SOMERSET.—One was shot on October 19th, 1903, at Axbridge (S. Lewis, *Zool.*, 1904, p. 461) ; said to be the fourth for Somerset (F. L. Blathwayt, *t.c.*, 1905, p. 36).

IRELAND.—An adult was caught on Clare Island, *co. Mayo*, on June 14th, 1906 (R. M. Barrington, *Irish Nat.*, 1906, p. 193).

GREAT AUK *Alca impennis* L. S. page 697.

The late Professor Newton in an interesting article on the " Orcadian Home of the Garefowl " (*Ibis*, 1898, pp. 587–592) explains that the breeding place of the Great Auk was on the Holm of Papa Westray, and not in Papa Westray itself. Professor Newton, in company with the late Henry Evans, Colonel Bolland, and Mr. Joseph Whitaker, landed on the Holm on June 27th, 1898, and visited the very spot which he thought must have been the " true home of the species whose extirpation, so far as Orkney is concerned, was compassed in 1813 by Bullock."

Bones of this species have been found in Antrim, Donegal, and Clare, in addition to Waterford (R. J. Ussher, *List of Irish Birds*, p. 51, and *Irish Nat.*, 1899, pp. 1–4, 1902, p. 188).

BRÜNNICH'S GUILLEMOT *Uria bruennichi* E. Sabine. S. page 701.

YORKSHIRE.—One was procured near Flamborough Head in November, 1899, and one was shot about two miles off Castle Foot on October 28th, 1902 (T. H. Nelson, *Birds of Yorks.*, p. 725).

[On June 14th, 1908, when off the Pinnacle Rocks, Farne Islands, in a boat, Messrs. H. B. Booth and Riley Fortune saw a bird which they identified as an example of this species. " . . . . it was not in full summer plumage, and it was the fact of having more white upon its neck and lower throat in contrast to its companions, the Common Guillemots, that first drew my attention to it, and it was rather darker on the upperparts. . . . . Its thicker, slightly shorter, and differently-shaped beak was quite distinct from that of the Common Guillemot. . . . . I could distinctly see (through my field-glasses) the white line along the edge of the basal half of the upper mandible " (H. B. Booth, *Nat.*, 1908, p. 289).]

BLACK GUILLEMOT *Uria grylle* (L.). S. page 703.

CORNWALL.—One was picked up dead near the St. Anthony Lighthouse, Falmouth, on March 12th, 1905, during very stormy weather. One of the rarest casuals in Cornwall (J. Clark, *Zool.*, 1907, p. 287).

NORFOLK.—Two were seen near Wells by Mr. C. Hamond, on January 8th, 1898 (J. H. Gurney, *t.c.*, 1899, p. 118).

LITTLE AUK *Mergulus alle* (L.). S. page 705.

A great irruption of Little Auks occurred during February and March, 1900, when numbers were washed up chiefly on the Norfolk coast, and many in Suffolk. Compared to the " invasion " in 1895 there were more if anything on this coast in 1900, but " the incursion expended itself in a space of about fifty miles extending from the Wash to Lowestoft, and reaching its maximum at Cley." Not so many were found inland as in 1895, and although the numbers were large, there appeared to be fewer on the Yorkshire and Lincolnshire coasts. Norfolk appears to have recorded five great irruptions of this bird, viz., October, 1841 (probably the greatest) ; December, 1848 ; November, 1861 ; January, 1895 ; February, 1900 (J. H. Gurney, *Zool.*, 1901, pp. 124–126 ; *cf.* also T. H. Nelson, *Birds of Yorks.*, p. 731). One bird recorded " about mid-winter, 1900," on St. Agnes, Scilly Isles (J. Clark and F. R. Rodd, *t.c.*, 1906, p. 345), perhaps was a straggler from this horde. On January 4th, 1900, a great number were reported from North Uist, and in February many along the Aberdeen coast and several in the Forth area (*Ann. S.N.H.*, 1901, p. 144). In the latter half of February, 1901, also over fifty were reported on the Norfolk coast (J. H. Gurney, *Zool.*, 1902, p. 87).

WHITE-BILLED NORTHERN DIVER *Colymbus adamsi* G. R. Gray. S. page 711.

[On December 31st, 1901, a Diver with the whole of lower and about two-thirds of upper mandibles white, was picked up at Caister, Norfolk, but Mr. Gurney does not think the bill was sufficiently upturned for this species. Nor does he consider the specimen figured in Babington's "Birds of Suffolk" a true *C. adamsi* (J. H. Gurney, *Zool.*, 1902, p. 99 ; *cf.* also W. R. Ogilvie-Grant, *antea*, Vol. I., p. 295).]

BLACK-THROATED DIVER *Colymbus arcticus* L.
S. page 713.

NORTH WALES.—One seen by Mr. T. A. Coward in Aberffraw Bay, Anglesey, on April 21st, 1905, was in summer plumage (H. E. Forrest, *Vert. F. N. Wales*, p. 406).

RED-THROATED DIVER *Colymbus septentrionalis* L.
S. page 715.

Moult.—At the end of September and beginning of October, 1898, Mr. W. Farren had several examples of this species sent to him, and the majority of the adults were entirely devoid of flight feathers, both primaries and secondaries being shed *en masse* (*Ann. S.N.H.*, 1899, p. 114).

GREAT CRESTED GREBE *Podicipes cristatus* (L.).
S. page 717.

SCOTLAND.—*Tiree.*—Two pairs in breeding plumage on a loch on May 22nd, 1900, were reported (*Ann. S.N.H.*, 1901, p. 145). "Is now (1904) a rapidly extending species in the nesting season, and nests freely in many parts both south and north of Forth and Clyde. One was seen on May 7th, 1903, in Assynt by Mr. F. L. Blathwayt, the first record for Sutherlandshire" (J. A. Harvie-Brown, *A Fauna of N.W. Highlands and Skye*, pp. 345–346). Three pairs nesting on Lake of Menteith (Perth), in 1905. Breeding range still slowly but surely extending (T. T. Mackeith. *Zool.*, 1905, p. 314).

From the published records there seems to be no doubt that in England the nesting birds are still increasing in numbers. This increase is very noticeable in the Midlands. Records of breeding from the northern counties (Cumberland, Durham, Northumberland, etc.) seem to be lacking, however.

RED-NECKED GREBE *Podicipes griseigena* (Bodd.).
S. page 719.

This species is rare on the west side of Great Britain and on the south coast of England.

JERSEY.—One is recorded without date (H. Mackay, *Zool.*, 1904, p. 382).

SARK.—One was seen in March, 1902 (H. E. Howard, *t.c.*, 1902, p. 422).

SURREY.—The adult male picked up on Farthing Down in 1890 (*Birds of Surrey*, p. 346) was in full breeding plumage (J. A. Bucknill, *Zool.*, 1901, p. 254).

KENT.—A male in full summer plumage was shot at sea off Dungeness on April 14th, 1907. (N. F. T.)

MID-WALES.—A pair was seen on the Dovey in October and November, 1899 (J. H. Salter, *t.c.*, 1902, p. 1).

SCOTLAND.—One was shot at Portmary on February 20th, 1900 (R. Service, *Ann. S.N.H.*, 1900, p. 120). Another was shot at Glencaple on October 6th, 1903 (*t.c.*, 1904, p. 217). One was shot on Spiggie on November 14th, 1901 (*Ann. S.N.H.*, 1902, p. 198), and another on Baltasound (Shetlands) on December 30th, 1901 (T. E. Saxby, *Zool.*, 1902, p. 113).

IRELAND.—Eleven or twelve have been taken at long intervals on or near the coasts (R. J. Ussher, *List of Irish Birds*, p. 52).

SLAVONIAN GREBE *Podicipes auritus* (L.). S. page 721.

Saunders says " its occurrence on the southern and western shores of England seems to be irregular even in winter."

JERSEY.—Frequent (H. Mackay, *t.c.*, 1904, p. 382).

SCILLY ISLES.—An autumn and winter casual chiefly on Tresco, by no means rare. The last was recorded in November, 1902 (J. Clark and F. R. Rodd, *Zool.*, 1906, p. 345).

DEVON.—" I have noticed one or two on the river [? Taw] for the past two winters, and I am inclined to think that they are regular winter visitors " (B. F. Cummings, *t.c.*, 1905, p. 469).

OXFORDSHIRE.—Mr. O. V. Aplin gives particulars of six winter occurrences previously unrecorded (*t.c.*, 1899, p. 441), and of a seventh (*t.c.*, 1907, p. 331).

NORTH WALES.—Occurs frequently in winter on the Merioneth coast (H. E. Forrest, *Vert. F. N. Wales*, p. 409).

SCOTLAND.—There is considerable but not conclusive evidence of its having bred in Benbecula (Outer Hebrides) in 1893. Two were shot in full summer plumage in April, 1898, in Barra (J. A. Harvie-Brown, *Ann. S.N.H.*, 1903, pp. 21–22). One at Arisaig (Inverness) in full summer plumage on May 13th, 1907 (*t.c.*, 1908, p. 207).

Food.—In the stomach of one shot in the winter were besides feathers, elytra of water-beetles and numbers of *larvæ* of the Crane-fly (*Tipula oleracea*) (O. Grabham, *Zool.*, 1899, p. 32). One examin ed by G. Sim from Bruckley Castle, Dee area, contained flies, beetles, grubs, and stickle-backs (*Vert. Fauna of Dee*, p. 190).

(*To be continued.*)

ON THE EGGS OF THE TREE-PIPIT.

BY

PERCY F. BUNYARD, F.Z.S., M.B.O.U.

IT is surely a little surprising that no one has yet seriously attempted to analyse and systematise the marvellous range of variation which the eggs of the Tree-Pipit (*Anthus trivialis*) present, in the matter of colour and arrangement of markings. How great is this range may be gathered from the extremely divergent descriptions which have from time to time been published by the various authors who have had occasion to refer to this subject.

It has been contended indeed that it is impossible to define the limits of this variation. But with this view I cannot agree. On the contrary, as I propose to show, the apparent medley of colour and markings here presented can be reduced to an orderly system comprising no less than seven distinct types. This result, I need hardly say, could never have been arrived at if I had not, through the kindness of many friends, been enabled to examine a very large number of specimens. These seven types (not *varieties*, be it noted) are, in my opinion, of sufficiently frequent occurrence, and so constant and well-defined as to justify this classification. They may be divided into two classes, namely, mottled, and spotted. I recognise three types of mottled eggs, two of which are *very* distinct; while in the spotted eggs I can distinguish four types, three of which have *very* strongly marked characteristics. It will be observed that I have endeavoured to describe the extreme, and less modified, forms of each of these; varieties I have not in this paper attempted to deal with; though they are of frequent occurrence they may, with a keen eye, and a little trouble, be traced to one or other of the types just referred to.

A few words in regard to the system upon which I have worked to obtain these results may be of interest, though I do not pretend that this system would be applicable to the eggs of all species. The work of most importance is the separation of the clutches into their respective types (by no means a difficult operation), keeping them separate by placing each type in a separate tray upon which white cotton wool has been carefully and evenly spread; if glass-lid boxes are used the lids should be removed before attempting to distinguish the colours; always use a magnifying glass of low power, which assists very materially in obtaining accuracy in regard to colour, shape, formation of the markings, texture of shell, etc. ; the stronger the light the better ; I prefer sun-light, but of course not *direct* sun-light. As each point is determined, it should at once be carefully noted down, thus : ground colour, colour of markings, position and arrangement of markings, shape, and, finally, the texture of the shell.

In the following descriptions it should be remarked that I have referred to " Eggs of the Birds of Europe," by H. E. Dresser, Parts VII. and VIII., plate 4, and also to " Eggs of British Birds," by Henry Seebohm, plate 58.* Mr. Dresser figures six types, Mr. Seebohm four only, and, curiously, the one type not figured by Mr. Dresser. To have done full justice to this article I should have preferred to have had plates specially drawn, however, I trust that I have made myself as clear as possible in referring to those mentioned. Some interesting points have been brought to light in connection with the description of these various types. Most noticeable among them is the slight variation in the thickness and texture of the shell. A fact which I think is pretty generally known is that some types occur much less frequently than others, as is also the case with the eggs of the Red-backed Shrike and others. Locality, or climatic conditions have apparently nothing whatever to do with these variations. I have received the whole seven distinct types, from as many different localities ; continental eggs exhibited precisely the same types.

MOTTLED TYPE (No. 1)—Brick-red, very distinct.

GROUND COLOUR.—White. The markings are so close as almost to obliterate the ground colour, though there is generally one or more eggs in a clutch in which the ground colour is fairly conspicuous.

MARKINGS, *normal.*— Rich brick-red to light red (Dresser, pl. 4, No. 15), mottlings very close. *Extreme type*—Mottlings obliterate ground colour. *Modified type*—Markings well defined, ground colour conspicuous, shell markings more or less absent (Dresser, pl. 4, No. 13). A rare variety of this type occurs in which the markings are bold and well defined (Dresser, pl. 4, No. 22) which is intermediate between the red mottled type (No. 1) and the red spotted, or blotched, type (No. 4). This variety has also a slight suspicion of shell markings of purplish grey. Fine hair-like scrollings at the broad end occur in this type.

---

* These works will throughout the rest of this paper be quoted simply as " Dresser, pl. 4, No. —," and " Seebohm, pl. 58."

SHAPE.—Normal, a short conical oval, sometimes fully rounded. Narrow pointed ovals occur less frequently.

SHELL.—Finely grained, glossy, sometimes dull, fairly thick and strong for the size.

MOTTLED TYPE (No. 2)—Purplish-red, distinct.

GROUND COLOUR.—White. The markings do not obliterate the ground colour so much as in type No. 1.

MARKINGS.—In general appearance similar to the mottled brick-red type (No. 1). The purplish tint is caused by the presence of minute purplish-grey shell markings, which are conspicuous, though the pigment markings are distinctly purplish in tinge. Very little variation occurs in this type, which is constant and well set, the fine hair-like scrollings do not occur so frequently as in No. 1 (Dresser, pl. 4, No. 14).

SHAPE.—Similar to No. 1, but the full rounded shape is more frequent. Narrow pointed ovals occur.

SHELL.—Similar in every respect to No. 1, except that there is less gloss.

MOTTLED TYPE (No. 3)—Sepia-brown, very distinct.

GROUND COLOUR.—Greyish-white to white; sometimes distinctly pale greenish blue. Compared with the two other mottled types, the ground colour is conspicuous, except in the extreme type in which the mottlings are so close as almost to obliterate it.

MARKINGS.—Precisely the same in general arrangement as in types Nos. 1 and 2. *Normal*—Mottled rich sepia-brown, shell markings brownish-grey, very inconspicuous, but in some cases sufficiently present to alter the general appearance to a greyish purple-brown (Dresser, pl. 4, Nos. 17 and 18). *Extreme type*—Appearance entirely altered by the running together of the mottlings, which form dark patches of colour, giving this type an intermediate appearance between the normal of this and the brown spotted, or blotched, type No. 5. *Modified type*—Precisely the same in appearance, but several shades paler in colour; fine hair-like markings occur, as in types 1 and 2. The normal of this type is often confused with the egg of the Meadow-Pipit, and in general appearance it somewhat resembles the egg of that species, except in shape, size, and texture. The eggs of the Meadow-Pipit, as a rule, do not show so much gloss.

SHELL.—Similar in every respect to type No. 1.

SHAPE.—Goes through the same variation as type No. 1.

SPOTTED, or BLOTCHED, TYPE (No. 4)—Red, very distinct. ¨

GROUND COLOUR.—Varies considerably from palest grey and red, to white tinged with purple or mauve. This great variation is brought about by very minute shell markings, spots, and cloudings of varying shades of red, purple, and reddish-brown, so closely conglomerated as to alter the entire appearance of the actual ground colour.

MARKINGS.—The variation in the markings is even more marked than in the ground colour. *Normal*—Rich reddish-brown cloudings, evenly distributed spots with very dark centres, marginated with paler shades with eye-spots of very dark brown, a few hair-like lines of the same colour as the margins of the eye-spots, shell markings few and very inconspicuous (Dresser, pl. 4, Nos. 19 and 20). *Modified type* —Markings more or less confined to the broad ends, similar in arrangement, except that the shell markings are more conspicuous, and of a purplish tinge (Dresser, pl. 4, No. 21). *Extreme type*—Markings take the form of short scrollings and cloudings. Other markings are present, but to a very slight extent, and are small and inconspicuous (Dresser, pl. 4, No. 16). This type is much more subject to variation than the others.

SHAPE.—Inclined towards pointed ovals, rather more than in the other six types.

SHELL.—Fragile compared with the mottled types, finely grained, moderately glossy, sometimes glossless.

SPOTTED, or BLOTCHED, TYPE (No. 5)—Brown, very distinct.

GROUND COLOUR.—From palest brown to brownish-grey. In some cases there is a slight suspicion of purplish-grey. In the modified type the ground colour is conspicuous, in the normal and extreme types it is almost obliterated by the markings.

MARKINGS.—Precisely the same in arrangement as in type No. 4 except that there is a tendency to form dark caps. *Normal*—Clouded rich brown, eye-spots black-brown, marginated with paler brown, small fine short scrollings of the same colour as the eye-spots. Shell-markings, rich brown-grey, inconspicuous. *Extreme type*—Rich brown spots and cloudings, ground colour almost obliterated ; shell-markings inconspicuous or totally absent. *Modified type*—Finely dotted and " short-scrolled " with rich brown ; ground colour conspicuous (Dresser, pl. 4, No. 23). This type is constant, well set, and subject to little variation.

SHAPE.—Broad ovals, sometimes slightly pointed.

SHELL.—Very fragile, finely grained, displaying more gloss than in the other types.

SPOTTED, or BLOTCHED, TYPE (No. 6)—Purplish brown, very distinct.

GROUND COLOUR.—Purplish-grey, giving the whole egg a distinctly purplish appearance. The ground colour in this type is very conspicuous and seldom obliterated by the markings.

MARKINGS.—Similar in arrangement and appearance to types Nos. 4 and 5, except that the markings are more prominent and better defined. *Normal*—Eye-spots and short scrollings, rich purplish-brown, marginated with paler shades, cloudings pale purple-grey ; shell markings, dark purplish-grey, few. *Extreme type*—Markings more abundant and richer in colour, ground colour is also darker by several shades. Shell markings almost absent. *Modified type*—Similar in general appearance, but the ground colour more conspicuous. Markings form zones, or caps ; shell-markings conspicuous (Dresser, pl. 4, No. 24). This type is constant, well fixed, and subject to less variation than No. 4.

SHAPE.—Goes through precisely the same variations as in types Nos. 4 and 5.

SHELL.—Thin and fragile, but less so than in types Nos. 4 and 5.

SPOTTED, or BLOTCHED, TYPE (No. 7)—Green ground, distinct.

GROUND COLOUR.—Very conspicuous, and distinctly greenish, giving the whole egg a green appearance, which separates it from types Nos. 4, 5 and 6.

MARKINGS.—In appearance similar to types Nos. 4, 5 and 6, but as a rule more evenly distributed and better defined. *Normal*—Eye-spots and short scrollings, rich umber to sepia-brown, marginated with paler shades, cloudings pale brown ; shell markings pale purplish-brown (Seebohm, pl. 58A, third from right). *Extreme type*—Eye-spots very dark brown, less marginated than in the normal ; ground colour inclined towards olive-brown ; shell markings almost absent. The markings sometimes form caps and zones, giving the egg a very rich appearance. *Modified type*—Similar, except that the short scrollings predominate ; ground colour more conspicuous ; shell markings dark grey-brown, few, sometimes quite absent. This type occurs much less frequently than any of the other six types, but is distinct and well fixed.

SHAPE.—Similar to that in types Nos. 4, 5 and 6, but there is a tendency towards a smaller size.

SHELL.—Very thin and fragile, rather more gloss than in the other types.

THE BIBLIOGRAPHY OF BRITISH BIRDS.

To compile a complete bibliography of a subject which has attracted so much attention for so many years as the ornithology of this country would be a task of great magnitude, and so far as we know no such bibliography has been attempted, although we have had most useful papers on the subject by Dr. Elliott Coues in the "Proceedings of the U.S. Museum" (1880), and in Mr. Miller Christy's "Catalogue of Local Lists" (1891), as well as by the late Professor Newton in the "Dictionary of Birds." A valuable contribution towards the subject has just reached us in the form of a pamphlet entitled "A List of Books relating to British Birds published before the Year 1815." These are from the library of our contributor, Mr. W. H. Mullens, and the pamphlet forms an "Occasional Publication No. 3," of the Hastings and St. Leonards Natural History Society. Seven plates giving facsimiles of rare and notable editions will be much appreciated, while the extremely carefully drawn up details of the works themselves cannot fail to be of the greatest value. We are glad to see the words "to be continued" at the end of the pamphlet, and we would suggest that if those who possess valuable ornithological libraries would co-operate with Mr. Mullens the task of forming a bibliography of British birds might be accomplished.

EDS.

COMPARATIVE LEGISLATION FOR THE PROTECTION OF BIRDS.

IN Vol. I., page 354, we called attention to an offer by the Royal Society for the Protection of Birds of a gold medal and a prize of twenty guineas for the best essay on the subject of "Comparative Legislation for the Protection of Birds." This prize has been awarded to Mr. A. H. Macpherson, while a second prize of ten guineas has been given to Lieut.-Colonel G. A. Momber.—EDS.

THE BRITISH SONG-THRUSH AND DARTFORD WARBLER.

DR. ERNST HARTERT has already described in these pages (Vol. I., pp. 208–222, Vol. II., pp. 130–131) a number of geographical races of birds which are peculiar to the British Islands. At the meeting of the British Ornithologists' Club held on January 20th last, he called attention to the differences between British and Continental examples of the Song-Thrush. He pointed out that the non-migratory race breeding in Great Britain and Ireland differed in the warmer, more rufous, colour of the upper surface, especially the rump. These parts are more olive-brown, generally paler, and with a faint greenish tinge, in the birds breeding on the Continent and migrating to the Mediterranean countries in winter. The underside of the British race was often more heavily spotted, and this was especially conspicuous in specimens from the Hebrides, while others from the same islands were in every way similar to English examples. For this reason Dr. Hartert did not, for the present, distinguish more than one British race, which he proposed to call TURDUS PHILOMELOS CLARKEI, in honour of Mr. Eagle Clarke, who had first called his attention to the dark coloration of the British race. The difference had also been noticed by other British ornithologists. Dr. Hartert mentioned that the correct name of the Song-Thrush was *Turdus philomelos*, the first description of *T. musicus* undoubtedly referring to the Redwing ; while the name *T. iliacus* was not available at all, as in the first instance it referred to three distinct species, viz., the Song-Thrush, Mistle-Thrush, and Redwing.

While we thoroughly agree with Dr. Hartert in his separation of these races, and applaud his good work, we think it only right to state that we cannot agree with him in abolishing old and well-known names and substituting for them names which are quite unknown to the average ornithologist. Dr. Hartert adheres most strictly to certain rules in order to secure stability in nomenclature, but in many cases, such for instance as the present, these rules act in our opinion in a directly opposite way to that which was intended, in that they disrupt the past. The Song-Thrush has been called " *Turdus musicus* " in countless books and papers, and if we now alter that name surely we show no regard for the past, while to the future ornithologist the innumerable references to this bird under the name of *T. musicus* will be obscured. With no wish to argue such an intricate question in these pages we can but state our firm conviction that to adhere strictly to a rule in such a case as this amounts to making the rule a fetish. Having no wish to be the blind slave of any rule, we are determined to call the British Song-Thrush *Turdus musicus clarkei*.

In part V. of Dr. Hartert's work (*Die Vög. der pal. Fauna*, p. 601) we note that he separates the Dartford Warbler of England and North-west France from the typical bird of the continent under the name of *Sylvia undata dartfordiensis* of Latham, by reason of its slightly smaller size, its dull, chocolate-brown, instead of slaty-grey, upperside, and by the flanks being washed with brown instead of grey.—EDS.

NORTHERN WILLOW-WREN IN NORFOLK.

IT may be of interest to record that a specimen of *Phylloscopus trochilus eversmanni* (*cf. antea*, Vol. II., p. 234) was shot on the Norfolk coast during the second week of May, 1908. Another specimen shot in the same locality during the month of September is of greater interest, because Dr. C. B. Ticehurst had not detected this bird in the autumn. Both specimens have been examined by Dr. Ticehurst, and the autumn bird exhibits in its plumage practically none of the green and yellow characteristic of the typical Willow-Wren.

CLIFFORD BORRER.

Another example of this race shot at Cley, Norfolk, in October, 1901, has been very kindly submitted to me by Mr. Ernest M. Connop, of Wroxham, in whose collection it now is. The bird, which has been examined by Dr. Ticehurst and myself, is greyish-brown on the upperside and greyish-white on the underside and has no green or yellow (except in the axillaries) in its plumage. The eyestripe is white. An additional interest attaches to this specimen in that it was examined by Howard Saunders, and I am indebted to Mr. Connop for a view of a letter regarding the bird which Howard Saunders wrote to Mr. Pashley, of Cley. Although it was not his practice to distinguish very closely allied forms by name, and although he makes no reference · to this race in his " Manual," it is clear from the letter that Howard Saunders fully recognised its characteristics. " Your bird," he wrote to Mr. Pashley, " is (in my opinion, of course) simply a Willow-Wren *Ph. trochilus*, but it is a very interesting example—and quite an *old* bird—of the northern form, which, as Seebohm says (*Cat. Birds B. M.*, V., p. 58) ' occasionally in high northern latitudes has all the green and yellow abraded and the general plumage earthy-brown, the eyestripe having faded to greyish-white and the underparts also to white.' The wing-formula is absolutely that of the Willow-Wren, and one of Seebohm's specimens from the Yenesei, Siberia, matches your bird exactly."

Seebohm's opinion that the brown and grey colouring was produced by fading and abrasion is now, of course, proved to be an error, since spring specimens exhibit the same characteristics.

As some confusion still exists in the minds of some of my correspondents with regard to the various races of Willow-Wrens and Chiffchaffs which have now been detected as occurring in this country, it may be well to summarise the information.

THE TYPICAL WILLOW-WREN (*P. trochilus trochilus*).

THE NORTHERN WILLOW-WREN (*P. trochilus eversmanni*).— Now found to occur both on the autumn and spring passage, but apparently much more frequently in the spring. Breeds in Northern Russia, from the Kolyma westwards to the Timan Hills, and possibly to Norway (*cf. supra*, p. 234). Except in the axillaries there is practically no green or yellow in its plumage.

THE TYPICAL CHIFFCHAFF (*P. rufus rufus*).

THE EAST EUROPEAN CHIFFCHAFF (*P. rufus abietina*).— So far has only once been detected in this country (*cf. supra*, p. 233). Breeds in Scandinavia, Russia (south of 65°), East Prussia, Austria, and the Balkans. It is of slightly larger size and paler coloration than the typical form.

THE SIBERIAN CHIFFCHAFF (*P. rufus tristis*).—Has been found on several occasions in winter in the Orkney and Shetland groups (*cf.* Vol. I., pp. 8 and 382). Breeds in Siberia from the Petchora to Lake Baikal. Easily distinguishable by its very brown upperside, grey underside, brownish flanks, and bright golden axillaries.

H. F. WITHERBY.

LESSER SPOTTED WOODPECKER BREEDING IN MERIONETH.

MR. R. J. LLOYD-PRICE, of Rhiwlas, Bala, informs me that last year a pair of Lesser Spotted Woodpeckers (*Dendrocopus minor*) nested in an oak-tree close to his head-keeper's house. They hatched and reared the young, but all left before winter. This is the first recorded instance of the species breeding in North Wales so far to the west as Merioneth, where, indeed, it has hitherto been met with only occasionally.

H. E. FORREST.

HOOPOE IN MERIONETH.

MR. LLOYD-PRICE writes me that a Hoopoe (*Upupa epops*) appeared on the lawn at Rhiwlas, Bala, one day in August, 1907. He watched it for a considerable time walking about, and every now and then erecting its crest. The Hoopoe is very rare in Wales, and has only once before been recorded in Merioneth.

H. E. FORREST.

LITTLE OWL IN WARWICKSHIRE AND WORCESTERSHIRE.

An adult female specimen of the Little Owl (*Athene noctua*) was shot at Barston, Warwickshire, on November 15th or 16th, 1908, by Mr. Russell. It is not the Sutton Coldfield specimen already recorded (*antea*, Vol. II., p. 240). The distance between the two places would be fifteen to twenty miles. No other specimen has been seen by Mr. Russell.

Another example was shot at King's Norton, Worcestershire, on October 14th, 1907. This bird rose from a ditch. and the gentleman who shot it mistook it for a Woodcock! Another has been seen there since, and I have urged the gentleman not to shoot it. Both my specimens are adult females.

F. COBURN.

THE FOOD OF THE COMMON EIDER.

That the Common Eider (*Somateria molissima*) feeds mainly on shell-fish is well known, yet the following summary of the results of a number of dissections which I have made may be of interest to the readers of BRITISH BIRDS.

On one occasion I found the remains of a crab in the gizzard, and of a crab and starfish in the crop.

"Periwinkles" seem to be very commonly eaten. I have taken as many as twenty of their shells from a single gizzard.

In many Eiders a bulge in the throat may often be seen, and on examination this proves to be caused by a "Razorshell" (*Ensis siliqua*), locally known as the "Spute-fish," and used by the fishermen as a bait. Sometimes one valve of the shell is missing. Examples as long as eight inches are sometimes swallowed, and often one end of the shell is broken, leaving a jagged edge. The dissolution of the contained animal evidently takes place in the crop, and the shell is, we may assume, ejected, as other birds eject pellets, since it could never pass through the intestines. It is curious that Razorshells are never found in birds killed in the early morning.

The shells of univalves are disintegrated, partly, apparently by the action of the gastric juices, and partly by the trituration of the gizzard.

The Eider is also fond of limpets. My boatman once reared an Eider drake, which was the terror of the limpet-pickers on the island, for it would steal the limpets as fast as they were detached from the rocks, and would attack the pickers with great spirit, using beak, wings, and feet, should they object to the levying of this toll! H. W. ROBINSON.

VELVET-SCOTER IN SHROPSHIRE.

HITHERTO the only known instance of the occurrence of *Œdemia fusca* in Shropshire was an adult male found exhausted near Whitchurch on November 23rd, 1866. It was preserved by John Shaw, of Shrewsbury, who recorded it at the time in the "Field." Mr. F. Coburn, of Birmingham, recently informed me of a second example which came into his hands—an immature female shot on December 12th, 1890, at Clungunford, near Ludlow, by Mr. Graham Williams.

H. E. FORREST.

INCREASE OF WOOD-PIGEONS IN ORKNEY.

IT is recorded in Howard Saunders' "Manual" that the Wood-Pigeon is pushing northwards, and breeds locally and sparingly in the Orkney Islands. It may be interesting to note that during the last two years—1907-08—I have found the bird breeding in increasing numbers in the Island of Shapinshay, Orkney Islands. I noticed in 1907 at least two pairs in the trees round Balfour Castle, and last year I shot two and picked up one young bird dead in the garden of Balfour Castle, and frequently saw eight or nine birds on the grass opposite the castle. I may add the bird is most destructive in the garden at Balfour, and already the damage done is considerable to the kitchen garden crops.

JAMES R. HALE.

RED VARIETY (*P. MONTANA*) OF THE COMMON PARTRIDGE.

As the note in the last number of BRITISH BIRDS (p. 311) conveys the impression that Lord Forester's specimens are the only examples of the rufous form of Partridge obtained in Shropshire, it may be of interest to state that it has been met with in several places. There is a specimen in the British Museum from Acton Reynald, near Shrewsbury. An example described in the "Field," November, 1902, was shot at Farmcote, near Bridgnorth. Earlier in the same year Mr. H. L. Horsfall obtained four Partridges at Gatacre Park, Bridgnorth, one of which he sent me for examination. It was of the same dark red hue as *P. montana* beneath, but the back was beautifully spangled with creamy-white, on a dark ground. It closely resembled the variety figured by Mr. Frohawk in the "Field," February 13th, 1897. A similar bird in the museum at Whitchurch, Salop, was shot near that place in the autumn of 1902, by Mr. J. M. Etches, who informed me that there

were several others like it in the covey. Three examples of
the typical *P. montana* were shot at Albrighton, near
Shrewsbury, on October 6th, 1905.

H. E. FORREST.

[Mr. J. R. B. Masefield kindly sends us a copy of a paper
on a number of occurrences of this variety in Staffordshire
which he contributed to the " Transactions of the North
Staffordshire Field Club " (1902, pp. 65–68, with Plate), and
he tells us that he has examined from time to time examples
showing almost every possible gradation between what may
be termed the true *P. montana* and the normal *P. cinerea*.
The erythristic variety of the Partridge, as is well-known,
constantly occurs and recurs in many parts of this country,
and the subject is of considerable interest in that no satisfactory reason, so far as we know, has as yet been adduced
to explain the much more persistent nature of erythrism in
this, than in apparently in any other, species.—EDS.]

* * *

RARE BIRDS ON THE ISLE OF MAY (FIRTH OF FORTH).—We
referred in our last volume (p. 295) to the results of a visit
to this island in 1907 by two energetic lady ornithologists.
In 1908 the island was again visited by Miss Evelyn V. Baxter,
from September 10th to October 9th, and we extract the most
important results from her paper in the " Annals of Scottish
Natural History " (1909, pp. 5–20). RED-SPOTTED BLUE-
THROAT (*Cyanecula suecica*).—Single birds were seen on
September 22nd and 23rd, two on the 24th, and several on
the 25th, and one on October 5th. YELLOW-BROWED
WARBLER (*Phylloscopus superciliosus*).—One on September
22nd, one on the 24th, one on the 25th, and another on October
3rd. BRITISH COAL TIT (*Parus ater britannicus*).—One procured October 1st. BRITISH BLUE TIT (*Parus cœruleus
obscurus*).—One on September 30th. [Both these records
are interesting as there is little proof that Tits are wanderers.
WHITE WAGTAIL (*Motacilla alba*).—Four or five adults on
September 20th. GREAT GREY SHRIKE (*Lanius excubitor*).—
One on October 25th. SCARLET GROSBEAK (*Pyrrhula ery-
thrina*).—An adult female on September 12th.

Miss Baxter also kindly informs us that the ROBINS and
GOLDCRESTS which she obtained have been examined at the
Royal Scottish Museum and pronounced to be of the British
race.

BLACK REDSTARTS IN CO. WATERFORD.—Mr. R. J. Ussher
caught a female or immature male *Ruticilla titys* at Cappagh

House, on November 4th, 1908, and before liberating it he saw another on the window sill. On the same date in 1907 he caught one in his bedroom, and on November 2nd of that year he saw another, while two were caught in his house in 1895, on October 29th and November 2nd (*Irish Nat.*, 1909, p. 26).

WOOD-WREN IN HEREFORDSHIRE IN WINTER.—A Warbler seen at close range by Mr. A. B. Farn near the River Wye on January 9th last, is said by him to have been without doubt an example of *Phylloscopus sibilatrix* (*Zool.*, 1909, p. 28).

DIPPER IN KENT.—Mr. A. H. Hardy writes to the " Field " (19, XII., 08, p. 1103) that he saw a Dipper (*Cinclus aquaticus*) on the River Stour on December 11th, 1908. The species is a rare straggler to Kent. Dr. N. F. Ticehurst tells us that he has notes of some dozen occurrences, and adds that the bird is *supposed* to have nested on one occasion at Chartham, not far from the locality of the present record.

SCARCITY OF THE LONG-TAILED TIT IN A YORKSHIRE DISTRICT.—Mr. H. B. Booth records the scarcity of the Long-tailed Tit in Upper Airedale and Upper Wharfedale (West Yorkshire). Only three occurrences of the bird in the breeding season are known during the last fifteen years, although a few years before it nested annually in these districts, and does so commonly in adjoining districts. No reason can be assigned for the desertion of the neighbourhood by the bird (*Nat.*, 1909, pp. 55–57).

COAL-TITMOUSE ON THE BASS ROCK.—Mr. W. Evans reports that two *Parus ater* occurred on the Bass Rock on September 28th, 1908. Only a wing and leg were sent to him, so that he could not say whether the birds were British or Continental (*Ann. S.N.H.*, 1909, p. 49).

LATE STAY OF SWALLOW IN IRELAND.—An immature *Hirundo rustica* was seen (and unfortunately shot) on December 9th, 1908, near Clondalkin, co. Dublin (W. J. Williams, *Irish Nat.*, 1909, p. 56).

BRAMBLING IN WEST SUTHERLAND.—A number of Bramblings (*Fringilla montifringilla*) were seen on October 25th, 1908, at Inchnadamph. The bird has not hitherto been identified in this area (J. T. Henderson, *Ann. S.N.H.*, 1909, p. 47).

SNOW-GEESE IN CO. MAYO.—A flock of four *Chen hyperboreus* were seen flying over Bartragh Island by Mr. Claud Kirkwood "a day or so after December 29th, 1908." They were easily recognised by their snow-white plumage and black-tipped wings (R. Warren, *Zool.*, 1909, p. 77). For previous records of this species in the same district see page 27 of the present volume.

GADWALL IN FIFESHIRE.—A young male *Chaulelasmus streperus* was shot near Tayport on November 14th, 1908. The bird is of irregular occurrence on the east coast of Scotland (W. Berry, *Ann. S.N.H.*, 1909, p. 49).

GARGANEY BREEDING IN EAST YORKSHIRE.—Mr. W. H. St. Quintin, of Scampston, East Yorkshire, writes to the "Naturalist" (1909, p. 38) that an entirely wild pair of Garganey (*Querquedula circia*) made a nest in May, 1908, near the River Derwent, the female laying some eight eggs. These, being in a dangerous place, were taken, and from them four drake and two duck Garganeys were reared.

ADULT LONG-TAILED DUCK INLAND.—An adult female *Harelda glacialis* was shot on the Spey forty miles from the sea in October, 1908 (J. R. Pelham Burn, *Ann. S.N.H.*, 1909, p. 49).

TURTLE-DOVE IN CO. DONEGAL IN WINTER.—An adult male *Turtur communis* (scarce at any time in Ireland) was shot among some Wood-Pigeons near Muff, co. Donegal, on November 30th, 1908 (D. C. Campbell, *Irish Nat.*, 1909, p. 56).

SUPPOSED GREAT BUSTARD IN YORKSHIRE.—Mr. J. Morley records in the "Zoologist" (1909, p. 78) that a Mr. Bennett shot a Great Bustard near Scarborough "about last Christmas-time," which he had cooked and found superior in delicacy to a Turkey! Although the skin was not preserved, we find on enquiry that two of the tail-feathers were, and Mr. Oxley Grabham informs us that these have been positively identified as those of a female Silver-Pheasant! It is well to make sure of the facts before putting into print the record of a rarity.

Erratum.—We regret that in the last number, on p. 310, the scientific name of Montagu's Harrier was given by a slip as *Circus æruginosus* instead of *C. cineraceus.*—EDS.

BRITISH BIRDS

AN·ILLUSTRATED·MAGAZINE
DEVOTED·TO·THE·BIRDS·ON
THE·BRITISH·LIST

APRIL 1,
1909.

Vol. II.
No. 11.

MONTHLY·ONE·SHILLING·NET
326·HIGH·HOLBORN·LONDON
WITHERBY & Cº

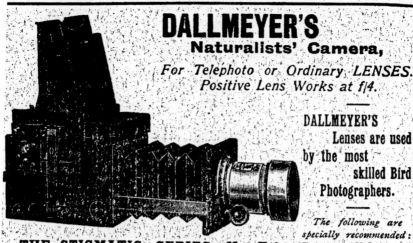

THOMAS BEWICK, from the engraving by F. Bacon, after the picture by James Ramsay.

BRITISH·BIRDS

EDITED BY H. F. WITHERBY, F.Z.S., M.B.O.U.
ASSISTED BY W. P. PYCRAFT, A.L.S., M.B.O.U.

SOME EARLY BRITISH ORNITHOLOGISTS AND THEIR WORKS.

BY
W. H. MULLENS, M.A., LL.M., M.B.O.U.

VIII.—THOMAS BEWICK (1753—1828) AND GEORGE MONTAGU (1751—1815).

IT is difficult to determine what position Thomas Bewick
holds among the principal British ornithologists ; it is
difficult indeed to determine whether he was, in the
strict sense of the word, an ornithologist at all. It was
by a series of entirely unforeseen events that Bewick

found himself called upon to write even a portion of the famous " History of British Birds " that bears his name, and it certainly cannot be said that the text of that work contains anything of much originality or importance. " It is respectable but no more," and would by itself, founded as it was on the style of Pennant, and admittedly deriving most of its information from his works,* in all probability have attracted but scant and passing attention. And yet this work of Bewick has met with extraordinary success, it has passed through edition after edition ; it has instructed and delighted thousands upon thousands of readers, and has in the opinion of one† who was fully competent to judge, done more than any other work in existence, Gilbert White's " Natural History of Selborne " alone excepted, to promote the study and pursuit of ornithology in this country.

This great popularity and widespread influence of Bewick's " History of British Birds " arose solely from the brilliance and fidelity of the wood-cuts, with which he was able to illustrate that work.

What Bewick and his fellow-author together entirely failed to do with the pen, he alone most successfully accomplished with the burin and the graver. Such was Bewick's skill, and so wonderful his power of transferring his impressions to paper, that his engravings of birds, especially of those which he was enabled to draw from life, or from freshly-killed specimens, remain even to this day amongst the finest black and white illustrations of the kind which we possess. Their effect therefore at the time of their appearance,‡ and for many years afterwards, may be easily understood, and this, coupled with the fact of Bewick's general renown as an artist and with the charm of the curious and often beautiful tail-pieces with which he and his pupils adorned his work, made

---

* " Memoir of Thomas Bewick," p. 162.

† Newton, " Dict. of Birds," Introd., p. 19.

‡ Pennant's fourth edition of the " British Zoology," which appeared in 1776, contained numerous plates of birds, but they were not very successful.

his name one to be ever associated with the study of British ornithology. Claims to be considered a scientific naturalist he had none, and yet his works will be remembered and revered, when those of far more erudite and accomplished writers have passed away.

Many books have been written about Thomas Bewick, his art, and his "life and times," but by far the best account of the artist and his work is to be derived from the "Memoir" which he compiled between the years 1822 and 1828, and on which he was still engaged at the time of his death.* It was written for the information of his daughter Jane and her brother and sisters, and is a bulky volume of some 316 pages. From it we learn that Thomas Bewick was born in August, 1753,† at his father's house of Cherryburn, near Eltringham, in Northumberland, and was baptized at the neighbouring church of Ovingham, on August 19th of that same year. Thomas was the eldest son of John Bewick, who farmed some eight acres of land at Cherryburn, and leased a small colliery at Mickley Bank.

Of Bewick's somewhat tempestuous youth it is here necessary to say but little ; he was educated first at Mickley School, and afterwards by the Rev. C. Gregson, of Ovingdean. At a very early age he developed a taste for drawing, and in spite of constant reproof for " misspending " his time, he tells us that " many of my evenings at home were spent in filling the flags of the floor and the hearthstone with my chalky designs." From this the transition to pen and ink, and brush and colour, was rapid ; and the young artist soon commenced to decorate the walls of his neighbours' houses with rude pictures, chiefly consisting of hunting scenes. At the age of fourteen young Bewick was apprenticed to Ralph Beilby, of Newcastle, an engraver in a considerable way of business. Under Beilby's tuition Bewick soon began

* The " Memoir " was first published in 1862 and again in 1887.

† Bewick kept his birthday on August 12th, but there is a doubt about the exact date.

to excel as an engraver, and the firm having been " applied
to by printers to execute wood-cuts for them,"* Beilby,
who had no liking for this branch of engraving, entrusted
the execution of the blocks to Bewick, who made so good
a job of it that henceforward orders for this particular
sort of work increased rapidly. Bewick's progress in
engraving was so rapid, and was so well thought of by
his master, that he sent some of his apprentice's cuts,
executed for " Select Fables," to the " Society for the
Encouragement of Arts," and for these Bewick received
a premium of seven guineas. In 1774 Bewick's apprentice-
ship came to an end, and he commenced to work on his
own account, chiefly for Newcastle printers, till the middle
of 1776. In the summer of that year he made an expedition to Scotland, travelling on foot, and afterwards
went to London, where he arrived in October, 1776.
Bewick disliked the Metropolis, and returning to
Newcastle next year, entered into partnership with his
former master, Ralph Beilby. For some years Bewick
continued to busy himself with the ordinary work of his
profession, but at length having come to the conclusion
that the figures of animals, as they were represented in
the children's books then available, were very inferior,
he resolved to try what he could do in that direction,
and on the advice of his friend, Solomon Hodgson, book-
seller and editor of the " Newcastle Chronicle," he
commenced on November 15th, 1785, to cut the figure of
the dromedary,† the first of a series of wood-cuts for the
" History of Quadrupeds," which was published in
1790.‡ While Bewick was engaged in drawing and cutting
the figures for the " History of Quadrupeds " his partner,
who was of " a bookish or reading turn, proposed to write
or compile the descriptions, but not knowing much about
natural history we got books on that subject to enable

* " Memoir," p. 59.

† Those animals which were not familiar to Bewick were copied from
Dr. Smellie's " Abridgment of Buffon."

‡ It reached an eighth edition in 1824.

HISTORY

OF

BRITISH BIRDS.

THE FIGURES ENGRAVED ON WOOD BY T. BEWICK.

VOL. I.

CONTAINING THE

HISTORY AND DESCRIPTION OF LAND BIRDS.

NEWCASTLE:

PRINTED BY SOL. HODGSON, FOR BEILBY & BEWICK: SOLD BY THEM,
AND C. C. & J. ROBINSON, LONDON.

[*Price* 1*l.* 1*s. in Boards.*]

1797.

GEORGE MONTAGU, from the Original Miniature in the possession
of the Linnæan Society, London.

him to form a better notion of these matters." These descriptions Bewick helped to revise and correct. When, however, the title page was in preparation, Beilby wished to appear as the author, and desired the book to be announced as being " by R. Beilby " ; but although this idea was abandoned through the influence of Mr. Hodgson, the foundation of the quarrel between Bewick and Beilby was commenced, which finally led to the dissolution of their partnership. The "History of Quadrupeds" proved so great a success, being appreciated by young and old alike, that Bewick began to turn his thoughts to a " History of British Birds."* For this purpose he commenced to study various works on the subject, and informs us that "in addition to Pennant's works, [he] perused ' Albin's History of Birds,' Belon's very old book,† Willoughby and Ray, etc. Mr. John Rotherham gave me ' Gesner's Natural History,' with some of these I was in raptures. Willoughby and Ray struck me as having led the way to truth and to British Ornithology. . . . . I was much pleased with ' White's History of Selborne.' Pennant, however, opened out the largest field of information, and on his works I bestowed the most attention. The last of our ornithologists, and one of the most indefatigible, was the late Col. George Montagu, author of the ' Ornithological Dictionary ' " (*Memoir*, pp. 161, 162.)

In addition to the time he devoted to the works above mentioned, Bewick, who at the beginning of his undertaking had made up his mind "to copy nothing from the works of others, but to stick to Nature as closely as I could," availed himself of an invitation from Mr. Constable, the owner of " Wycliffe,"

---

* This project was, however, in full consideration in 1790, *vide* letter from John Bewick (1760-1795). Robinson's " Thomas Bewick : his Life and Times," p. 94.

† L'Histoire de la Nature des Oyseaux, avec leurs descriptions, & naifs portraicts. . . . Par Pierre Belon du Mans, Paris, 1555. 1 vol. folio. This work of Belon's, though not so diffuse as Conrad Gesner's " Historia Avium " of the same date, is nevertheless the most trustworthy authority of that period.

to visit the museum there, which contained the collection
of birds formed by Marmaduke Tunstall.* For nearly
two months Bewick remained at "Wycliffe," making
drawings from the specimens there (some of these being
in water-colour) and commenced to engrave from them
as soon as he returned to Newcastle. Finding, however,
"the very great difference between preserved specimens
and those from Nature . . . . I never felt satisfied with
them . . . . and was driven to wait for birds newly shot
or brought to me alive." All this, of course, involved
considerable delay, but "after working many a late hour
upon the cuts" the first volume of "British Birds,"
entitled "Land Birds," appeared in 1797. "Mr. Beilby,"
as Bewick tells us (*Memoir*, p. 171), "undertook the
writing or compilation of this (the first) volume, in which
I assisted him a great deal more than I had done with the
' Quadrupeds.' " Bewick was therefore surprised to
find that Beilby was determined on being recognised as
the sole author of the book. To this claim Bewick
strongly objected, and although through the intervention
of mutual friends, the title-page of the first volume
merely bore the legend "Printed . . . for Beilby and
Bewick," neither of them being named as authors,†
they found it impossible to work in harmony any longer,
and their partnership was dissolved, Bewick buying up
Beilby's share in the "Quadrupeds" and the first volume
of the "Birds."

Bewick was now thrown upon his own resources as an
author, and by consulting all the available authorities,
and making use of his own knowledge and observations,
he composed the text of the second volume, entitled
"Water-Birds." This appeared in 1804, and in the
preface Bewick states that "owing to a separation of

---

* Marmaduke Tunstall (1743-1790), the anonymous author of the
"Ornithologica Britannica," London, 1771, 1 vol. folio. For an account
of his life, *vide* Fox's "Synopsis of the Newcastle Museum," where his
collection now is. It was for this same Marmaduke Tunstall that
Bewick had in 1789 executed his famous wood-cut of the "Chillingham
Bull."

† *cf.* conclusion of Preface to 1st vol. "British Birds."

interests between the editors . . . . the compilation and completion of the present work devolved upon one alone." He also acknowledges his obligations to the Rev. H. Coates, the vicar of Bedlington, for "literary corrections."

A facsimile title-page of the first volume of the first edition of "British Birds" is here given, that of the second volume is somewhat similar, but Beilby's name does not appear in it.

The collation of the book is as follows :—

Vol. 1 ; pp. XXX., title, preface, introduction and contents, + pp. 335, + 117 figures of birds, and 91 tail-pieces.

Vol. 2 ; pp. XX. + pp. 400, + 101 figures of birds, and 139 vignettes.

The first edition was printed on paper of three different sizes, viz., imperial, royal, and demy 8vo, that of the latter size being of two qualities, thick and thin. The publishing prices were 21s., 18s., 13s., and 10s. 6d. respectively, and of the imperial paper copies (of the first issue) only twenty-four were printed. The prices of the second volume being 24s., 18s., and 12s. Of the first volume of the "British Birds" there were two issues, both bearing the same date, viz., 1797 ; the second issue being, however, printed in 1798. The first issue may be determined from the fact that on the reverse of page 335 the *third* edition of "Bewick's Quadrupeds" is announced, while in the second, the *fourth* is advertised.*

The success of the "History of British Birds" was immediate and complete, six editions were issued in Bewick's lifetime, and in the year 1847, an eighth,† edited by John Hancock with great skill, and containing some twenty extra tail-pieces, which Bewick had executed for a projected "History of British Fishes,"

---

* For further particulars, *cf.* Newton, "Dict. Birds," Introd., p. 20.

† Dates of the eight editions of Bewick's "Birds" are as follows :— 1st, 1797-1804 ; 2nd, 1805 ; 3rd, 1809 ; 4th, 1816 ; 5th, 1821 (with Supplement) ; 6th, 1826 ; 7th, 1832 ; 8th and last, 1847.

appeared, this edition being in many respects the best. The " Birds " marked Bewick's high-water mark as an artist, the only book of any real importance which he subsequently produced being " Æsop's Fables," in 1818.

As has above been mentioned, the value of the " History of British Birds " rests on its wood-cuts alone, and although it has been frequently stated that Bewick had from his youth upwards a great leaning towards the study of birds, a careful investigation seems to show that he only possessed the ordinary interest in Nature common to most intelligent boys brought up in the country; indeed, on his own showing his chief delight as a youth consisted in joining the local " hunting parties," and in observing the habits of the various " beasts of the chase." It is true that in his " Memoir " he makes some not infrequent mentions of his early observations and interest in ornithology, and he further enlarges on this subject in the preface to the sixth edition of his " Birds " ; but it was only in human nature that a man who had seen edition after edition of his ornithological writings eagerly absorbed by the public, should come to consider himself as a zoologist, both by inclination as well as study. Be this as it may, the excellence of his wood-cuts* stands out beyond all doubt or question, and the debt we owe to the memory of Thomas Bewick is great and lasting.

Of the remainder of Thomas Bewick's life we can here make but the briefest mention. His wife (Isabella Elliot, of Ovingdean), whom he had married in 1786, died in 1826, and in November, 1828, at the ripe old age of seventy-five, he followed her to the grave, and lies buried by her side in Ovingham churchyard, " at the west end of the church near the steeple." He continued working

---

* Although Bewick seems to have been the first engraver to use wood-blocks for the representation of birds with any signal success, the process had, of course, been made use of on the Continent for that purpose, while in this country it had already been employed in 1743 to illustrate a work entitled " Ornithologia Nova: or a new General History of Birds," a second edition of which, with a somewhat different title, appeared in 1745.

to the close of his busy life, and when seized with his fatal illness was engaged on a large block entitled, " The Old Horse waiting for Death."

George Montagu, whose " Ornithological Dictionary " has already been referred to, as having been issued while Bewick was engaged on the compilation of the second volume of his " Birds," was born at Lackham, in Wiltshire, in 1751. He entered the army at an early age, and served as a captain in the 15th Regiment of Foot during the American War. He afterwards settled down at Easton Grey, in Wilts, and became acting colonel of the County Militia. He died at Kingsbridge, in Devonshire, in August, 1815. Montagu was a prolific writer,* but his reputation rests on his " Ornithological Dictionary," a work so able and so well-known that it is only necessary to say that its merits have been as widely acknowledged abroad as at home ; and to quote Coues' *dictum* " It is one of the most notable treatises on British Birds, as a *vade mecum* which has held its place at a thousand elbows for three-quarters of a century."

The full title of the book is as follows :—

Ornithological Dictionary ; / or, / Alphabetical Synopsis / of / British Birds. / By / George Montagu, F.L.S. / In two volumes, / Vol. I. [Vol. II.]. London : / Printed for J. White, Fleet Street, / by T. Bensly, Bolt Court. / 1802.

Two Vols. 8vo. Collation : Vol. I. pp. 2 un. + pp. XLIII., + pp. un. (being sheets B–Y) + Slip of Errata. Plate of Cirl Bunting. Vol. II., Title + pp. un. (sheets B–Y) + Slip of Errata.

A supplement (unpaged) exceeding in bulk the two volumes of the original edition, with 24 plates, was issued by Col. Montagu in 1813 ; and there were numerous editions and re-issues after his death.

---

* For a list of his works on Natural History *vide* " Agassiz," Vol. III., p. 614.

MARKING BIRDS: NOTES ON THE WORK AT THE ROSSITTEN STATION.

BY

A. LANDSBOROUGH THOMSON.

FROM time to time references have been made in the pages of BRITISH BIRDS * to the work of the various investigators who are endeavouring to obtain fuller and more accurate data with regard to migration, by liberating birds marked with metal foot-rings. It may be of interest, however, to give a fuller account of the methods employed, and of these I was able to gain some knowledge during a couple of weeks' stay last autumn (1908) at Rossitten, on the Baltic. Some details of the results obtained there may indicate what may be looked for by following similar lines of research.

A word about the situation of Rossitten : at the very south-eastern corner of the Baltic, the River Niemen (or Memel) flows through many mouths into a large lagoon—the Kurisches Haff. This lagoon is connected with the sea by a narrow channel at one end, and for the remainder of its length is separated only by a tongue of land, or Nehrung, about sixty miles long by from less than half a mile to more than two miles broad. It is among the " wandering " dunes—the highest in Europe— on the Kurische Nehrung, that the little out-of-the-way fishing village of Rossitten lies. And it is there that the German Ornithological Society has established its permanent Vogelwarte, or ornithological station. Lying in the midst of a large tract of uninviting country, the neighbourhood of Rossitten, combining as it does within a small area, examples of many different types of country —woods, meadows, sandy wastes, ponds, marshes, reed-beds, open shore, and cultivated land—may be regarded as a sort of oasis where vast numbers of resting mi-

* Vol. I., pp. 58, 298, 326; Vol. II., pp. 35, 171, 245, and 246.

grants of widely different needs and habits congregate. So great is the number of migrants passing along the Nehrung, and so large the proportion that break their journey at Rossitten, that, as a station from which to observe migration, it is now regarded as rivalling, if not surpassing, the more famous Heligoland, being inferior only in that it is less easily searched.

For five years now Dr. Thienemann, director of the Vogelwarte, has been actively engaged in marking birds at Rossitten. The mark employed consists of a strip of aluminium bent into the form of a ring, with the two ends projecting outwards together where they meet, and fastened by folding one of these ends over the other. The inscription engraved on these rings varies with the size of the ring. On the Crow and Gull sizes it reads : " Vogelwarte, Rossitten," followed by a number. On the larger sizes, for Storks, etc., " Germania," or even " Ost-Preussen Germania," is added. On the smaller sizes, for Terns, small Waders, small Passerine birds, etc., there is no room for anything except the number, and consequently very little success has been obtained with these birds. The difficulty lies entirely with the finder of the bird—once sent in to the Vogelwarte it is easy to tell whether the ring is a Rossitten one or not. This indicates the need for a new pattern of ring for small birds. Dr. Thienemann has rejected the idea of a metal label attached to the ring ; such a label would interfere with the bird too much, and would create a doubt as to whether results thus obtained could be regarded as normal.

Birds are procured for marking at Rossitten in two ways. A large number of birds are marked as nestlings, not only at Rossitten, but also in other parts of East Prussia. The other method is to capture resting migrants, mark them, and then let them continue their journey. Large numbers of Hooded Crows are marked in this way every autumn. Huge flocks of these birds pass along the Nehrung at this season, and large numbers are ingeniously netted—and *bitten* to death (!)— by the Rossitten natives, who preserve them for winter food. Many of the birds thus caught uninjured are not killed, but are sold alive to the Vogelwarte for marking and liberation. About 8 per cent. of these marked Crows are killed or recaptured by persons who send them, or the rings and feet, or at least notice of the capture, to Rossitten. This proportion of returns is far larger than was originally expected, and it will probably come as a surprise to many. The only other figures I have seen are those quoted on p. 246 of this volume of BRITISH BIRDS : the returns for a species so much shot as the Woodcock are shown to be scarcely more than 5 per cent. The insufficient " address " given on the rings in that case must, however, be taken into account. It must also be remembered that on the Continent the Hooded Crow has almost the status of a " game bird," shooting Crows decoyed by a captive Eagle-Owl being a recognised and popular form of sport.

To show the value of bird-marking I conclude by giving short summaries of the results obtained at Rossitten in the case of a few species, beginning with the Hooded Crow (*Corvus cornix*).

The places from which these Crows, marked while stopping on migration at Rossitten, have been again recorded, lie within a broad belt of country extending from southern Finland and the St. Petersburg district of Russia, southwards through Livonia and Courland to Rossitten, and then westwards, still bounded on one side by the Baltic, through northern Germany, and terminating in the north-eastern corner of France (Solesmes). The most northerly point from which one of these marked birds has been recorded is Wiisala, in the Government of St. Michel, Finland (April 20th, 1907 : liberated October 12th, 1905). From a large number of records I select, as fairly representative of the whole series, those of birds which were liberated—along with nearly a hundred others not heard of again—at Rossitten on the same day, October 4th, 1906.

DATE AND PLACE OF RECAPTURE.

Feb., 1907.	Friedland, Mecklenburg-Strelitz,
9th April, 1907.	Agilla, East Prussia.
12th April, 1907.	Lalendorf, Mecklenburg-Schwerin.
14th May, 1907.	Wainoden, Courland.
9th June, 1907.	Rossitten.
12th Oct., 1907.	Pernau, Livonia.
7th Dec., 1907.	Crefeld, Rheinland.

There are some interesting records of birds marked about the same time and again recorded about the same time from the same place :—

MARKED AT ROSSITTEN.	RECORDED AT AGILLA, EAST PRUSSIA.
4th Oct., 1906.	9th April, 1907.
8th Oct., 1906.	14th April, 1907.

	RECORDED AT SARKAU, KURISCHE NEHRUNG.
9th Oct., 1904.	12th Oct., 1905.
20th Oct., 1904.	16th Oct., 1905.

The second couple of these records also shows that the birds passed along the Nehrung at almost the same time in the autumn of 1905 as in the previous season.

Three records which bear upon the time taken upon migration, give rather different results :—

MARKED AT ROSSITTEN.	DATE AND PLACE OF RECAPTURE.
28th March, 1904.	31st March, 1904, Pillkoppen, Kur. Neh.
16th April, 1904. (6.30 p.m.)	17th April, 1904 (morning), Pillkoppen (ca. six miles north-easterly from Rossitten.)
18th April, 1904.	26th April, 1904, Peterhof, St. Petersb.

The fact that it was later in the season may account for the much greater distance in proportion to the time, in the last case than in the first.

The longest time, so far, between the liberation of a marked Crow and its recapture, is four years and a week : liberated at Rossitten, October 12th, 1903 ; shot at the mouth of the Vistula, October 20th, 1907.

The proportion of marked birds again recorded is even greater among the Gulls than among the Crows, but this is partly explained by the commonness of gull-shooting as a form of " sport " on the Continent. Moreover, one of these larger-sized rings would be visible on a bird at some distance. The proportions, as reckoned about eighteen months ago, were 12.5 per cent. and 16.6 per cent. for Herring-Gulls (*Larus argentatus*), and Common Gulls (*L. canus*) respectively. One of the latter species marked at Rossitten was obtained in the Faröes, so that it is probable that some Rossitten birds may reach the British Isles.

Although the proportion is smaller among the Black-headed Gulls (*L. ridibundus*) the total number of returns is greater as these birds are marked in large numbers as nestlings in a colony at Rossitten. The records have shown that, on the approach of winter the birds of this colony cross Europe by two routes. One leads south-wards, following the Vistula at first it is supposed, over Vienna and Trieste to the Adriatic, where quite a number have been recorded near the mouth of the Po. One, also, has been obtained in the south of Italy, and a bird marked on July 26th, 1907, was obtained near Tunis on January 12th, 1908. The other route follows the Baltic coast westwards, crosses to the North Sea, follows the Rhine upwards, and reaches the Mediterranean by the Lake of Geneva and the valley of the Rhone.

Storks (*Ciconia alba*) have been marked in considerable numbers, as nestlings, in East Prussia and elsewhere. Among those returned are a few from different parts of Africa. These include one of a brood of three marked by Dr. Thienemann near Königsberg on June 21st, 1906, and one of a brood of three marked near Köslin, in Pomerania, on July 5th, 1907. The ringed foot of the first of these was brought by a native to a French officer near Lake Chad in October, 1906. The other, having left for the south on the 25th or 26th of August, when it had been about a fortnight out of the nest, was obtained that winter near Fort Jameson, in north-eastern Rhodesia : the record came to Dr. Thienemann's notice through a note in the " Field " for January 25th, 1908 (p. 150).

For reasons already explained the returns for the smaller species are disappointing, but there are a few isolated records of interest. A Dunlin (*Tringa alpina*), for instance, was marked at Rossitten on September 5th, 1904, and recorded on the 22nd of the same month, from the Arenholzer See in Schleswig-Holstein.

The director of the Vogelwarte earnestly requests that anyone finding one of his marked birds, will send him the ring and foot, or at least the ring, with full particulars as to date and place of capture.*

* The Editors will be glad to forward any information · to Dr. Thienemann, and to publish in these pages the data relating to the capture of any marked birds in the British Islands.

ON THE MORE IMPORTANT ADDITIONS TO OUR KNOWLEDGE OF BRITISH BIRDS SINCE 1899.

BY

H. F. WITHERBY AND N. F. TICEHURST.

PART XIX.

(*Continued from page* 334.)

BLACK-NECKED GREBE *Podicipes nigricollis* C. L. Brehm.
S. page 723.

OXFORDSHIRE.—A pair shot on a large pond near Bloxham on September 19th, 1899, were thought to have bred, or attempted to breed, somewhere in the district (O. V. Aplin, *Ibis*, 1902, p. 165, and *Zool.*, 1903, p. 10).

It was reported by Mr. Aplin in 1904 that some well-known ornithologists, who wished to remain anonymous, had discovered that several pairs of these birds nested and reared their young that year in Britain. The birds were on a shallow lake surrounded with marshy ground. Early in June four pairs were seen with one, two, two, and three young respectively, in one part of the lake, while further off was a fifth pair with rather larger young and two unattached adults (O. V. Aplin, *Zool.*, 1904, pp. 417–420). In 1906 Mr. Aplin announced that he had himself been able to pay a visit to the place, and had seen four or five adult birds in full breeding plumage, but they had not at that time, he thought, yet hatched their young (*t.c.*, 1906, p. 315).

MIDDLESEX.—The plate in Sowerby's " British Miscellany " of a male and female Grebe with nest and eggs taken on a pond on Chelsea Common in 1805, and ascribed by Mr. Harting (*Birds of Middlesex*, p. 244, and *Handbook*, p. 269) to this species, is stated by Mr. Aplin to represent Little Grebes in summer plumage (*Zool.*, 1904, p. 266).

CHESHIRE.—One was shot on Dee Marshes, near Chester, in November, 1906 (A. Newstead, *t.c.*, 1907, p. 153).

LANCASHIRE.—An adult male in full summer plumage was caught alive on a pond at Middleton, near Lancaster, on July 28th, 1904 (H. W. Robinson, *t.c.*, 1904, p. 350).

NORTHUMBERLAND.—Two seen in the middle of June were in *winter* plumage (A. Chapman, *Birdlife of the Borders*, 2nd ed., p. 94).

SCOTLAND.—One was shot at Lendalfoot (Ayr) on January 27th, 1906. The species had not previously been recorded in the Clyde area (*Ann. S.N.H.*, 1907, p. 207).

IRELAND.—A male " coming into summer plumage " was shot on Belfast Lough on February 28th, 1907 (W. H. Workman, *Zool.*, 1907, p. 111). Has been obtained in twenty-four instances (R. J. Ussher, *List of Irish Birds*, p. 53).

Food.—In the stomach of one shot March 2nd, 1898, at Strathbeg (Dee area), were found many feathers, amongst which were numbers of a stalk-eyed crustacean (*Mysis vulgaris*), showing that though killed on a loch, it had shortly before been feeding in the sea (G. Sim, *Vert. Fauna of Dee*, p. 190).

STORM-PETREL *Procellaria pelagica* L. S. page 727.

BREEDING ON THE EAST COAST.—A pair, at first thought to be Fork-tailed Petrels, but afterwards identified as of this species, were found breeding on the Bass Rock in 1904 (H. N. Bonar, *Field*, 1904, pp. 908 and 983 ; W. E. Clarke, *Ann. S.N.H.*, 1905, p. 55).

LEACH'S FORK-TAILED PETREL *Oceanodroma leucorrhoa* (*Vieill.*). S. page 729.

FLANNAN ISLANDS (OUTER HEBRIDES).—Regarded as the chief breeding stations of the species in the British Isles. On Eilean Mor they are more plentiful than the Storm-Petrel. They lay earlier—the first eggs being found on May 29th, but their nesting habits are similar. The chicks are sooty-black and much darker than those of the Storm-Petrel (W. Eagle Clarke, *Ann. S.N.H.*, 1905, p. 86).

IRELAND.—A very few have been found breeding on islands off Mayo and Kerry (R. J. Ussher, *List of I. Birds*, p. 53).

MADEIRAN FORK-TAILED PETREL *Oceanodroma castro* (Harcourt). S. page 731.

A female (the second British example) was shot near Hythe, Kent, on November 8th, 1906, while flitting along the shore in a tired manner after a heavy south-westerly gale (N. F. Ticehurst, *Bull. B.O.C.*, XIX., p. 20).

WILSON'S PETREL *Oceanites oceanicus* (Kuhl). S. page 733.

[SURREY.—Four specimens in the Charterhouse Collection are *said* to have been killed on Godalming Pease Marsh, after a very severe storm, but no dates are given, and Mr. Bucknill does not consider them to be sufficiently authenticated (J. A. Bucknill, *B. of Surrey*, p. 352).]

GREAT SHEARWATER `Puffinus gravis` (O'Reilly).
S. page 737.

SCILLY ISLES.—A fairly regular visitor in flocks during autumn and winter to the seas around the islands. Never seen among the islands (J. Clark and F. R. Rodd, *t.c.*, 1906, p. 346).

SUFFOLK.—One was obtained off Lowestoft in November, 1898 (T. Southwell, *Knowledge*, 1899, p. 41).

LINCOLNSHIRE.—About November 27th, 1902, a male was shot near the mouth of the River Welland (F. L. Blathwayt, *Zool.*, 1903, p. 30).

YORKSHIRE.—Autumn, 1904, a female at Scarborough. A number of examples obtained prior to 1899 are also detailed (T. H. Nelson, *B. of Yorks.*, p. 754).

SCOTLAND.—On June 27th, 1894, between the *Butt of Lewis* and *North Rona* forty to sixty pairs were seen, " nearly each pair sitting [on the water] lovingly together." On June 24th, 1895, between *Barra Head* and *St. Kilda* over fifty pairs were seen sitting on the sea in pairs. A specimen was killed by a fisherman on August 7th, 1897, and two others in the fourth week of July, 1899, near *St. Kilda* (Alfred Newton, *Ann. S.N.H.*, 1900, pp. 142–147). One was obtained from a small flock in the Summer Islands, *Loch Broom*, on October 31st, 1897 (J. T. Henderson, *t.c.*, 1906, p. 114). A few were seen off the *Flannans* on September 21st, 1904 (W. E. Clarke, *t.c.*, 1905, p. 86).

IRELAND.—In September, 1900, when cruising off the coasts of *Kerry*, *Cork*, and *Waterford*, Mr. H. Becher found this species surprisingly numerous ; on several days he saw eight or ten (R. J. Ussher, *I. Nat.*, 1901, pp. 42–43). On September 9th, 1901, the same observer sailed into a flock of two to three hundred of these birds between *Cape Clear* and *Mizen Head* and shot four. On September 13th he again saw large numbers, both off *Valentia* and between the *Blaskets* and *Skelligs* (*id.*, *t.c.*, 1905, p. 43). In 1906 Mr. G. P. Farran saw many in August off *co. Kerry*, and on November 1st off *co. Cork ;* and several on November 6th off *co. Kerry* (*id.*, *t.c.*, 1907, pp. 163 and 184). In 1907 the same observer saw off the same coast many in August, a few in September, several in November, and in 1908 two in August and many in November (*id.*, *t.c.*, 1909, p. 80).

Moult and Habits.—In the specimens recorded above under Scotland, as taken in July and August, Professor Newton found that the primaries were all new and only partially grown, and he concludes that the birds were practically incapable of flight (*loc. cit.*). The birds observed in June, 1894, were subsequently seen by Mr. H. L. Popham, who reports that "there were no young birds amongst them, but the old birds could scarcely fly, having apparently moulted out their primaries" (*cf. Trans. Roy. Irish Acad.*, XXXI., Pt. III., p. 72). Howard Saunders, in replying to a question of Newton's as to the statement in the "Manual" that this bird strikes the water with great violence on alighting, gives as his authority Captain J. W. Collins, who had had remarkable opportunities for observing this species on the American fishing banks. The habit had also been observed by Mr. R. Warren (*cf. Zool.*, 1894, p. 22). Mr. Saunders further remarks that Captain Collins stated that the primaries and other flight-feathers of this species were shed and renewed somewhat abruptly from the end of June to the latter part of July, and that Baron d'Hamonville had drawn attention to the rapid moult of the flight-feathers in the Manx Shearwater (Howard Saunders, *t.c.*, 1901, pp. 15–18).

SOOTY SHEARWATER *Puffinus griseus* (J. F. Gm.). S. page 739.

CORNWALL.—One was shot near Looe on August 21st, 1899 (J. Clark, *Zool.*, 1907, p. 287).

YORKSHIRE.—"Now known to be a fairly regular visitant to the Yorkshire coast in autumn and winter" (T. H. Nelson, *B. of Yorks.*, p. 756). Records since 1899 :—A male and female October 2nd, 1901, one October 1st, one October 4th, 1904, all off Scarborough (W. J. Clarke, *Zool.*, 1901, p. 477, 1905, p. 74). Others reported in 1904 off Flamborough and Bridlington (T. H. Nelson, *t.c.*, p. 758). Mr. Clarke also informs us that he obtained another example also off the coast of Scarborough on October 6th, 1908.

KENT AND SUSSEX, *cf. supra*, p. 243.—Three recent occurrences.

SCOTLAND.—A female was captured in *Stromness Harbour* on October 16th, 1902. Of extreme rareness in Scottish seas, this bird is new to the fauna of the Orkneys (W. E. Clarke, *Ann. S.N.H.*, 1903, p. 25). In the mouth of the *Firth of Forth* Mr. William Evans is disposed to consider it "a fairly regular, though usually far from common, autumn visitant." It appears that only two specimens have been preserved, but

Mr. Evans bases his opinion on his own personal observations and those of fishermen who know the bird well, and have frequently seen it, though usually in small numbers. In 1902 it appears to have been specially numerous, about a dozen being seen at one time (W. Evans, *t.c.*, 1903, pp. 26–28).

IRELAND.—A good many were seen during September, 1900, by Mr. H. Becher when cruising off the coasts of *Kerry, Cork,* and *Waterford* (R. J. Ussher, *I. Nat.*, 1901, pp. 42–43). Great numbers were seen by the same observer in September, 1901, and four were shot. " The observations of Mr. Becher in 1892, 1899, 1900 and 1901, go to show that both [Great and Sooty] these oceanic species may be met with in August and September off the south-west extremity of Ireland, and some-times in considerable numbers." One was shot off *Achill Island* on May 22nd, 1901 (*id.*, *t.c.*, 1905, p. 43).

MANX SHEARWATER *Puffinus anglorum* (Temm.).
S. page 741.

SCILLY ISLANDS.—Breeds on Annett in " prodigious numbers " (J. Clark and F. R. Rodd, *Zool.*, 1906, p. 346).

NORTH WALES.—*Bardsey Island.*—Mr. O. V. Aplin in 1901 found a considerable breeding colony on the north-east end of the island (*t.c.*, 1902, p. 16), and they undoubtedly breed *on the mainland* of West Carnarvonshire (*idem, t.c.*, 1900, p. 505).

IRELAND.—On June 18th, 1904, when crossing from Liverpool to Belfast, Mr. R. Lloyd Patterson saw a large assemblage of between 150 and 200 in the early morning a few miles off the *Skullmartin Lightship,* near the coast of co. Down (*Irish Nat.*, 1904, p. 171).

The September migrations of this species are deserving of closer study than they appear to have received. There are a number of scattered records referring to these migrations but they are not sufficiently continuous to allow of any conclusion being drawn from them.

LEVANTINE SHEARWATER *Puffinus yelkouanus* (Acerbi).
S. page 741.

The occurrences of this species have already been dealt with (*antea*, pp. 138, 206–208, 313). Mr. T. H. Nelson points out that in the " Birds of Yorkshire " it is recorded that three examples were obtained in 1904 (only two were mentioned on p. 207), but no details are given.

LITTLE DUSKY SHEARWATER *Puffinus assimilis* Gould.
S. page 743.

Puffinus obscurus bailloni, Bp., Rothschild and Hartert, Nov. Zool., VI. (1899), p. 196.

Puffinus bailloni, Bp., Godman, Monograph Petrels, pp. 138–141.

Messrs. Rothschild and Hartert separate the Australian form *P. assimilis* from the African form *P. bailloni*, and Dr. Godman, who coincides in this view, states that in the latter bird the quill-lining is greyish, or ashy-white, and not of such a pure white as in *P. assimilis*, while the lateral feathers of the under tail-coverts are more or less black along their outer webs and not entirely white as in *P. assimilis*. Dr. Godman considers that the British examples are referable to the Madeiran form.

The third British example was a female picked up exhausted on the beach near Bexhill during the severe gale from the W.S.W. on December 28th, 1900. The bird was shown to belong to the form *P. obscurus bailloni* (W. R. Butterfield, *Bull. B.O.C.*, XI., p. 45).

The fourth example—a male—was caught alive near Lydd, Kent, after the disastrous south-westerly gale of November 26th–27th, 1905 (N. F. Ticehurst, *t.c.*, XVI., p. 38).

BULWER'S PETREL *Bulweria bulweri* (Jard. and Selby).
S. page 749.

The second British example was picked up dead near Beachy Head, Sussex, on February 3rd, 1903, after a succession of strong south-westerly gales (N. F. Ticehurst, *Bull. B.O.C.*, XIII., p. 51).

The third—a female—was found dead on the shore near St. Leonard's-on-Sea, Sussex, on February 4th, 1904, also after prolonged south-westerly gales (W. R. Butterfield, *t.c.*, XIV., p. 49).

The fourth, *vide antea*, p. 282.

FULMAR *Fulmarus glacialis* (L.). S. page 751.

. #### *Breeding Stations.*

FLANNAN ISLES.—Reported as breeding (J. A. Harvie-Brown, *Ann. S.N.H.*, 1903, p. 19). A few pairs have bred on the outer islands for several years, and in 1904 two couples had nests on Eilean Mor (W. E. Clarke, *t.c.*, 1905, p. 86).

BARRA.—Birds were seen in 1899, and in 1902 eggs were actually seen, while in 1906 there were from eight to twelve pairs breeding (N. B. Kinnear, *t.c.*, 1907, p. 85).

SUTHERLANDSHIRE.—In 1897 Fulmars were seen on June 19th and 30th by Mr. Eagle Clarke about a mile to the east of Cape Wrath, and again at the same place on July 10th, 1900, by Mr. Howard Saunders, and they both considered that the birds were then nesting there (*Ann. S.N.H.*, 1897, p. 254, 1901, p. 50). A colony was established during 1901 (or possibly a year sooner ?), 1902, and 1903 on Handa (*cf.* J. A. Harvie-Brown, *Fauna N.W. Highlands and Skye*, pp. 355–361, where a very full account of the extension of this bird's range in Scotland and its status up till 1904 will be found).

CAITHNESS.—First observed at Dunnet Head in 1900. Have gradually increased in numbers since. About thirty pairs there now (J. A. Harvie-Brown, *Ann. S.N.H.*, 1907, p. 118).

SHETLANDS.—In 1903 there were eight or nine actually occupied nesting sites (J. A. Harvie-Brown, *Fauna N.W. Highlands and Skye*, p. 359). *Fair Isle.*—In 1902 it was present during the summer : in 1903 about a dozen pairs bred, since which it has thoroughly established itself (W. E. Clarke, *t.c.*, 1906, p. 80). *Whalsay and Yell.*—Found breeding in 1906 (J. S. Tulloch, *t.c.*, 1906, p. 240). *Fitful Head.*—A pair or two first seen in 1900, now (1905) about thirty pairs nesting (N. B. Kinnear, *t.c.*, 1905, p. 246).

ORKNEYS.—A number building nests June 8th, 1901 (*t.c.*, 1902, p. 199). Since 1891 two localities in Orkney, one of which is Hoy Head, have been occupied (J. A. Harvie-Brown, *Fauna N.W. Highlands and Skye*). Thirty or forty nests in 1901 at Hoy Head ; over fifty in 1902 (*Ann. S.N.H.*, 1904, p. 94). Several pairs were discovered during the summer of 1907 frequenting the cliffs between Stromness and the Bay of Skaill, in Orkney. This is believed to be the first record from this locality (J. Walpole-Bond, *Country Side*, 7, XII., 07).

In connection with the very marked and steady increase of this bird as a breeding species in Scotland, it should be remarked that a great increase has taken place in its numbers in the St. Kilda group, and this may partly account for the establishing of new nesting colonies (*cf.* J. A. Harvie-Brown, *Ann. S.N.H.*, 1903, p. 19).

IRELAND.—"Frequently met with at all seasons on the Atlantic . . . . rarely comes to land " (R. J. Ussher, *List of I. Birds*, p. 54). In 1906 Mr. G. P. Farran saw a few off *co. Cork* on November 1st, off *co. Kerry* on November 6th (*id., t.c.*, 1907, p. 163), while he also noticed them in May and August of the same year, and in 1907 in February (*t.c.*, p. 184), May (a few), August (many), September (a few), November (many), and in 1908, in January and February (many), August (many), November (a few), (*t.c.*, 1909, p. 80).

_**[*] This instalment takes us to the end of the " Manual," and our task of collecting the more important additions to the second edition will be completed next month by an article detailing the omissions from, and corrections of, our " additions." *We shall be extremely obliged for notes of such omissions and corrections which may have been detected by our readers, and these should be sent in not later than April 12th.*

(*To be continued.*)

NOTES FROM SUSSEX.

GREY WAGTAIL (*Motacilla melanope*).—This species nests regularly in north and west Sussex, probably in the east of the county, where Dr. C. B. Ticehurst found a brood in 1906, and probably also in other parts of Sussex, though very locally, and in no great numbers. For three years in succession, 1906–07–08, I have found the nest (two in 1908) in the two former areas.

HOBBY (*Falco subbuteo*).—Twice certainly within the past few years has the Hobby bred in Sussex. At the time of writing " Sussex Hobbies," which was published in " Country Side " on February 1st, 1908, detailing the finding of an eyrie on June 15th of the previous year, I was under the impression that this was actually the first record. So it is as far as the notifying of the fact goes, but careful inquiries have elicited the information that in 1906 a brood was taken off in a totally different part of the county to where I found my eyrie. In 1907 (the year in which I found it) I also located a second pair, but both birds vanished entirely from the wood they were frequenting. Probably they were destroyed. In 1906 a pair meant settling down in a Heron's old nest, but before the eggs were laid the male was mercilessly shot. On June 17th, 1908, I saw a single bird in a certain district in the north of the county.

COMMON SHELD-DUCK (*Tadorna cornuta*).—The Sheld-Duck nests in one spot in Sussex for certain, in the extreme south-west coast corner of the county. I visited the place on May 6th, 1908, and saw one pair of birds. The young were seen in 1906 and 1907 by Mr. Padwick—a capital observer.

SHOVELER (*Spatula clypeata*).—On April 18th, 1908, I discovered the Shoveler's nest in the north of Sussex. This is the first record for the county. The following is an extract from my diary for that day :—" As I left a withy bed at the tail of the big mill-pond a pair of Shovelers, easily recognised as such by the drake's plumage, chanced to be flying towards me over the water. This was the first year I had ever seen them here. Making a circuit they both settled by the second withy bed, half-way up the reach, where, by the aid of glasses and a careful stalk, I could study them to perfection.*

* A full description of the birds, both in flight and at rest, here follows but has been omitted for want of space. – EDS.

" Enlisting the services of the keeper I proceeded to hunt for the nest, which I was confident was somewhere near. The area of this mill-pond is a wide one, and there is much possible ground encircling it. But bearing in mind my experiences of the Shoveler in the north Kent marshes, where I had studied it very closely, I at once looked upon a stretch of rough grass adjoining the water and lying between the two withy beds as the most likely place for success. Taking the piece in beats the keeper and myself worked the place carefully. Suddenly the keeper stopped and held up his hand. I knew he had something ; and he had—the Shoveler's nest. It was about thirty paces from the margin of the pond, and placed between three tufts of ordinary grass, and then only held two eggs partially covered with bits of grass. As it happens so often at this stage of laying, there was not a shred of down, though of course the size and colour of the eggs, as well as the size of the nest, betokened the Shoveler." Five days later, however, there was down in the nest, and this with the feathers amongst it set the matter beyond dispute. Thirteen eggs were ultimately laid, but unfortunately they were deserted owing to the heavy snowstorm at the end of April.

FERRUGINOUS DUCK (*Fuligula nyroca*).—For nearly the whole afternoon of March 20th, 1908, I watched three Ferruginous Ducks on a certain mill-pond in the north of Sussex. Luck is with me over this species, because in 1903, on April 19th, Mr. Gwynne-Vaughan and myself identified three on the Wye at Builth Wells, Breconshire, at really close quarters.

J. WALPOLE-BOND.

RARE BIRDS IN PEMBROKESHIRE.

GREY PHALAROPE (*Phalaropus fulicarius*).—I shot a bird of this species that was swimming about a duck-pond within fifty yards of a private house in the neighbourhood of Haverfordwest on December 6th, 1908. The pond is about one and a half miles from the sea, and there are no sand or mud-flats within about ten miles. This is quite an uncommon bird in this county, and I only know of two stuffed specimens.

COMMON BITTERN (*Botaurus stellaris*).—Mr. Jeffery, taxidermist, of Haverfordwest, informs me that he had one to stuff this winter, shot near St. David's.

On January 23rd I was one of a party of seven guns who had a splendid view of a Bittern standing quite motionless, with head and beak at an angle of about 45°, in some tall

yellow rushes on the lake at Stackpole Court, a residence of
Earl Cawdor. The head-keeper there informed me that
one or two appear regularly every winter. •

NUTHATCH (*Sitta cæsia*).—This bird is reputed to be exceedingly rare in this county, but has either been overlooked
or has lately become commoner. At the end of November
I saw a bird of this species in the grounds of Picton Castle,
and on March 7th I saw one in the grounds of Hean Castle,
Saundersfoot. At both of these places there are a good
number of old and large trees, the exception in this exposed
and windswept county.

W. MAITLAND CONGREVE.

BIRD PROTECTION IN YORKSHIRE.

THE Wild Birds and Eggs Protection Committee of the
Yorkshire Naturalists' Union are this season placing a special
watcher at Hornsea Mere to protect the rarer birds nesting
there. They have for several years employed a watcher on
Spurn Point during the nesting season, with good effect, as
was shown in Mr. Oxley Grabham's article in our last number.
—EDS.

THE BIRDS OF KENT.

As some of our readers may be aware Dr. Norman F.
Ticehurst has for many years been studying the birds of Kent
with a view to writing a history of the avifauna of the county.
Dr. Ticehurst informs us that his manuscript is now complete,
and that he intends to publish the work forthwith. The book
is to be offered to subscribers, and the edition is to be limited.
For many reasons Kent is an extremely important county
ornithologically, and an adequate history of the Kent avifauna
has long been needed. We have every confidence that Dr.
Ticehurst's work will be one of exceptional merit, and will
take an honourable place in the splendid roll of the local
avifaunas of our islands.—EDS.

BLACK-THROATED THRUSH IN KENT.

ON February 1st, 1909, Mr. G. Bristow, taxidermist, of St.
Leonard's-on-Sea, brought to me in the flesh a male specimen
of the Black-throated Thrush (*Turdus atrigularis* Temm.),
which had been shot by a man named Fuller on the previous
Saturday (January 30th) at Newenden, in Kent. The bird was
killed on the Kentish side of the River Rother, which separates

the two counties of Kent and Sussex. This, so far as we know, is the third example of this Siberian Thrush which has been obtained in Great Britain.

The first was killed near Lewes, in Sussex, on December 23rd, 1869, and was bought by the late Mr. T. J. Monk from the man, a bricklayer, who had just shot it, and was at the time Mr. Monk met him, carrying the bird in his hand. I have often had the story from Mr. Monk's own lips. After

Male Black-throated Thrush shot at Newenden, Kent, on
January 30th, 1909.

Mr. Monk's death, the rarest of the birds in his collection were, through Mr. A. F. Griffith, obtained for the Booth Museum at Brighton, and amongst them was this specimen of *T. atrigularis*.

For the second recorded specimen obtained in Great Britain, Mr. J. A. Harvie-Brown kindly draws my attention to the one now in the Perth Museum, which was shot in February, 1879, on the banks of the Tay, and originally recorded by Col.

Drummond-Hay (*cf. Trans. Perth Soc. Nat. Sciences*, Vol. I., pp. 135–138 ; see also *Ibis*, 1889, p. 579).

Mr. H. E. Dresser in his " Manual of Palæarctic Birds " gives the habitat of this species as Asia, north to the Obi and northern Yenesei, south to the Altai and Turkestan, east to Lake Baikal ; in winter migrating south to Assam, northern India, Baluchistan, and Afghanistan ; has occurred in Europe as a rare straggler in the Caucasus, Hungary, Austria, Germany, Denmark, Belgium, and France and Great Britain. Mr. Dresser also informs us that " it has been found breeding in the Altai Range, and at Imbatskaya, on the Yenesei River," and that it lays four to six eggs, which vary considerably, some resembling the ordinary type of the Blackbird, whereas others more resemble those of the Mistle-Thrush, but have the ground-colour of a deeper blue.

THOMAS PARKIN.

CURIOUS NESTING SITE OF A WOOD-WARBLER.

WHILE walking along a road on June 5th, 1908, I saw a Wood-Warbler (*Phylloscopus sibilatrix*) with a mouthful of green grass. Being curious to know what it would do with it, I watched a few seconds. While looking, another Wood-Warbler came up with a white insect in its beak and, strange to say, entered a rabbit-hole on a perpendicular bank, not five yards from me, and in full view. It remained in the hole out of sight for several seconds, and then came out without the insect. I went nearer to see into the rabbit-hole, when both birds flew within a foot of my head, fluttering and tumbling about, and uttering the usual alarm note. On looking into the hole I could see nothing, so put in my hand, and out flew six little Wood-Warblers and joined their parents. The nest was exactly twelve inches down the hole, and was quite invisible from its mouth. The nest was not domed as usual. It may be interesting to note also that this particular bird was first heard by me near the spot where I subsequently found the nest, on May 6th, and I think it arrived on that day. The young flew out of the nest on June 8th : this seems quick work. Another peculiarity about this pair of birds was that the cock omitted the preliminary " chit, chit," and uttered only the second part of the ordinary shivering song—the trill, which was very loud and very prolonged, in some cases lasting fifteen seconds.

Subsequent observation revealed the fact that the bird with the mouthful of grass referred to above was building a second nest, while the other bird of the pair was feeding six young in the first nest. The second nest was also in the hole, not so far in as the first (only about six inches), but quite invisible when one looked on to the face of the bank. This nest was also not domed.

W. S. MEDLICOTT.

[Mr. Medlicott very kindly sent us the nest in question, and we have submitted it and the note to the Rev. F. C. R. Jourdain, who mentions the following abnormal sites :—

Nest under shelter of a root of a tree (*cf. Zool.*, 1896, p. 375).

Nest in a diagonal cleft in the perpendicular face of a big square boulder found in North Wales, May 21st, 1904 (O. V. Aplin, *in litt.* to F. C. R. J.).

Mr. Jourdain adds :—" The date (June 5th) is a very early one for young to be able to leave the nest. The eggs are usually laid about May 16th to 26th, often not till the end of May. This record seems to imply that the birds would have reared a second brood. Of this I have no previous evidence, and should consider it unlikely, as the Wood-Warbler is a late breeder."—EDS.]

CHAFFINCH BREEDING IN WINTER.

I WAS surprised to hear that a pair of Chaffinches (*Fringilla cœlebs*) had nested and reared a brood during the past winter at Churchstoke, Montgomeryshire. Being sceptical I inquired into the matter. I found the report quite correct ; the nest—undoubtedly a Chaffinch's—being now in my possession. It was in a sycamore, twelve feet above the ground, and is made mainly of wool, with a few bits of lichen outside. The Chaffinches built during the mild weather, in December, and both parents were seen repeatedly at or about the nest, and were observed feeding the young on February 20th. Mr. G. Mountford, the master, and one of the boys in Churchstoke School, kindly furnished me with the above details.

H. E. FORREST.

ALPINE SWIFT IN PEMBROKESHIRE.

AN Alpine Swift (*Cypselus melba*) was shot on November 20th, 1908, on the land of Colonel Mirehouse on the east side of Angle Bay, Pembrokeshire. The gamekeeper said that he had seen a pair of them in the neighbourhood for some time previously.

CHARLES J. P. CAVE.

[In connection with the above record we have received further interesting particulars from Lieut. W. Maitland Congreve, R.A., who writes as follows :—" The bird was shot by one of a number of guns (who nearly all fired at the bird, thinking it was a Hawk), the guests of Colonel Mirehouse. The bird was sent to Mr. W. E. de Winton, of Orielton, who at once pronounced it to be an Alpine Swift. It is now stuffed and in the possession of Colonel Mirehouse, and I saw it some weeks ago. The bird is particularly remarkable for the enormous span of the wings. The back is of a dull brown colour ; throat white, then a brown band and belly white. It is not in the least like an ordinary Swift, owing to the white, its size, and the span of its wings."—EDS.]

DOWNY WOODPECKER (DENDROCOPUS PUBESCENS) IN GLOUCESTERSHIRE.

ON January 14th, 1908, a friend who occasionally shoots birds for me brought me in a little Woodpecker that he had shot that day at Frampton Cotterel, near Bristol. It was climbing up the trunk of an old apple tree some five feet from the ground when shot. I supposed it to be simply a Lesser Spotted Woodpecker, and so labelled it, and it was only after sending the skin to Mr. Marsden, of Tunbridge Wells, last month, that the bird was discovered to be a specimen of the North American Downy Woodpecker (D. pubescens).

WM. A. SMALLCOMBE.

[In connection with this record we have received the following letter from Mr. H. W. Marsden :—" Amongst some Woodpeckers I received from Mr. Smallcombe there were a male and female, supposed to be Dendrocopus minor. The day I got them I was very busy, and sent on the two skins to the Hon. N. C. Rothschild. He handed them, without examination, to Messrs. Rowland Ward, to be remade, and it was by them the bird was identified as Dendrocopus pubescens. Mr. Smallcombe is quite a young ornithologist, and had probably never seen a foreign skin of D. pubescens."

Both Mr. Marsden and Messrs. Rowland Ward have satisfied us that this skin was undoubtedly not of American origin (we had suggested that the label might have been inadvertently changed), and that the bird was in fact shot in Gloucestershire. The record is an interesting one, but we cannot believe that this North American Woodpecker crossed the Atlantic unaided, and we think that the bird must have escaped from captivity.—EDS].

WHITE-TAILED EAGLE IN ESSEX.

WE could distinctly notice a White-tailed or Sea-Eagle (*Haliaëtus albicilla*) soaring over this park (Weald Hall, Brentwood) at midday on Saturday, February 6th. It was high up and being mobbed by a smaller bird, which I could not distinguish. I could see the Eagle quite clearly through field-glasses. It kept wheeling quietly round for nearly half-an-hour, and then disappeared.

CHRISTOPHER J. H. TOWER.

OSPREY IN ESSEX.

AN Osprey (*Pandion haliaetus*) appeared in this park (Weald Hall, Brentwood) from October 11th till the 24th, 1903. When it first came it was very tame, coming and taking some golden carp out of a pond in the garden, where some gardeners were at work. Afterwards it generally took up a position on the dead bough of a tree on an island in the lake, where it was generally mobbed by rooks, for whom, however, it seemed to have a supreme contempt. There is absolutely no doubt about its identity. It' was of course protected, and notice was given about so that it should not be shot.

An Osprey, presumably the same bird with more mature plumage, came again the following year, staying about a week.

CHRISTOPHER J. H. TOWER.

POCHARD NESTING IN NORTH KENT.

ON April 29th, 1907, I found a nest of the Pochard (*Fuligula ferina*) on the marshes in the north of Kent, in a district which need not be precisely specified. This is the first authenticated nest found in the county, and the only other that I am aware of was discovered a year afterwards by Major R. Sparrow in the south-west of Kent (*cf. antea*, p. 96). For several days before the actual discovery of the nest I had seen and watched closely two pairs of Pochards. One afternoon as a small tongue of rough ground infringing on one of the " fleets " (as all dykes are termed in Kent) was being worked for a Shoveler's nest, a duck Pochard clattered cumbrously from a thick screen of reeds. The fact that she was alone suggested the possibility of a nest ; and next morning on wading into the reed-bed, a duck Pochard again rose, not more than ten paces from the bank, from a swampy ridge of soil plastered with aquatic plants partitioning the " fleet." On this natural groyne was the nest, which, although very exposed, was not visible from the "mainland," but I marvelled greatly that it had escaped the prying eyes of the Crows. Facing the reeds growing in the deep backwater opposite, it was secured under the lea of a large spread-eagled tuft of extremely coarse sword-like grass, and was built up from the ooze beneath to a height of five inches. It was a moist affair of freshly-plucked green grass, flat shreds of dried grass, fragments of reed, sedge and water-weeds, finished off with a few wisps of green grass. To some extent it resembled a Coot's nest, though it was not so utterly exposed as most nests of that species. There was a well-trodden sloping platform, or "slide," of vegetation about a foot long and seven inches wide leading up to the nest from the water. It was then, as above cited, only April 29th, far too early for a full clutch of Pochard's eggs, and there were but two in the nest. These were uncovered and cold, for, of course, their owner had merely been standing by them. But their shape, size, and coloration, not to mention the presence of the bird, the disposition of the nest, and a few tufts of down, settled their identity beyond quibble.

J. WALPOLE-BOND.

THE FOOD OF THE EIDER.

IN a note in our last issue (p. 344) on this subject we made Mr. Robinson say that it was *curious* that Razor-shells were never found in birds killed in the early morning. Mr. Robinson points out that he wrote that these shells "may often be found in Eiders shot at any time except early morning." By this, he now tells us, he merely meant that the shells were not present before the birds had breakfasted, and not, as we inferred, that the birds do not feed on these shell-fish in the early morning.—EDS.

GOOSANDER IN BEDFORDSHIRE.

ON February 27th, 1909, I saw a female Goosander (*Mergus merganser*) amongst the ducks on one of our ponds at Woburn. The bird has often been recorded in Bedfordshire, but it is perhaps sufficiently rare to be worthy of mention.

M. BEDFORD.

RED GROUSE AND BLACK GROUSE HYBRIDS.

IT would be natural to suppose that species which are closely allied and which frequent the same ground would often interbreed; but such is not the case. It is well known that

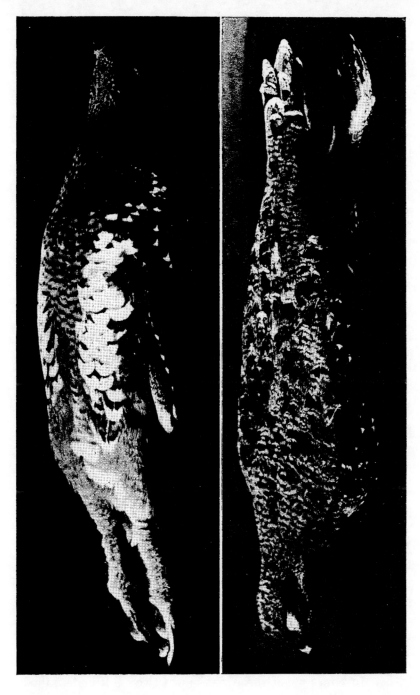

Immature Female Black and Red Grouse Hybrid, shot near Bala, North Wales, August, 1908.
Now in the collection of J. G. Millais.

Black Grouse and Capercaillie frequently interbreed, and there are four or more known instances of hybrids between such diverse species as the Pheasant and the Capercaillie, but crosses between Black Grouse and Red Grouse, or Red Grouse and Ptarmigan, are extremely rare. Mr. J. A. Jones spoke to me one day concerning some Grouse which he and his son had killed at Llanerch bog, near Bala, North Wales, in August, 1908. On examination they proved to be undoubted hybrids between Black and Red Grouse. All the seven young birds in the covey were killed, but only two were preserved; these exhibit very clearly the characteristics of both parents. The back, wings, and scapulars are similar to those of the immature Black Grouse, whilst the new plumage, coming in on the breast and flanks, is like that of the " White " form of the Red Grouse, being deep chestnut and black widely tipped with white. The feet, legs, and forked tail are similar to those of the larger parent. Neither of the parent birds was seen. It seems a great pity that the whole covey was destroyed, for had any reached maturity they would have been very beautiful and interesting birds, quite dissimilar to the other specimens of this hybrid that I have seen.

Mr. H. E. Forrest, in " The Vertebrate Fauna of North Wales," states (p. 107) that he has seen what appears to be a hybrid between the Black and the Red Grouse in the possession of Mr. Foster of Bettws-y-Coed; the specimen was shot at Yspythy Moor on the 20th of November, 1897. Mr. Foster also procured a similar specimen on the 9th of December, 1895.

By the kindness of Mr. J. A. Jones I was enabled to exhibit the two specimens referred to above—a male and female—at the meeting of the British Ornithologists' Club, held on January 20th last.

J. G. MILLAIS.

AN immature male of the hybrid between the Red Grouse and Black Grouse was killed on October 6th, 1908, by Mr. F. W. Stobart, in Glen Troot, Kirkcudbrightshire, where Black Grouse are extremely plentiful. It was shot during a drive when flying in company with four Red Grouse. The bird is exactly of the same type as the two adult males already in the British Museum (one of these was one of two adult males killed at Millden, Forfarshire, on October 1st, 1900, by Mr. J. L. Cadwalader, while the other, presented by Lord Tweedmouth, bears no particulars regarding its capture), but it still retains a considerable amount of the first plumage, particularly on the sides of the head and neck, where the feathers are mostly light reddish-buff barred with black. The bill is large and rather coarse, and the basal portions of the toes are feathered as in the Red Grouse, while the terminal portions are naked and pectinate on the sides as in the Black Grouse. Mr. Stobart has kindly presented the specimen to the British Museum, and I had the pleasure of exhibiting it at the February meeting of the British Ornithologists' Club. I have a further communication from Mr. Stobart saying that his keeper in Kirkcudbright has examined at close range a second example of this hybrid on the same ground with some Red Grouse.

Another male example of this rare hybrid has also been offered to the British Museum by Mr. G. Ashley Dodd, but has not yet been received.

W. R. OGILVIE-GRANT.

THE BILL OF THE GREAT NORTHERN DIVER.

IN the March number of BRITISH BIRDS there is a reference to the White-billed Northern Diver. For the last three years I have had opportunities of watching large numbers of Great Northern Divers (*Colymbus glacialis*) in the Outer Hebrides in the end of October and beginning of November. At that time none of the birds had attained their full winter plumage, and the neck bands were in every case easily detected, but in nearly the whole of them the lower mandible was ivory-coloured, and the upper mandible partially so. The bill of *C. adamsi* is so remarkably " up-turned " that it would be a far safer guide in winter than the colour.

M. BEDFORD.

[The plumages of the Great Northern Diver are very little known, but it would seem from the Duchess of Bedford's observations that it is not only the young that have light-coloured bills in autumn. Mr. Ogilvie-Grant has noted (Vol. I., p. 295) that in the young *C. adamsi* the up-curved character of the lower mandible is much less marked, and " mistakes may easily be made," but by the end of October it is possible that birds of the year would have attained this characteristic. We fancy that the purer white colour of the bill of *C. adamsi* would make it distinguishable even at a distance from *C. glacialis*.—EDS.]

FULMAR PETREL IN LANCASHIRE.

As Mitchell, in his " Birds of Lancashire," only mentions the Fulmar Petrel (*Fulmarus glacialis*) as having occurred four times in Lancashire, perhaps the occurrence of a fifth example at Galgate, near Lancaster, on April 3rd, 1904, may be of interest, especially as it was picked up alive in a field quite three miles from the sea.

H. W. ROBINSON.

✳ ✳ ✳

LITTLE RINGED PLOVER IN NORTH UIST.—Mr. J. E. Harting reports in the " Field " (20, XI., 09, p. 329), that he has received word from Mr. H. E. Beveridge, of Kelso, of a small Plover which he shot in North Uist, in October, 1908. By means of a sketch, drawn to the natural size, and a description of the bird, Mr. Harting comes to the conclusion that it was undoubtedly a specimen of *Ægia'itis curonica*. So far as we know an authenticated example of the Little Ringed Plover has not been obtained in this country for very many years, and the bird has never before been recorded for Scotland.

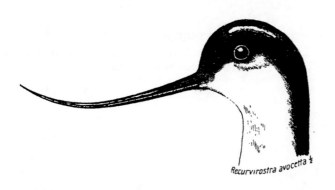

Recurvirostra avocetta ♀

RTISH
BIRDS

JSTRATED·MAGAZINE
TED·TO·THE·BIRDS·ON
THE·BRITISH·LIST

I,

Vol. II.
No. 12.

HLY·ONE·SHILLING·NET
HIGH·HOLBORN·LONDON
WITHERBY & CO

BRITISH·BIRDS

EDITED BY H. F. WITHERBY, F.Z.S., M.B.O.U.

ASSISTED BY W. P. PYCRAFT, A.L.S., M.B.O.U.

SOME EARLY BRITISH ORNITHOLOGISTS AND THEIR WORKS.

BY

W. H. MULLENS, M.A., LL.M., M.B.O.U.

IX.—WILLIAM MACGILLIVRAY (1796—1852) AND WILLIAM YARRELL (1784—1853).

AMONG the many famous names which adorn the long
roll of British ornithologists, that of William Macgillivray
stands forth as *facile princeps*. His work was not only
far superior to that of his predecessors and contemporaries,
but it remains to this day with but few, if any, serious
rivals, and will·probably continue to do so for many years
to come.

Macgillivray's great book, " The History of British

Birds," which was completed in 1852, has nevertheless failed to receive the appreciation which it deserves, and although it is probably far more widely read to-day than it has been hitherto, it has taken upwards of sixty years for the student and compiler to recognise its transcendent merit.

The causes of this neglect are somewhat difficult to understand ; and probably they arose from a variety of circumstances. Macgillivray's personal character was no doubt an obstacle to his success. One of his warmest admirers, the great American bibliographer and naturalist, Elliott Coues, describes him in these words :—

"Macgillivray appears to have been of an irritable, highly sensitized temperament, fired with enthusiasm and ambition, yet contending, for some time at least, with poverty ; ill-health and a perhaps not well-founded, though not therefore the less acutely-felt, sense of neglect ; thus ceaselessly nerved to accomplish yet as continually haunted with the dread of failure. . . . . This author was undoubtedly unwise in his frankness ; but diplomacy is a stranger to such characters. . . . . If he never hesitated to differ sharply with anyone, or to express his own views pointedly—if he scarcely disguised his contempt for triflers, blockheads, pedants, compilers and theorizers . . . . he was nevertheless a lover of Nature, an original thinker, a hard student, and finally an ornithologist of large practical experience, who wrote down what he knew or believed to be true with great regard for accuracy of statement and in a very agreeable manner."

To this must be added the curious coincidence that in the same year as the first volume of Macgillivray's "History of British Birds" was published (that is in 1837), another very famous work on the same subject, and bearing a precisely similar title, made its appearance. This was the well-known work of William Yarrell, which, from the clearness of its descriptions, the skill of its illustrations and the useful conciseness of its information, speedily became recognised as the standard authority on British ornithology. The success of Macgillivray's masterpiece was undoubtedly retarded by the simultaneous appearance of Yarrell's work, and it was further hampered by the fact that while Yarrell completed his task in 1843, it was not until nine years later that Macgillivray's was brought to a conclusion, twelve years having been allowed to elapse between the publication of the first three and the last two volumes. The matter of nationality had also perhaps some bearing on the question ; the English public naturally preferring the work of a fellow countryman to that of a Scotsman, however able.* But all these circumstances, much as they tended to prevent the due appreciation of Macgillivray's labours, were but trivial in comparison with the predominating cause of his comparative failure. The failure of the " History of British Birds " lay in the intrinsic value of the book itself.

To understand how this arose it is necessary to consider not only the scope of Macgillivray's book itself, but also the state and condition of ornithology in this country at that time. The increasing study of ornithology had produced in that science, in common with many others, specialists; i.e., students and writers who devoted themselves to some particular department or branch of their favourite science. These had gradually formed themselves into three distinct groups : the anatomists or morphologists, the chamber-naturalists, and the field-naturalists. The first named carried out their work in the dissecting room and the laboratory, the second devoted their attention to the study of the skins of birds in the museum, and of the labours of others in the ornithological library ; the third gave their time to the observation and study of living birds in their natural surroundings. The labours of the chamber-naturalists were chiefly

* In much the same way—to compare small things to great—the undoubted merit of Fleming's " History of British Animals," 1828, had been injuriously affected by the greater popularity accorded to a similar undertaking by an Englishman, viz., Jenyns' "Manual of British Vertebrate Animals," which appeared in 1835.

directed towards the manufacture of new genera and the subdivision of existing ones ; to proclaiming the superiority of one system of nomenclature over another ; to the endless alteration and confusion of the classification of species, to the disparagement of each other's labours and the laudation of their own. On one point and on one only were they agreed, much and bitterly as they differed on most other matters : they united in a common hatred and contempt for the field-naturalists.

In the opinion of the chamber-naturalists the existence of this third group of ornithologists was only justified by the fact that their observations and investigations provided fresh material for the use and advancement of the very men who decried their labours. It is true that most of the really important contributions to the literature of ornithology had come from the pen of the field-naturalists, but these works were not deemed " scientific " and the chamber-naturalist regarded them as but of small account.

And now suddenly all this was changed, the pedants and the pundits were threatened with a new and unconsidered danger and driven by it to seek their common safety in united action. A Scotsman who had spent his youth in observing and collecting birds, both in the distant islands of the Hebrides and on the mainland of his native country, had in due course of time become professor of Civil and Natural History in a northern university, had devoted his acute and highly trained intellect to the study, not of a single branch but of the whole science of ornithology, and had produced a book which not only recorded the most careful and accurate investigations in the field, but also proposed to create a new scientific classification of birds, founded on the consideration of their digestive organs, which, from the fact that his skill as an anatomist was unassailable and that the proposed scheme of classification had the further disadvantage of being original, constituted in the opinion of the chamber-naturalists a pressing and immediate peril. Presumption combined

WILLIAM MACGILLIVRAY, from the engraving in "A Vertebrate
Fauna of the Outer Hebrides," by J. A. Harvie-Brown and
T. E. Buckley, 1888. (By permission of Messrs. Harvie-Brown
and David Douglas.)

WILLIAM YARRELL, after the frontispiece by F. A. Heath, to
the Third Edition of Yarrell's " History of British Fishes," 1859.

with merit must be crushed, and crushed it was in a speedy and most effective manner. The word went forth that Macgillivray's work was "choked with anatomical details." The half-truth repelled the public, the "History of British Birds" was doomed to oblivion and the chamber-naturalists returned to their discussions in triumph. That they had incidentally broken the heart of the greatest ornithologist this country has ever possessed, that they had nearly prevented the completion of one of the greatest books on British birds, was to them of course, not a matter of the least importance.

Fortunate indeed it is that at the present day all this is changed, and that the "chamber-naturalist" is now as able in the field as in the museum.

From this combination of adverse circumstances Macgillivray's work has never completely recovered, and probably never will. Although the copyright has long expired and it now commands a price in the auction rooms which places it beyond the reach of many who would gladly possess it, yet the fact remains that in these days of constant re-issues and new editions of ornithological books, many of which are more or less worthless, Macgillivray's great work has never been reprinted and brought up to date.

No adequate account of the life and work of William Macgillivray has yet been published; some knowledge of his character and career can however be derived from a privately printed book, written by a namesake of the great ornithologist, and entitled "A Memorial Tribute to William Macgillivray" (Edinburgh, 1901, 1 vol., 4to). The preface to Macgillivray's "Rapacious Birds of Great Britain" and that to the fourth volume of his "British Birds" may also be consulted to advantage.

William Macgillivray was born in Old Aberdeen in 1796. He left Aberdeen when a child of three, and lived with his two uncles in the island of Harris—his father, who was an army surgeon, being absent with his regiment—till he was eleven years of age, when he returned to Aberdeen to complete his education. At the age of twelve Macgillivray entered King's College, and one year later, in 1809, lost his father, who fell on the stricken field of Coruña. Macgillivray, as he himself informs us,* "Commenced the study of zoology in 1817 while qualifying for the medical profession." . . . . "My only guides were Linnæus and Pennant," but a fellow student, William Craigie, evinced an equal interest in Nature, and the two together undertook a series of "pleasant and successful excursions in quest of plants and animals " . . . . "and most zealously strove to add to our common store of knowledge both in zoology and botany." . . . . "The fascinations of these pursuits were such that, after studying medicine for nearly five years, during part of which time I officiated as dissector to the lecturer on anatomy at Marischal College, I resolved to relinquish it and devote my attention exclusively to natural history." In pursuance of the project Macgillivray now commenced to wander over most parts of Scotland ; he explored the " desolate isles of the west " and walked from Aberdeen to London for the purpose of visiting the British Museum. He afterwards went to Edinburgh and attended Professor Jameson's natural history lectures. He then again betook himself to the Outer Hebrides, "where he hammered at the gneiss rocks, gathered gulls' eggs and shot plovers and pigeons " till finding this dull he returned to the mainland and became assistant and secretary to Professor Jameson, under whose supervision he took charge of the museum at Edinburgh University. Having held this post for several years he retired, and renewed his " observations in the fields," supporting himself meanwhile by his labours with the pen. In 1830–1831 he was unanimously elected as Conservator of the museum of the Edinburgh College of Surgeons, and this position he held till 1841, but meanwhile he in no way relaxed his ornithological labours, save, as he writes, for " about a year when hope seemed almost to have deserted me."

* Preface to "Rapacious Birds," p. 2.

In 1833, with a view to re-arranging the catalogue of his museum, he paid a series of visits to some of the more notable collections in this country, including the museums of Glasgow, Liverpool, Dublin, Bristol and London. In 1834 Macgillivray commenced to give lectures on natural history, and in 1835 he finished the new catalogue. In 1841 his connection with the museum came to an end, he having been appointed to the professorship of " Civil and Natural History " in Marischal College, Aberdeen. During the ten years he was at Edinburgh, Macgillivray in addition to his other work, published in 1836 " Descriptions of the Rapacious Birds of Great Britain " (1 vol., 8vo), the first volume of the first edition of " A Manual of British Ornithology " (2 vols., 8vo, London, 1840–1842 : the second edition appeared in 1846), and the first three volumes of his " Great Work,'" as he rightly termed it, " A History of British Birds, Indigenous and Migratory." Besides the above he contributed a " History of British Quadrupeds " to Jardine's " Naturalist's Library " (40 vols., 1838–1843 ; 2nd edition, 1844–1855), and compiled the scientific part of Audubon's " Ornithological Biographies." There is no need to deal at any length with Macgillivray's "Great Work " here, his object in writing it was "to lay before the public, descriptions of the birds of Great Britain, more extended and if possible more correct than any previously offered," and this he most ably succeeded in doing, but the illustrations, the anatomical plates excepted, can hardly be called worthy of the text. Macgillivray occupied his chair at Marischal College for eleven years, but in 1850–1851 he was attacked by a serious illness, the result it is said of a pedestrian excursion undertaken in the Upper Valley of the Dee, to study that locality for his last written and posthumously printed book, " The Natural History of Deeside and Braemar " (1 vol., 8vo, 1855). In the autumn of 1851 he removed to the milder climate of Torquay, and while still at that place he in March, 1852, published the fourth volume of his " Great Work." The fifth appeared in July after his return to

Aberdeen, and on September the 8th of that same year he died at his residence in Crown Street in that city. In the pathetic " Conclusion " to the fifth volume he states, " I have finished one of the many difficult and laborious tasks which I had imposed on myself." . . . . " Commenced in hope and carried on with zeal, though ended in sorrow and sickness, I can look upon my work without much regard to the opinions which contemporary writers may form of it, assured that what is useful in it will not be forgotten . . . . and knowing that it will powerfully influence the next generation of our home-ornithologists."

If Macgillivray was not " the most eminent ornithologist in Europe," as he has been designated by his admirers, and perhaps that description better applies to Naumann, he certainly was by far the greatest ornithological genius that this island has produced, and as such we have every reason to honour his memory.

William Yarrell, Macgillivray's great contemporary, was born on the 3rd of June, 1784, in the parish of St. James', London. His father carried on the trade of a newspaper agent in Duke Street, and to this business Yarrell succeeded in due course. He was educated at Ealing, and in his eighteenth year entered the banking-house of Herries, Farquhar & Co., as a clerk, but soon left to assist his father in business. Yarrell seems to have turned his attention to the study of ornithology while engaged on the fishing and shooting expeditions with which he varied the monotony of business. As he neared middle age his love for natural history increased, and he abandoned field sports, and henceforward devoted himself to the systematic study of zoology. In 1823 he commenced to note the appearance of rare and interesting birds, and is said to have aided Bewick by sending him rare specimens. He became a fellow of the Linnæan Society in 1825, and was one of the original members of the Zoological Society. In 1836 he completed a " History of British Fishes," and in July, 1837, published the first part of his well-known "History of British Birds." This was completed in May, 1843, and the first supplement was printed in 1845. The "History of British Birds," which originally appeared in three volumes, proved a great success. Yarrell, besides being an accomplished ornithologist, knew exactly what the general public wanted in a popular text-book, and, moreover, possessed the skill of presenting his knowledge in a concise and agreeable manner. A second edition of the book appeared in 1845, and a third, incorporating the second supplement, in 1856. In 1871 a fourth edition was commenced; this was finished in 1885, and consisted of four volumes, the original text being almost entirely rewritten, Professor Newton undertaking that of the first two volumes and Mr. Howard Saunders that of the remainder; the latter of these two—both, alas, recently deceased—further condensed the whole into a single volume, illustrated with the same figures as the larger work, and entitled "An Illustrated Manual of British Birds" (1st edition, 1889 ; 2nd, 1899).

Yarrell was a man of unbounded energy, and in addition to his business labours was the author of many and various writings on natural history.*, He was also a zealous supporter of several learned societies. After a long and busy life he was seized with a sudden illness while on a visit to Yarmouth, where he died on September 1st, 1853. He was buried at Bayford in Hertfordshire in a spot which had been selected by himself, and a medallion tablet at the west end of the south aisle in St. James' Church records his memory in his native parish.

* For a list of these, 81 in number, cf. "Memoir," third edition, " British Fishes."

NOTES ON THE NESTING OF THE GOOSANDER.

BY

NORMAN GILROY, M.B.O.U.

FOR the last eight years or so it has been my custom to spend a portion of the spring in a remote part of Sutherlandshire, and although each May I have observed a duck Goosander (*Mergus merganser*) with a brood of newly-hatched young ones on the large loch near which I stayed, it was not until the spring of 1908 that I actually came across the nesting site, or rather, sites, for the main object of these notes is to show that with the Goosander there is a slight inclination to sociability.

My previous experience of the Goosander as a nesting species had been slight—confined, in fact, to the finding, or rather, to the assisted discovery of a nest on April 21st, 1905, in a deep cavity on the steep, rocky, and sparsely-wooded bank of a river in Ross-shire (the sides of the ravine were in places almost inaccessible), and to the discovery, after a long and interesting watch, on April 25th of the same year, of a second nest on a wooded hillside in Sutherlandshire.

In the first-mentioned case I saw little of the sitting bird, for she at once scuttled out of the cavity containing the nest, and flew rapidly down the gorge to the main river. There were thirteen eggs and a profusion of down, and incubation had commenced, although at the time the ground was white with snow. I saw no sign of the drake anywhere, and I am credibly informed that as soon as the clutch is complete the male Goosanders leave the neighbourhood and repair probably to the sea. My subsequent observations tended to confirm this.

My experience with the second pair was considerably more interesting. I was sitting in a sheltered spot on the wooded hillside above mentioned (which overlooks a loch of considerable size) watching a couple of Eagles hunting, when my attention was arrested by the movements of a

pair of ducks, which suddenly rose from the loch and
flew rapidly towards me. I was at first unable to determine
accurately the species. But fortunately the sun was
shining brightly ; still more fortunately the birds came
quite close to me before they turned, so that the characteristic plumage of the male was easily discernible. After
turning once they flew round at varying heights in wide
ellipses, the duck leading, whilst both birds uttered a
curiously muffled, but harsh, quacking noise. I noticed
that the duck invariably dived down over a particular
spot on the hillside, and it at once struck me that the nest
was not far from this point, so that when they finally
flew down to the loch again I started to explore the hillside carefully in the immediate neighbourhood of the spot
mentioned. The ground was sparingly strewn with
boulders of considerable size, most of them half buried
in the soil, but at the base of the second one, which I
examined, was a wide dry cavity containing a lot of
withered grass which had evidently not been blown there
by the wind. I could not possibly reach it or get at it
from the front in any way, but found that by the removal
of some small stones from behind the whole cavity could
be comfortably examined, and in it was the Goosander's
nest, containing one fresh egg. The nest itself was composed of masses of white, withered grass, and at this time
I saw only one or two down tufts. I visited the nest
twice afterwards, but had no opportunity of seeing the
duck sitting, as I had to come south before the clutch
was completed. An egg appeared to be laid every
other day, and I afterwards heard from a gillie, to whom
I showed the nest, that this bird ultimately laid eight.

I had no further experience of the Goosander until
1908 ; for, although each spring up till then I carefully.
searched a heavily wooded hillside hanging right over a
Sutherland loch, I could never discover a nest, in spite of
the fact that after perhaps a fortnight's hunting I invariably saw the duck Goosander with her young on the
water.

That year I arrived in Sutherlandshire on May 12th. The weather was beautifully fine and warm, but this was only of recent occurrence, as apparently the heavy snows of late April had but just melted away. On the evening of my arrival I had a short conversation with the keeper, in the course of which I asked him if he had seen anything of the Goosander. He at once replied that a few mornings previously as he was coming down the road which runs parallel with a small stream that flows from the hills through a deep gorge down into the loch, he had seen a Goosander flying rapidly up-stream, and that at a certain point it had appeared to dive into the bank. There was of course no doubt in my mind then that I was at last on the right track and that the explanation of my previous years' fruitless search was at hand.

The keeper had described the spot where he had seen the Goosander disappear so fully that I did not think it necessary to take him with me next morning, and I accordingly started away at an early hour to explore the gorge, the banks of which are in places very steep and rocky—in others less precipitous, but thickly grown with heather, with here and there a mountain ash, or birch, or an aspen, now just bursting into leaf. Although I was perfectly familiar with the stream, having often tramped it from mouth to source in search of the Ring-Ousel, curiously enough it had never struck me before as an ideal place for the nest haunt of the Goosander, and I naturally was full of excitement at the prospect before me. On reaching the spot which I imagined the keeper to have described—a high and somewhat bare hummock, forming almost an island in the stream, with a solitary tree and thin growth of heather on the top, the whole overlooking a beautiful waterfall, I at once commenced searching the holes and rifts in the peat, some of which are fringed with heather. In about ten minutes I came across a deepish cavity with a well-worn track leading in to it, and two tell-tale down-tufts clinging to the heather at the entrance. I could by no means reach the nest, and as I was anxious to catch the sitting duck if possible I broke away a piece of the bank. As soon as I did so, however, the duck escaped by another hole which I had not previously noticed, and flew down towards the loch. Her plumage was very bright, and she appeared to be in perfect condition. The ·nest was perhaps four feet from the entrance—the cavity being dry and warm—and it contained ten eggs and a profusion of down mixed with good-sized bunches of heather, and a very few birch leaves, evidently taken there by the bird. The site had

A Nesting Haunt of the Goosander.

been used before as was amply demonstrated by the presence of old eggshells. The eggs were in an advanced state of incubation, and should have hatched out in a week or less. Both entrances to the nest were quite open and unprotected, and both were apparently used regularly.

I told the keeper on my return that I had found the Goosander's nest and the matter dropped. But on the 21st I happened to be rambling along the same stream early in the morning looking at a Kestrel's eyrie, when to my astonishment, on suddenly turning a corner, I saw a Goosander flying rapidly towards me. I concealed myself hurriedly, and the duck passed me at a distance of a few feet. She was quacking in the same harsh but curiously muffled way as I have before mentioned, but unfortunately she disappeared before I could gain a spot commanding a view of the entire gorge. This place was only about three hundred yards from the nest described above ; but I could not think that the duck belonging to that one would be here alone after her eggs had been taken, so I determined to come out at four o'clock on the following morning and take up a suitable position to watch for her. A wait of five hours, however, was unavailing ; I saw no sign of the Goosander at all. My search of the banks, a very difficult matter at this point, seeing that they were heavily fringed with heather, and that the rocks were very sheer, proved equally fruitless. I got back almost exhausted, but arranged to go out with the keeper early next morning to clear up the mystery.

The morning broke bright and clear, and we started off at 4.30. On reaching the point at which I had taken the Goosander's nest a few days previously, the keeper passed it unconcernedly, so I at once knew there was a second nest close by. We crossed the stream almost exactly where I had seen my bird disappear, and the keeper then remarked, " This is the spot." We climbed up the bank with considerable difficulty, but after a short search came upon a large and very deep hole almost concealed by a heavy fringe of heather. There were half-a-dozen pieces of down scattered about, and immediately I raised the heather fringe I heard the Goosander hissing inside. The nest itself was about seven feet in, and I caught the sitting duck, which was in perfect plumage and condition, although the ten eggs were within a few days of hatching, and the quantity of down was considerable. The nesting hole was damp and filthy, and had evidently been used for years, so that I was surprised to find the duck so beautifully clean, the rose tints being particularly fine. She was very fat, but did not struggle much, and when I released her, flew off down the stream quacking quietly.

I was, of course, greatly astonished to find two Goosanders nesting in such close proximity; and the fact that they must have been for years inhabiting a stream with which I was perfectly familiar goes far to prove how easily the species may be overlooked.

I saw no sign of the drakes during the whole of this, or any previous, visit to the district.

Squatarola helvetica ½

B

ON THE MORE IMPORTANT ADDITIONS TO OUR KNOWLEDGE OF BRITISH BIRDS SINCE 1899.

BY

H. F. WITHERBY AND N. F. TICEHURST.

PART XX.

(Continued from page 375.)

Corrigenda et Addenda.

IN concluding this series of articles we would express our great indebtedness to the Rev. F. C. R. Jourdain, who has given us generous and continual assistance ; to Mr. R. J. Ussher, who has taken the greatest pains to make our record of Irish birds complete, and has added much new information; to Mr. H. E. Forrest, who has given us much help in Welsh and Shropshire records ; to Mr. T. H. Nelson, who has most kindly assisted us in Yorkshire records ; and to Messrs. W. Evans and J. A. Harvie-Brown, who have given us much advice in Scottish records.

In making use of this series of articles on the additions to the second edition of Howard Saunders' "Manual," the reader is warned that it is necessary to consult also the indices of the two volumes of the Magazine now complete, since many observations have been recorded while these articles have been in progress.*

WHITE'S THRUSH (Vol. I., p. 53).—Saunders describes it as " probably " breeding in Japan, and also describes Swinhoe's eggs. These are now known to have belonged to some other species. Many authenticated nests and eggs have now been taken in Japan (*cf.* Heatley Noble, *Bull. B.O.C.*, X., p. 47, Collingwood Ingram, *Ibis*, 1908, pp. 132 and 386, Plate IV., figs. 2 and 3 (eggs)).

REDSTART (Vol. I., p. 54).—*Scotland.*—Quoting from the " Manual " we stated that this bird had not previously been recorded from the " Hebrides." Mr. D. Macdonald, of Tobermory, kindly writes that this should read *Outer* Hebrides, as the Redstart is common in Mull. Moreover, Mr. Harvie-Brown has noted (*Ann. S.N.H.*, 1902, p. 140) that it was met with once by Finlayson, of Mingulay, on August 6th, 1889,

---

* Some corrections of Irish records not mentioned in this article are made on pp. 248 and 276 of this volume.

and another was recorded at Barra Head on May 15th, 1894. Saunders says : " In Scotland it has of late years spread northwards; now breeding freely in the Moray basin, and only less so in Sutherland, Caithness and West Ross." But, as far back as 1839 Jardine (*Brit. Birds*, Vol. II., p. 119) wrote, " It extends, . . . , to the northern parts of Scotland." In the sixties, Booth noted their abundance in the Highlands : in the Catalogue of the Cases in his Museum, at Brighton (p. 121), he says : " I have noticed them particularly abundant in the wooded glens in the Highlands, where the old stone dykes and rugged, weather-beaten trees afford ample choice for the selection of a nursery." Was also noted in Moidart prior to 1865 (Mrs. Blackburn, *Birds drawn from Nature*), and in Ross-shire prior to 1872 (Bateson, *Proc. Glasgow N. H. Soc.*, II., p. 182) (W. Evans, *in litt.*).

BLACK REDSTART (Vol. I., p. 54).—*Ireland.*—One was seen near Courtown Harbour, co. Wexford, in February, 1909 (M. D. Haviland, *Field*, 27, II., 09). February is an unusual month for its occurrence in Ireland, October and November being the usual months.

NIGHTINGALE (Vol. I., p. 55).—We omitted some information with regard to its range in Yorkshire, but this is not now included as it has been decided to open an inquiry into the exact range of the Nightingale in England. This inquiry will be organised by the Rev. F. C. R. Jourdain and N. F. Ticehurst, and will commence in the next volume. Meanwhile those who are able to make observations at any point on the outskirts of the normal range of the Nightingale, would greatly assist the inquiry by keeping full and careful notes of occurrences, and especially of instances of nesting which come under their notice.

WHITETHROAT (Vol. I., p. 55).—*Scotland.*—One was shot in June, 1897, and a pair nested at Eoligary, Barra, in 1900. They had been seen at Barra in May for several years (J. A. Harvie-Brown, *Ann. S.N.H.*, 1902, p. 140).

LESSER WHITETHROAT (Vol. I., p. 55).—*Scotland.*—One was shot on October 24th, 1898, on Barra (W. E. Clarke, *Ann. S.N.H.*, 1899, p. 109). An adult male was killed at the Suleskerry Lighthouse on September 17th, 1902 (*id.*, *t.c.*, 1903, p. 24). Both this and the Common Whitethroat were observed by Dr. Hamilton at Traigh, Loch Morar, in autumn of 1880 (*cf. Zool.*, 1880, p. 503) (W. Evans, *in litt.*). *Ireland.*—The second example was taken on October 10th, 1899, at the Innishtrahull Lighthouse (most northerly Irish light) (R. M. Barrington, *Mig. B. Irish Lt. Stations*, p. 72).

GARDEN-WARBLER (Vol. I., p. 56).—*Scotland.*—One was shot at Barra on October 24th, 1898 (W. E. Clarke, *Ann. S.N.H.*, 1899, p. 110). It is, in my experience, more plentiful and more generally distributed than the Blackcap in the Forth District, and also in Perthshire (W. Evans, *in litt.*).

WOOD-WREN (Vol. I., p. 83).—*Scotland.*—Saunders says : " In Scotland it is fairly distributed, and has apparently spread northward of late years, being recorded by Messrs. Harvie-Brown and Buckley as breeding in the north-east of Sutherlandshire, and as having been identified at Dunbeath, in Caithness, and in West Ross." According to Booth (*Catalogue*, p. 107), it was abundant in the north of Scotland in the sixties. " I have," he says, " noticed this bird as being particularly numerous in the wildest glens of Perthshire, Ross-shire, and Caithness." Was also noted in Moidart prior to 1865 as seemingly a regular summer visitor, but " less common with us than the Willow-Warbler " (Mrs. Blackburn, *t.c.*). " It is of general diffusion through the kingdom " (Selby, *Brit. Ornithology*, Vol. I., 1st ed., 1825, p. 189) (W. Evans, *in litt.*).

GREENISH WILLOW-WARBLER (Vol. I., p. 82).—The specimen taken at the Suleskerry Lighthouse in 1902 now proves to be an example of Eversmann's Warbler (*Phylloscopus borealis*). Mr. Eagle Clarke obtained a similar bird on Fair Isle in 1908, and found that, although it only had a single wing-bar, it was a specimen of *P. borealis.* The fact that this species sometimes exhibited this character had escaped the attention of Mr. Howard Saunders and himself when they identified the Suleskerry bird. It is possible that the only other British record for *P. viridanus* may also prove an error, and that the species may have to come off the British list (W. Eagle Clarke, *Ann. S.N.H.*, 1909, p. 114).

HYPOLAIS ? sp. (Vol. I., p. 83).—Mr. F. C. Selous writes : " My friend Major Mangles when at school in Croydon took a nest with four eggs in an osier-bed in 1884. Two of these eggs were broken and the other two I have in my collection. They are undoubtedly eggs of either the Melodious or the Icterine Warbler. Howard Saunders and Mr. E. Bidwell both thought they belonged to the former species."

REED-WARBLER (Vol. I., p. 84).—The first authentic example for Ireland was killed by striking the Rockabill Lighthouse on October 20th, 1908. Mr. A. H. Evans stated that he heard this species singing in a reed-bed on the Shannon, near Portumna, on July 23rd, 1904 (R. M. Barrington, *Scient. Proc. R. Dublin Soc.*, XII., p. 19).

GREAT REED-WARBLER (Vol. I., p. 84).—One was shot in a

reed-bed at one of the meres at Ellesmere, in Shropshire, about 1886. It had been noticed singing, and was supposed to be a Nightingale. It was stuffed by C. W. Lloyd, and purchased by H. Shaw ; subsequently it passed through the hands of G. Cooke and G. F. Fox, and is now in Mr. W. S. Brocklehurst's collection. It was examined by Mr. Forrest soon after Cooke bought it (H. E. Forrest, *in litt.*, and *Fauna of Shropshire*, p. 111).

GRASSHOPPER-WARBLER (S. page 89).—*Scotland.*—Breeding in Morayshire (near Elgin) in 1896-7-8, and not included in Messrs. Harvie-Brown and Buckley's " Fauna of Moray " (R. H. Mackessach, *Ann. S.N.H.*, 1900, p. 48).

ALPINE ACCENTOR (Vol. I., p. 109).—One was shot " a few years since " (1904) at Ettington, near Stratford-on-Avon, on the borders of Warwick and Worcester (R. F. Tomes, *Vict. Hist. Warwick*, I., p. 191).

BEARDED TITMOUSE (Vol. I., p. 109).—One was seen by Captain Henneker, who knew the bird well, in a reed-bed near Sudbury, Derbyshire, in the summer of 1896 (F. C. R. Jourdain, *Vict. Hist. Derby*, Vol. I., p. 126).

CRESTED TITMOUSE (Vol. I., p. 110).—The bird observed by Baron von Hügel was at Torquay, as already pointed out by Mr. W. S. M. D'Urban (*Vict. Hist. Devon*, p. 301).

TREE- PIPIT (Vol. I., p. 112).—*Scotland.*—Booth (*Catalogue*, p. 17) says : " Forest of Glenmore, in Inverness-shire, where in the summer of 1869 I found it breeding in considerable numbers " (W. Evans, *in litt.*).

WATER-PIPIT (Vol. I., p. 113).—Mr. O. V. Aplin writes that the square brackets enclosing his Oxfordshire record should be removed. Our only reason for inserting them was because Mr. Aplin considers there is no distinction between *A. s. spipoletta* and *A. s. rupestris*, but we consider that the two forms are quite distinct, and did not know to which his record referred.

A third example from Merioneth was obtained by Mr. Caton Haigh on February 21st, 1898 (H. E. Forrest, *Vert. Fauna N. Wales*, p. 123).

GOLDEN ORIOLE (S. page 145).—*Ireland.*—A female was found dead at the Skelligs Lighthouse, co. Kerry, on May 23rd, 1899 (R. M. Barrington, *Mig. B. Irish Lt. Stations*, p. 11).

RED-BACKED SHRIKE (Vol. I., p. 148).—*Ireland.*—An immature bird (the second Irish specimen) was received by Mr. R. M. Barrington from the Wicklow Head Lighthouse, where it had been caught at the lantern on the night of September 1st, 1908 (R. M. Barrington, *in litt.*). *Scotland.*—Saunders says: " In the south-east of Scotland it has occasionally been known to breed, as well as at Cambuslang, Lanarkshire, in 1893 [1892 in *Ann. S.N.H.*, 1893, p. 182] ; but beyond the Forth it is rare," etc. The first part of the above statement is too strong, no nesting having ever been proved ; only birds seen in " the breeding season," which may have been passing migrants, while the Cambuslang record is apparently of doubtful value ; Mr. J. Paterson, who sent it to the " Annals," refers to it in his list of Clyde birds (*Brit. Assocn. Handbook Fauna, etc., Clyde Area*, 1901, p. 161) thus : " Red-backed Shrike ; recorded by the writer as having nested in Lanark (*Ann. S.N.H.*, 1893, p. 183), but I fear deception somewhere " (W. Evans, *in litt.*).

WAXWING (Vol. I., p. 148).—*Scotland.*—In connection with the occurrence of the example in Unst in November, 1903, it is of interest to note that many were obtained in the south-east of Scotland (and elsewhere) about the same time (W. Evans, *Ann. S.N.H.*, 1904, p. 54, etc.).

PIED FLYCATCHER (Vol. I., p. 148).—*Ireland.*—The eighth Irish specimen was obtained at the Fastnet, co. Cork, on October 9th, 1899 (R. M. Barrington, *Mig. B. Irish Lt. Stations*, p. *11*), and the ninth at the Tuskar Rock on September 11th, 1901 (*id., in litt.*). *Scotland.*—To the counties in which it has nested should be added Midlothian (W. Evans, *in litt.*).

RED-BREASTED FLYCATCHER (Vol. I., p. 149).—*Ireland.*— The *data* of the two last Irish specimens (*cf. supra*, p. 248) are : Leg and wing sent from the Blackwater Bank Lightship, co. Wexford, on September 24th, 1898 (R. M. Barrington, *t.c.*, p. *10*) ; a female, or immature bird, was obtained on the Bull Rock, co. Cork, on November 18th, 1903, and is now in Mr. Barrington's collection (R. M. Barrington, *in litt.*).

HAWFINCH (Vol. I., p. 152).—*North Wales.*—One was shot in a garden at Trescawen in June, 1906--the first record for the county (H. E. Forrest, *in litt.*). *Ireland.*—The following have been taken at Irish lights: Hook Tower, co. Wexford, one, October 25th, 1897, and one, November 4th, 1897 ; Tuskar, co. Wexford, one, November 1st, 1897 ; Mine Head, co. Waterford, one, November 10th, 1898 (R. M. Barrington, *t.c.*, p. *127*). *Scotland.*—" So many occurrences of both old and young birds in the south-east of Scotland have of late come to my knowledge, that I now regard the species as widely distributed, and not rare in the district " (W. Evans, *in litt.*).

SISKIN (Vol. I., p. 180).—The record of its breeding in Derbyshire should be enclosed by square brackets.

TREE-SPARROW (Vol. I., p. 181). — *Scotland.* — To the counties in which it is now known to nest, add Linlithgowshire. (W. Evans, *in litt.*).

[NORTHERN BULLFINCH (Vol. I., p. 246).—One was shot on Caister Denes, Norfolk, on January 22nd, 1893 (*cf. Zool.*, 1894, p. 85). Mr. J. H. Fleming, of Toronto, has kindly sent for our inspection two skins (collectors unknown, but the labels seem quite genuine), a male taken at Bolton on June 18th, 1894, and a female at Gloucester on May 1st, 1889.]

CROSSBILL (Vol. I., p. 247).—" I have recently got satisfactory evidence that several pairs nested on the borders of Shropshire and Herefordshire in the spring of 1895. Crossbills were remarkably numerous here at that time " (H. E. Forrest, *in litt.*). *Ireland.*—In 1907 Mr. C. Langham reported that he had Crossbills in several places at Tempo Manor, co. Fermanagh. They had been scarce for a year or two (R. J. Ussher, *in litt.*). A nest was found with fresh eggs on April 17th, 1907, in co. Wicklow (R. Hamilton-Hunter, *Irish Nat.*, 1907, p. 208).

SNOW-BUNTING (Vol. I., p. 250).—*Ireland.*—An adult male was sent to Mr. Barrington from Aranmore, co. Donegal, on July 28th, 1898 (*Birds of Ireland*, p. 78).

NUTCRACKER (Vol. I., p. 254).—One was seen near Scotton Common, Lincolnshire, on August 14th, 1900 (F. M. Burton, *Nat.*, 1900, p. 319).

KINGFISHER (S. page 279).—One on River Broom, West Ross, September, 1898 (*Ann. S.N.H.*, 1899, p. 47).

WRYNECK (Vol. I., p. 280).—*Ireland.*—One was found dead on the Fastnet, co. Cork, on September 17th, 1898 (R. M. Barrington, *Mig. B. Irish Lt. Stations*, p. 181).

GREAT SPOTTED WOODPECKER (Vol. I., p. 281).—*Scotland.* —The Wells and Minto woods are in Roxburghshire. *Ireland.* —Two bones found by Mr. Ussher in separate caves in co. Clare and identified by Mr. E. T. Newton as belonging to this species, point to its being formerly a resident in Ireland (*cf. List of I. Birds*, p. 24).

ROLLER (Vol. I., p. 281).—*Ireland.*—Ten have been obtained —the last in co. Donegal on October 10th, 1891 (R. J. Ussher, *List of I. Birds*, p. 24).

BEE-EATER (Vol. I., p. 282).—The Yorkshire record should

have been ascribed to Mr. G. W. Murdoch (*cf. Birds of Yorks.*, p. 284, and *Yorks. Weekly Post*, 23, IX., 05).

HOOPOE (Vol. I., p. 282).—One shot near Brackley, Northampton, in May, 1908 (O. V. Aplin, *Zool.*, 1908, p. 312).

LITTLE OWL (Vol. I., pp. 315 and 335).—Several were seen in the summer of 1908, and five were liberated in Essex ("A. W.," *Field*, 15, VIII., 08). One was shot early in 1909 at Burton-on-Trent (F. C. R. Jourdain, *in litt.*). Of the one recorded from Scotland (p. 315) Mr. G. Sim stated that he had since heard of facts which led him to suspect that the bird was an escape (*Vert. Fauna Dee, addenda*).

SNOWY OWL (Vol. I., p. 315).—*Ireland.*—One was seen by Captain Kirkwood in December, 1906, at Bartragh, co. Mayo (R. Warren, *Zool.*, 1907, p. 73). An immature female was obtained in co. Mayo about the beginning of December, 1906, and an immature male (?) was shot near Ardagh, co. Kerry, and was received on March 6th, 1907 (R. J. Ussher (*fide* W. J. Williams), *Irish Nat.,* 1909, p. 100).

MARSH-HARRIER (Vol. I., p. 316).—The following have been shot at Hickling, Norfolk :—Adult ♂ May 9th, 1905 ; adult ♂ May 17th, 1906 ; and adult ♀ June 25th, 1906 (F. Smalley, *in litt.*).

MONTAGU'S HARRIER (Vol. I., p. 317).—The following have been shot in Norfolk :—Adult female, Hickling, May 11th, 1906 ; adult female, June 1st, and adult male, June 17th, 1907, near Lynn (F. Smalley, *in litt.*).

All these records point to the fact that the birds would have bred in these districts had they not been shot. No words are too strong to condemn this wanton destruction, and it may be pointed out (since it is not generally realised) that, as long as keepers know that they can dispose of such birds, the more inclined will they be to destroy them.

ROUGH-LEGGED BUZZARD (Vol. I., p. 319).—*Ireland.*—Only twelve (eleven obtained, one seen) are recorded in the " Birds of Ireland," and of these the one shot on October 4th, 1899, is mentioned in the Appendix. The second example referred to (p. 319) was received by Mr. Williams on November 5th, 1902, and not " in the early part of 1903," as recorded in the " Irish Naturalist." Of the two seen in December, 1906, the second—a female—frequented the moors during the winter, and was eventually poisoned and received by Mr. Williams on February 26th, 1907 (R. J. Ussher, *in litt.*, and *Irish Nat.*, 1909, p. 100). There are thus seventeen records from Ireland, sixteen obtained, and one seen.

BLACK KITE (Vol. I., p. 319).—The date given for the Aberdeen specimen in Sim's "Vert. Fauna of Dee" is April *18th*, 1901; and he says it was shot *within* the city boundary (W. Evans, *in litt.*).

GREENLAND FALCON (Vol. I., p. 320).—*North Wales.*— Mr. A. Heneage Cox writes that there is a specimen at Voelas Hall, Denbighshire, which was trapped by the keeper there. This is the second record for North Wales (H. E. Forrest, *in litt.*).—*Ireland.*—Several mistakes occur in this summary, and several more birds have been obtained and not recorded. Mr. Ussher provides us with the following list, which it seems better to print in full, and it must be taken to cancel that on p. 320 :—

ULSTER.
juv., Horn Head, Donegal, shot end December, 1903 (*Zool.*, 1904, p. 115).
♀ juv., Horn Head, Donegal, trapped March 21st, 1905 (*I. Nat.*, 1905, p. 119).
one, Owey Island, Donegal, seen March 14th, 1905 (*I. Nat.*, 1905, p. 201).
♂ Glenties, Donegal, shot October 25th, 1905 (*I. Nat.*, 1905, p. 263).
♂ Carrickfergus, Antrim, shot February 12th, 1906 (*I. Nat.*, 1906, p. 77).

CONNAUGHT.
♀ nearly adult, Crossmolina, Mayo, trapped April 9th, 1905 (*I. Nat.*, 1905, p. 202).
one, Clare Island, Mayo, seen March 10th, 1905 (*I. Nat.*, 1905, p. 201).
♀ juv., Belmullet, Mayo, shot March 29th, 1905 (*I. Nat.*, 1905, p. 201).
adult (?) Belmullet, Mayo, shot March 31st, 1905 (*I. Nat.*, 1905, p. 201).
adult (?) Belmullet, Mayo, shot April 2nd, 1905 (*I. Nat.*, 1905, p. 201).
one Belmullet, Mayo, captured and escaped, March 28th, 1905 (*I. Nat.*, 1905, p. 202).
two, Belmullet, Mayo, seen March and April, 1905 (G. Wallace, *in litt.*).
♂ juv., Castlegore, Mayo, shot March 30th, 1906 (in coll. of C. J. Carroll).
♀ juv., Ballysodare, Sligo, shot December 29th-30th, 1906 (*fide* Williams and Sons).
♂ juv., Westport, Mayo, shot April 10th-12th, 1907 (*fide* Williams and Sons).

MUNSTER.
♀ juv., Mizen Head, Cork, shot March, 1905 (*I. Nat.*, 1905, p. 202).
♂ Skelligs, Kerry, shot March, 1905 (*I. Nat*, 1905, p. 202).
♀ very white, Skelligs, Kerry, shot March, 1905 (*I. Nat.*, 1905, p. 202).
♀ juv., Skelligs, Kerry, shot March, 1905 (*I. Nat.*, 1905, p. 202).

This list comprises details of fifteen birds preserved, a six-teenth caught and escaped, and four or more others seen between 1903 and 1907. The Greenland Falcon seems to visit Ireland more frequently in March and April than any other months.

ICELAND FALCON (Vol. I., p. 321).—*Ireland.*—The examples recorded were seen and both obtained in the early part of 1905 (see R. J. Ussher, *List of I. Birds*, p. 29).

HOBBY (Vol. I., p. 321).—Found breeding in June, 1894, near Goyts Bridge, on the Derbyshire side of the River Goyt (Coward and Oldham, *B. of Cheshire*, p. 255). *Ireland.*—The tenth example from Ireland was picked up under a telephone wire at Loftus Hall, Fethard, co. Wexford, on April 16th, 1899 (R. M. Barrington, *Mig. B. Irish Lt. Stations*, p. *3*). The eleventh included in Mr. Ussher's " List of Irish Birds " rests upon insufficient evidence (R. J. Ussher, *in litt.*).

OSPREY (S. p. 359).—We did not mention the visits of this species to Great Britain owing to their regularity, but the following records from Ireland may be noted:—One, co. Sligo, May 3rd, 1901; ♂ juv., co. Kerry, September 30th, 1903; ♂ ad., co. Louth, end of April, 1907; ♂ juv. and one juv., co. Sligo, October 15th and 29th, 1907 (R. J. Ussher, *in litt.*). The two last have already been referred to (*cf.* Vol. I., p. 327), but the month was wrongly given as November.

NIGHT-HERON (Vol. I., p. 348).—*Ireland.*—An adult female was obtained at Ardee, co. Louth, on May 10th, 1900 (R. J. Ussher, *fide* Williams and Sons, *in litt.*). An immature bird taken on the Dodder, co. Dublin, on March 31st, 1904, is in the National Museum, Dublin (*id.*). At the Belfast Nat. Field Club meeting on October 25th, 1907, Mr. S. M. Stears exhibited a specimen of this bird (*Irish Nat.*, 1908, p. 65). Twenty-four records since 1834 (R. J. Ussher, *List of I. Birds*, p. 31).

LITTLE BITTERN (Vol. I., p. 349).—An adult male shot at Claverley, near Bridgnorth, in September, 1897, is the fifth record for Shropshire (H. E. Forrest, *in litt.*). *Ireland.*— About thirty have been obtained (R. J. Ussher, *List of I. Birds*, p. 31).

AMERICAN BITTERN (Vol. I., p. 349).—*Ireland.*—The four obtained since the publication of the " Birds of Ireland " (*cf. antea*, Vol. II., p. 276) are as follows :—Tralee, co. Kerry, November 2nd, 1901 ; Carlow, January, 1902 ; Moorstown, co. Tipperary, November 30th, 1904 ; near Colligan, co. Waterford, December 24th, 1904 (R. J. Ussher, *in litt.*).

GLOSSY IBIS (Vol. I., p. 350).—*Ireland*.—The "Manual" gives twenty from Ireland, the "Birds of Ireland" twenty-two, "List of Irish Birds" (1908), thirty-six. Of the fourteen extra to the "Birds of Ireland" three were obtained in each of the counties of Cork, Waterford, and Wexford, two in Clare, and one each in Dublin, Down, and Galway. Seven of these occurred in September, five in October, and one in November (R. J. Ussher, *in litt.*).

SPOONBILL (Vol. I., p. 350).—*Ireland*.—One was obtained in co. Galway on December 16th, 1890 (R. J. Ussher, *fide* Williams and Sons, *in litt.*). One was obtained at Tralee, co. Kerry, in September or October, 1904 (*id.*, *fide* Rohu and Sons).

GREY LAG-GOOSE (Vol. II., p. 24).—With reference to the young bird obtained in the Tay area in 1906 Mr. Harvie-Brown writes :—" Semi-domesticated Grey Lags have bred for some years close to and even within the watershed of Tay, near the southern boundaries, and truly wild Grey Lags have never been recorded as nesting anywhere within forty miles of the north-west boundary of the Tay area."

[RUDDY SHELD-DUCK (Vol. II., p. 51).—A pair was shot on the Essex coast early in January, 1908 (J. C. F. Fryer, *Field*, 1, II., 08).]

GADWALL (Vol. II., p. 52).—*Ireland*.—One, of a pair, shot on the Barrow, near Stradbally, Queens co., February 7th, 1908 (John W. Young, *Field*, 15, II., 08). One was shot in November, 1907, near Wexford Harbour (W. Rocke, *Field*, 22, II., 08). Not so rare in Ireland (Sligo, Roscommon and Leitrim) as might be supposed from records given. It is a regular visitor in small numbers to Lough Arrow : three or four couples were on the lough during the first half of April, 1909. This information comes to me from Messrs. J. Henderson, senr. and junr., both of whom are well acquainted with the bird (F. C. R. Jourdain, *in litt.*).

SHOVELER (Vol. II., p. 52).—A few pairs are said to breed on the moors of Somersetshire (F. L. Blathwayt, *Vict. Hist. Somerset*, Vol. I., p. 155). A brood was hatched both in 1904 and 1905 on a pool at Patshull, Staffs. (Lord Dartmouth, *Field*, 15, VIII., 08). *Ireland*.—Increasing as a breeding species in all the provinces (R. J. Ussher, *in litt.*).

PINTAIL (Vol. II., p. 54).—Hawick is in Roxburgh, not Berwick. But the lochs at which *both* nests are said to have been found are in Selkirk (W. Evans, *in litt.*).

GARGANEY (Vol. II., p. 54).—A pair was seen on Ellesmere, Shropshire, in April, 1906 (H. E. Forrest, *in litt.*). *Ireland*.—

One was shot near the Curragh of Kildare on September 21st, 1899 (R. J. Ussher, *List of I. Birds*, p. 34). *Scotland.*— Seven killed at Pitfour, Aberdeenshire, October 22nd, 1898 (*Ann S.N.H.*, 1899, p. 50).

WIGEON and POCHARD (Vol. II., pp. 55, 56.—*Scotland.*— "The Wigeon and Pochard have both been found nesting in Ross-shire [Gairloch (?)], and the eggs obtained " (Bateson, 1872, *Proc. N. H. Soc., Glasgow*, II., p. 182). In the last line of page 55 (*supra*) for " 1902 " read " 1899."

RED-CRESTED POCHARD (Vol. II., p. 56).—The second record mentioned under Yorkshire refers to the same bird as the first, and is wrongly dated (see *Nat.*, 1900, p. 322).

FERRUGINOUS DUCK (Vol. II., p. 57).—*Ireland.*—" Mr. F. Dyer, of Ramsgate, preserved one of two shot at Cruiserath, co. Meath, in December, 1889, as I am informed by Mr. J. E. Harting, *in litt.*" (R. J. Ussher, *in litt.*).

TUFTED DUCK (Vol. II., p. 83).—*Scotland.*—Along with Mr. Harvie-Brown's two papers in the " Ann. S.N.H." and " Proceedings of the Royal Physical Society" should be read Mr. W. Evans' paper in "Ann. S.N.H.," 1896, pp. 148 to 155.

SCAUP-DUCK (Vol. II., p. 85).—*Ireland.*—" I have seen several on a lake near the coast in Mayo in June, 1907 " (R. J. Ussher, *in litt.*). *Scotland.*—For correction of Stark's nesting record see Vol. II., p. 132.

GOLDEN-EYE (S. page 451).—Flocks, partly composed of males in adult plumage, seen on two lochs in the Forth area in summer of 1908 (W. Evans, *Ann. S.N.H.*, 1909, p. 49).

LONG-TAILED DUCK (S. page 455).—*Ireland.*—Occurs in winter on the north and west coasts down to Kerry (*Zool.*, 1907, p. 159). Several have been met with on Lough Corrib, and one of them (a male in breeding plumage) was shot in April, 1900, and is in the Nat. Museum in Dublin. A flock of five was seen on Lough Beg, and specimens were obtained (*Brit. Assoc. Guide to Belfast*, 1902). Three mature birds in Belfast Lough in May, 1898 (*ibid.*) (R. J. Ussher, *in litt.*).

EIDER DUCK (Vol. II., p. 86).—Line 3, for " Hansa " read " Handa."

GOOSANDER (Vol. II., p. 87).—*Ireland.*—" On January 16th, 1909, W. J. Williams wrote ' Adult male Goosander from Wicklow. I have not handled one for at least seven years ' " (R. J. Ussher, *in litt.*).

STOCK-DOVE (Vol. II., p. 125).—*Scotland.*—An old record for the south-west of Scotland (Ecclefechan, 1858) is given in the " Zoologist " for 1859 (p. 6378), as pointed out in my note in " Ann. S.N.H.," 1896, p. 254 (W. Evans, *in litt.*).

TURTLE-DOVE (Vol. II., p. 126).—*Yorkshire.*—Nests annually, and is by no means rare in the Scarborough district. Also found at Wetherby, and nests regularly near Harrogate and Driffield (W. Gyngell, *Nat.*, 1908, p. 464).

BLACK GROUSE (Vol. II., p. 127).—Two hybrids, apparently between Black and Red Grouse, were shot by Mr. A. Foster, of Bettws-y-coed, one on December 9th, 1895, and the other on November 20th, 1897, on Yspytty Moor, Carnarvon (H. E. Forrest, *Vert. Faun. N. Wales*, p. 307). *Ireland.*—Note the correction, Vol. II., p. 167.

PTARMIGAN (Vol. II., p. 128).—Note the correction Vol. II., p. 167.

SPOTTED CRAKE (Vol. II., p. 129).—*Ireland.*—In addition to those mentioned Mr. R. J. Ussher provides us with particulars of the following examples :—One, Castlerea, co. Roscommon, October 20th, 1900 ; two, King's co., October 7th, 1904 ; one, co. Fermanagh, October 13th, 1904 ; one, Balbriggan, co. Dublin, November 26th, 1906 ; one, Drogheda, co. Louth, December 4th, 1906 (*fide* Williams and Sons) ; one, co. Dublin, October 6th, 1902 (*fide* R. M. Barrington) ; one, Buttevant, co. Cork, January 4th, 1904 (Rohu and Son, *I. Nat.*, 1904, p. 98).

[CRANE (Vol. II., p. 147).—Two were shot on December 1st, 1903, at Knowle, Warwickshire, and were exhibited at a meeting of the Birmingham Nat. Hist. and Phil. Society (A. H. Duncalfe, *in litt.*). These seem likely to have been escaped birds.]

TURNSTONE (S. p. 557).—Has now been definitely recorded for *Derbyshire* (F. C. R. Jourdain, *Zool.*, 1909, p. 111) and also *Staffordshire* (J. R. B. Masefield, *R. and Tr. N. Staffs. F. Club*, 1909). *Ireland.*—Found regularly throughout the year along the Dublin coast. A female with ripe *ova* was obtained on July 18th, 1900, but was without a mate (C. J. Patten, *Nat.*, 1909, p. 51).

AVOCET (Vol. II., p. 228).—The east Sussex coast should have been included after Kent, as it forms a continuous coastline, but the records seem to be few. Mr. H. G. Alexander reminds us that one was seen on several days in March, 1906, by his brother and himself (*cf. Zool.*, 1906, p. 152).

GREY PHALAROPE (Vol. II., p. 229).—Five are recorded from the Isle of Man (P. G. Ralfe, *Birds I. of M.*, p. 213).

One was shot at Carno, Montgomeryshire, on October 25th, 1907 (H. E. Forrest, *in litt.*). One was shot near Hilbre Island in November, 1898 (Coward and Oldham, *B. of Cheshire*, p. 256). *Ireland.*—One was shot on Lough Foyle, near Eglinton, co. Londonderry, on September 18th, 1899 (D. C. Campbell, *I. Nat.*, 1900, p. 81). Mr. Ussher informs us of two others—one near Fethard, co. Tipperary, on November 30th, 1906 (*fide* C. J. Carroll), and one at Moyvally, co. Meath, on October 21st, 1902 (*fide* Williams and Sons).

RED-NECKED PHALAROPE (Vol. II., p. 229).—An adult in winter plumage (the first authentic record for Shropshire) was shot near Shrewsbury on November 1st, 1904 (H. E. Forrest, *in litt.*).

GREAT SNIPE (Vol. II., p. 229).—*Ireland.*—The thirteenth for Ireland was a male obtained in co. Antrim in October, 1901, and is in the National Museum, Dublin (R. J. Ussher, *in litt.*). *Scotland.*—One shot near Elgin, October 15th, 1898 (*Ann. S.N.H.*, 1899, p. 51).

LITTLE STINT (S. p. 585).—The third recorded specimen for Derbyshire was shot out of a trip of a dozen on the sewage farm at Egginton, September 26th, 1908, and is now in the possession of Mr. T. E. Auden (F. C. R. Jourdain, *in litt.*).

CURLEW-SANDPIPER (Vol. II., p. 268).—Recorded for the first time for Derbyshire (*Zool.*, 1906, p. 141). *Ireland.*—Flocks of considerable size (as many as 200 to 300) have been seen exceptionally in autumn on the Dublin coast (C. J. Patten, *Aquatic Birds*, p. 302).

PURPLE SANDPIPER (S. p. 593).—*Ireland.*—Frequently seen on the Dublin coast in nuptial plumage as late as the middle of May (C. J. Patten, *Aquatic Birds*, p. 306). A pair found by Witherby on a small island off the coast of Galway on May 30th, 1895, although in nuptial plumage had the sexual organs still undeveloped.

KNOT (Vol. II., p. 268).—*Ireland.*—One in summer plumage taken in July, 1904, at Belmullet, co. Mayo, is in the National Museum, Dublin (R. J. Ussher, *in litt.*).

SANDERLING (S. p. 597).—*Ireland.*—Has been observed on the Dublin coast in every month of the year. Even in July flocks of fifty have been seen, but the condition of their genital organs has not been examined (C. J. Patten, *Nat.*, pp. 83–85). A large flock was observed by Mr. Ussher on the shore at Cross, co. Mayo, on June 3rd, 1907, and one was shot there in the beginning of August, 1907 (*in litt.*).

RUFF (Vol. II., p. 269).—*Ireland.*—Mr. Ussher sends us a list of eighteen occurrences, briefly as follows :—1901, May, a male in breeding dress, co. Down ; one, August, co. Wexford ; one, September, co. Wicklow ; 1902, one, August, co. Mayo ; one, September, co. Donegal ; 1904, one, September, co. Westmeath ; one, October, co. Wexford ; 1905, two, August, and three September, co. Kildare ; two, September, King's co. ; one, October, co. Cavan ; 1908, two, September, co. Clare ; one from co. Limerick, date unknown.

WOOD-SANDPIPER (Vol. II., p. 269).—Only three appear to be recorded for North Wales, the last being shot by Mr. Caton Haigh on May 3rd, 1898, in Carnarvonshire (H. E. Forrest, *Vert. F. N. Wales*, p. 358).

SPOTTED REDSHANK (Vol. II., p. 270).—*Ireland.*—An immature female was obtained on Great Island, Cork Harbour, on December 26th, 1898 (W. B. Barrington, *in litt.*, to R. J. Ussher).

BAR-TAILED GODWIT (S. p. 623).—*Ireland.*—Flock of several hundred on Dublin coast on June 7th, 1899 (C. J. Patten, *Aquatic Birds*).

BLACK-TAILED GODWIT (Vol. II., p. 270).—*Scotland.*—In " Forth " (*Ann. S.N.H.*, 1903, p. 22, and 1904, p. 57), and Tiree (*Ibis*, 1903, p. 50).

BLACK TERN (Vol. II., p. 305).—An immature bird was shot near Broseley, Shropshire, on September 18th, 1901. In 1904 a pair visited Ellesmere from June 8th–11th, and two immature birds were seen on September 1st, 1904 (H. E. Forrest, *in litt.*).   Under *Derby*, for " Etwell," read " Etwall." *Ireland.*—Mr. Ussher gives us information of the following from Messrs. Williams and Sons' books :—One, co. Limerick, September 30th, 1901 (in National Museum) ; one (immature), Athlone, September 28th, 1903 ; one (immature), co. Cavan, October 7th, 1903.

WHITE-WINGED BLACK TERN (Vol. II., p. 306).—*Ireland.*— Six are recorded in the " Birds of Ireland " (p. 213) against five in the " Manual."

SANDWICH TERN (Vol. II., p. 306).—A male (the first recorded for Shropshire) was found dead near Shrewsbury in August, 1897 (H. E. Forrest, *in litt.*).

LITTLE TERN (Vol. II., p. 308).—*Scotland.*—A few pairs have again nested in the Forth area for a year or two past. This is of some importance in view of Saunders' statement that this Tern had ceased to nest in Haddingtonshire, which was quite correct (W. Evans, *in litt.*).

Sooty Tern (Vol. II., p. 308).—The specimen figured in the " Manual " was not shot as there stated (p. 653) but killed with a stone (F. C. R. Jourdain, *in litt.*).

Little Gull (Vol. II., p. 328).—*Ireland.*—An adult in winter plumage was killed about March 7th, 1909, near Laytown, co. Meath (R. M. Barrington, *Irish Nat.*, 1909, p. 99).

Great and Lesser Black-backed Gulls (S. pages 675 and 677).—*Scotland.*—Mr. Evans brings forward much evidence to prove that the naturalists who visited the Bass Rock during the first half of last century were unanimous in regarding the Black-backs that then bred there as *Larus marinus*, and that since about 1860, *L. fuscus* alone has been ascertained to nest there. The early ornithologists, however, left behind no *conclusive* evidence that their identification was correct (W. Evans, *Proc. R. Phys. Soc. Edin.*, Vol. XVI., pp. 42-51).

Glaucous Gull (Vol. II., p. 328).—*Ireland.*—In addition to those mentioned, the following occurrences supplied by Mr. Ussher do not appear to have been recorded :—1900, November 16th, co. Kerry ; 1901, March 19th, co. Mayo ; April 2nd, co. Kerry ; 1904, January 25th, co. Donegal ; February 9th, co. Wicklow ; March 3rd, co. Donegal ; March 15th, co. Donegal ; December 2nd, co. Mayo ; 1905, December 15th, co. Mayo ; 1906, December 27th, locality uncertain (*fide* Williams and Sons) ; 1907, March 31st, co. Donegal (*fide* W. A. Hamilton) ; January 11th, co. Mayo (in National Museum). This last may be the same bird as that recorded from Bartragh on December 8th, 1906 (*cf. supra*, p. 329), since the date here given is the date of the Museum register.

Iceland-Gull (Vol. II., p. 329).—*Ireland.*—Mr. Ussher gives us particulars of eleven obtained (many others have been seen), and amongst them we may note one from co. Galway on April 21st, 1906, and one from co. Londonderry on April 20th, 1903 (*I. Nat.*, 1903, p. 198), in addition to the late occurrences already mentioned.

Kittiwake Gull (S. page 683).—*Scotland.*—There are now considerably over 100 pairs breeding on the St. Abb's Cliffs. First noticed there about a dozen years ago (W. Evans, *in litt.*).

Great Skua (Vol. II., p. 330).—Holyhead record should have been under North Wales, not Ireland. An adult obtained on the River Shannon at Portumna, in October, 1906, is in Mr. C. J. Carroll's collection (R. J. Ussher, *in litt.*).

POMATORHINE SKUA (Vol. II., p. 330).—*Ireland.*—In addition to those mentioned, Mr. Ussher informs us of the following unrecorded examples :—Ballynakill, co. Galway, September 16th, 1902 ; co. Tipperary, September 19th, 1902 (*fide* Williams and Sons) ; off Cork Harbour, November 9th, 1903 ; off co. Galway, May 9th, 1904 ; off south-west coast, May, 1904.

LONG-TAILED SKUA (Vol. II., p. 331).—*Ireland.*—September 22nd, 1899 (? locality) ; co. Mayo, August 22nd, 1903 (R. J. Ussher, *in litt., fide* Williams and Sons). The following were seen by Mr. G. P. Farran : one fifty miles west of Tearaght, August 7th, 1906 ; one forty-five miles west of Skelligs, August 8th, 1906 ; one sixty-two miles south-west by west of Bull Rock, September 11th, 1907 (*id., in litt.*).

BLACK-THROATED DIVER (Vol. II., p. 333).—*Ireland.*— An adult and an immature bird are recorded in Williams and Sons' books for July 28th, 1906 (R. J. Ussher, *in litt.*).

GREAT-CRESTED GREBE (Vol. II., p. 333).—*Scotland.*—It may be recalled that Selby in 1838 (*British Ornithology*, Vol. II., p. 394) stated that it bred annually on a few of the northern Scottish lakes (W. Evans, *in litt.*).

SLAVONIAN GREBE (Vol. II., p. 334).—*Ireland.*—One in summer plumage was obtained at Belmullet, co. Mayo, in April, 1907 (R. J. Ussher, *in litt.*).

BLACK-NECKED GREBE (Vol. II., p. 369).—*Ireland.*—Three have been obtained since the publication of the " Birds of Ireland " ; one has been mentioned, the other two Mr. Ussher informs us are : Mullingar, December 30th, 1901 (*fide* Williams and Sons) ; King's co., July 4th, 1907 (H. E. Joly, *in litt.*).

MANX SHEARWATER (Vol. II., p. 372).—*Scotland.*—Present in the Firth of Forth every year from May to October ; at first, only a few, but in hundreds during August and September. Odd birds have occasionally been seen at other times (February, etc.), (W. Evans, *in litt.*).

FULMAR (Vol. II., p. 373).—One was shot by Mr. Wise on February 23rd, 1908, off Kingsgate, Thanet, Kent (C. Ingram, *Zool.*, 1908, p. 272). One was picked up dead at Canty Bay, near North Berwick, on July 16th, 1908. A pair is *said* by the lighthouse-keepers on the Bass Rock to have nested there in 1906 (W. and T. Malloch, *t.c.*, 1908, p. 396), but we can give no credence to this last statement. Mr. W. Evans writes :—" The principal lighthouse-keeper on the Bass tells me he never heard it suggested that the Fulmar had bred on the Rock. He has no doubt the recorded nesting of the Storm-Petrel there in June, 1904, is what is meant."

NOTES

LIFE OF THE LATE PROFESSOR ALFRED NEWTON.

I HAVE been invited to write a Life of the late Professor Alfred Newton. If any of your readers who have letters or reminiscences, or any other interesting information about Professor Newton, will be kind enough to communicate with me at the Savile Club, 107, Piccadilly, W., I shall be exceedingly grateful. I will, of course, undertake to return all letters, etc., to the senders.

A. F. R. WOLLASTON.

THE BIRDS OF FAIR ISLE.

IN the last issue of " The Annals of Scottish Natural History " (1909, pp. 69–75) Mr. W. Eagle Clarke gives a report on the observations made on this now well-known island during 1908. The results are even more extraordinary than in previous years (cf. Vol. I., pp. 233 and 381), and this may be due to the fact that by the generosity of friends Mr. Clarke has been enabled to instal a regular observer in the person of George Stout, a youthful inhabitant of the island, who had already shown himself an apt pupil. Beyond this important arrangement Mr. Clarke himself spent six weeks on the island in the autumn. Mr. Clarke's report is this year confined to those species which are additions to the fauna of the island, and he tells us that a great mass of information is reserved for publication in a further contribution. In those occurrences which are referred to, however, there is a most unfortunate lack of detail, which greatly lessens their interest. The list of Fair Island birds is now brought up to the remarkable total of 185. We learn that Mr. Eagle Clarke has had the good fortune to secure the interest of the proprietor of the island in the investigations, and we understand that Mr. Clarke has now been granted the sole right to shoot on the island. The following is a brief summary of the most notable items :—

BARRED WARBLER (*Sylvia nisoria*).—Several occurred in autumn and were identified beyond doubt.

SUBALPINE WARBLER (*Sylvia subalpina*).—This is one of the most interesting of these remarkable records, but only the bare fact that a bird of this species occurred during the year is chronicled. It will be remembered that the only other known occurrence of this species was at St. Kilda on June 13th and 14th, 1894 (cf. Saunders' *Manual*, p. 53). Now that Dr. Hartert has distinguished between the various races of this species (*Vög. pal. Fauna*, pp. 596–7), it would be interesting to discover the region of the origin of this specimen (we presume the bird·was secured) by a careful comparison.

ICTERINE WARBLER (*Hypolais icterina*).—The occurrence of this species can only be inferred by its inclusion in the list and by the remark that it has not been previously recorded from Scotland !

SAVI'S WARBLER (*Locustella luscinioides*).—The occurrence of this species in the spring is truly, as Mr. Clarke says, one of the most interesting events in British ornithology for many years. Since it became extinct as a breeding species in 1856 it has never been identified with certainty in England, and it has never before been known to visit Scotland.

ALPINE ACCENTOR (*Accentor collaris*).—One was seen at close quarters by Mr. Clarke in the autumn resting on the face of one of the great cliffs on the west side of the island. This species is new to Scotland.

BLUE-HEADED WAGTAIL (*Motacilla flava*).—This species occurred, but no details are given.

RED-THROATED PIPIT (*Anthus cervinus*).—This species occurred on two occasions during Mr. Clarke's visit in the autumn. Mr. Nicoll has shown (*antea*, p. 278) that there are very few reliable records of this bird's occurrence in the British Isles.

RICHARD'S PIPIT (*Anthus richardi*).—Several appeared in the autumn. Only once before recorded for Scotland.

GOLDEN ORIOLE (*Oriolus galbula*).—Observed both in spring and autumn.

HAWFINCH (*Coccothraustes vulgaris*).—A male in spring. There being no trees or shrubs it lived on the ground and fed on the dung of ponies.

TWO-BARRED CROSSBILL (*Loxia bifasciata*).—One in spring ; lived much the same as the Hawfinch. Only once previously recorded for Scotland.

RUSTIC BUNTING (*Emberiza rustica*).—Single birds on both passages. Mr. Clarke remarks that it has only once before been known to visit Scotland, but we may remind him that besides the bird recorded by himself from Cape Wrath on May 11th, 1906, a pair was reported as obtained at Torphins, Aberdeenshire, in March, 1905 (*cf.* Vol. I., p. 249).

ROSE-COLOURED STARLING (*Pastor roseus*).—An adult male in spring. A similar bird was reported on good evidence in 1907.

TUFTED DUCK (*Fuligula cristata*).—One or two appeared on migration, but whether in spring or autumn is not stated.

TEMMINCK'S STINT (*Tringa temmincki*).—Occurred in autumn ; very rare visitor to Scotland.

WOOD-SANDPIPER (*Totanus glareola*).—This bird also occurred, but no details are given. It seems curiously rare in Orkney and Shetland.

BLACK-TAILED GODWIT (*Limosa belgica*).—One visited the island in mid-winter.

EGGS OF THE CUCKOO.

IT may be of interest to put on record the following particulars of Cuckoos' eggs which I had the good fortune to find last year :—

Date.	Place.	Foster Parent.	Cuckoo's Egg.
May 18th.	Sussex.	Robin, 3 eggs.	(No. 1.) Colour like Robin's egg.
May 30th.	Surrey.	Tree-Pipit, 4 eggs, red spotted form.	(No. 2.) Dark grey-brown.
June 1st.	Same place.	Tree-Pipit, 3 eggs, grey blotched form.	(No. 3.) Pale pink - brown, similar to a Robin's egg. Found six inches from the cup of the nest, on the " platform."
June 7th.	Same place.	Tree-Pipit, 1 egg, red spotted form.	(No. 4.) Pale pink-brown, like No. 3.
			(No. 5.) Another egg outside the nest on the " platform." This one larger than No. 4, and grey-brown like No. 2. This egg had a small hole in it, and the two in the nest were stuck together by the contents of another egg, which had probably been eaten. No more eggs laid, and the nest deserted.
June 15th.	Hampshire.	Meadow-Pipit, 4 eggs.	(No. 6.) Green-yellow.
June 24th.	Surrey.	Hedge-Sparrow, 3 eggs.	(No. 7.) Like a Pied-Wagtail's egg.
July 3rd.	Same place.	Hedge-Sparrow, 2 eggs.	(No. 8.) As No. 7.

C. W. COLTHRUP.

CHAFFINCH NESTING IN WINTER.

WITH reference to the nest described in BRITISH BIRDS (*antea*, p. 381), I regret to say that on further investigation I found that the statement that the birds were seen feeding the young was incorrect. The position of the nest did not permit a view of its contents, and the nest itself contains no evidence of its having contained young. It can be proved that the Chaffinches built the nest in December, and were constantly about it during January and February, and that the hen brooded upon it. Beyond that nothing is certainly known, and it is extremely doubtful if a brood could have been reared successfully for lack of suitable food.

H. E. FORREST.

RED GROUSE AND BLACK GROUSE HYBRIDS.

IN the third line of Mr. Ogilvie-Grant's note on this subject (*supra*, p. 386) the name " Glen Troot " should be " Glen Trool." Two specimens of a similar hybrid obtained on the borders of Dumfriesshire and Kirkcudbrightshire, recorded in the " Field " at the time, may be seen in the Tullie House Museum, Carlisle. Another specimen, obtained near Kirkconnel (Dumfriesshire), is in the possession of the gentleman who shot it, and is now in Glasgow. These examples will be duly referred to in my book on the " Birds of Dumfriesshire," which it is hoped will shortly be published.

HUGH S. GLADSTONE.

NESTING RECORDS OF THE KITTIWAKE IN THE ISLE OF WIGHT.

MR. P. W. MUNN and I, in our " Birds of Hampshire," recorded a Kittiwake's egg, as picked up under the Culver Cliffs, Isle of Wight, in 1903. Mr. R. H. Fox, of Shanklin, now writes to me that Mr. G. T. Woods, the finder of the egg, after consultation with Mr. H. F. Poole, considers it to be a dwarfed egg of the Herring-Gull.

This puts back the nesting of the Kittiwake in the Isle of Wight for many years, and it would be interesting to know the date of any authentic specimens in local collections.

J. E. KELSALL.

BRÜNNICH'S GUILLEMOT IN THE FIRTH OF FORTH.

A FEMALE specimen of Brünnich's Guillemot (*Uria bruennichi*) was picked up dead on the shore at Craigielaw Point, on the Haddingtonshire coast of the Firth of Forth, on December 11th, 1908, and was sent to the Royal Scottish Museum by Mr. Valentine Knight. Judging by the size of the bill, which measures along the curve of the culmen only 1·2 inches, Mr. Clarke considers the specimen a bird of the year (W. Eagle Clarke, *Ann. S.N.H.*, 1909, pp. 75 and 76). Mr. Clarke is, however, mistaken in stating that since 1895 " no other specimen has been detected either in British waters or on our shores," for two have since been procured off the Yorkshire coast, and another, if correctly identified, has been seen off the Farne Islands (*cf. supra*, p. 331).

SLAVONIAN GREBE AND BLACK-NECKED GREBE IN HERTFORDSHIRE.

ON the 14th, and again on the 21st of March, I watched a Slavonian Grebe (*Podicipes auritus*) on Wilstone Reservoir, near Tring. The bird was in winter plumage, but the approaching change into breeding dress was heralded by a rufous tinge on the feathers of the flanks. On April 18th there was a Black-necked Grebe (*Podicipes nigricollis*) in full summer plumage on the same water. The slender, slightly recurved bill of this bird was in striking contrast with the thick, straight bill of the Slavonian Grebe.

CHAS. OLDHAM.

EARLY ARRIVAL OF THE SWIFT IN IRELAND.—Mr. Nevin H. Foster writes to us from Hillsborough, co. Down, Ireland, that he saw a Swift (*Cypselus apus*) on April 20th—a very early date for the appearance of the bird in that locality.

ROLLER IN CUMBERLAND.—An adult *Coracias garrulus* is recorded as having been shot by a keeper at Knorren, near Brampton, on June 17th, 1907 (L. E. Hope, *Zool.*, 1909, p. 156).

LONG-EARED OWL IN SHETLAND.—Three *Asio otus* were seen at Hayfield, near Lerwick, in February, 1909 (J. S. Tulloch, *Ann. S.N.H.*, 1909, p. 115).

LITTLE OWL IN NOTTINGHAMSHIRE.—An example of *Athene noctua* is recorded at Widmerpool on December 10th, 1907, and another near Clifton Grove on March 14th, 1908 (J. W. Carr, *Zool.*, 1909, p. 113).

GADWALLS IN FIFESHIRE AND ORKNEY.—A pair of *Anas strepera* was seen and one obtained on January 25th, 1909, on Morton Loch, near Tayport, and on the 29th a flock of thirty, of which three were shot and proved to be of this species, appeared on the same loch (W. Berry, *Ann. S.N.H.*, 1909, p. 116; *cf.* also *supra*, p. 348). An adult drake was shot out of a pack of Wigeon on March 8th, 1904, on Loch Stenness. Two days afterwards a female was seen on the same loch, and an adult male was seen on December 14th, 1906 (H. W. Robinson, *loc. cit.*).

STOCK-DOVE NESTING IN LANARKSHIRE.—In connection with the spread of the Stock-Dove (*Columba œnas*) as a breeding species in Scotland it is interesting to note that it " is now becoming quite established as a breeding species " in the Blantyre district of the Clyde Valley (W. Stewart, *Ann. S.N.H.*, 1909, p. 115).

REVIEWS

The Hastings and East Sussex Naturalist (Vol. I., No. 4, February 25th, 1909).

THIS number of the Journal of the Hastings and St. Leonards Natural History Society contains plenty to interest the ornithologist. The Society is much to be congratulated upon its vigour, and especially upon its strength in energetic and capable ornithological members—we believe it can boast of more M.B.O.U.'s among its members than any other local natural history society. The most important paper (pp. 153-173, plates XVIII.-XXIV.) in this number is one by our own contributor, Mr. W. H. Mullens, on " Gilbert White and Sussex." In this paper, which originally took the form of a lecture delivered before the 12th Congress of the South-Eastern Union of Scientific Societies, Mr. Mullens traces, with great care and thoroughness, Gilbert White's intimate connection with Sussex, and especially with the villages of Harting, near Petersfield, and Ringmer, near Lewes. He used to journey into Sussex frequently, and he greatly loved the Downs, of which he wrote : " I still investigate that chain of majestic mountains with fresh admiration every time I traverse it." There he saw Great Bustards and Kites, while along the chalky cliffs of the Sussex shore " the Cornish Chough builds, I know," he writes to Barrington. A careful paper is that by Mr. M. J. Nicoll on the Pipits which occur in the Hastings district. Here is recorded the fact, which we do not remember to have seen in print before, that a pair of *Tawny Pipits* " undoubtedly bred in Sussex in 1905, and again, possibly, the following year," while in 1906 Mr. Nicoll saw an adult bird collecting nesting materials (p. 183). Amongst the " Annual Notes," by the Rev. E. N. Bloomfield, we may note the following interesting records, which we do not think have been previously referred to :—*Red-footed Falcon*, Ashford, June 10th, 1908 ; *Night-Heron*, Lydd, October 3rd, 1906 ; *Spoonbill*, two, Romney Marsh, April 1st, 1908 (p. 187).

INDEX.

D

Falcon, Greenland (additions), 413.
—— Iceland, in Scotland, 310;
(additions), 414.
—— Red-footed, in Norfolk, 244;
in Kent, 427.
familiaris, Certhia. See Creeper,
Tree.
FEILDEN, COL. H. W., Notes on
Climbing Movements of the
Green Woodpecker, 93; Some
Sussex Ravens, 279.
ferina, Fuligula. See Pochard,
Common.
Finch. See Bullfinch, Chaffinch,
Greenfinch, Hawfinch, Serin.
Flamingo (additions), 24.
flammea, Strix. See Owl, Barn.
flava, Motacilla. See Wagtail, Blue-
headed.
fluviatilis, Sterna. See Tern,
Common.
Flycatcher, Pied, Nesting in Ayrshire, 139; (additions), 410.
—— Red-breasted, in Norfolk, 34,
200; Irish Records of, 248,
410; in Barra and at the Butt
of Lewis, 313; (additions),
410.
—— Spotted (nestling), 196.
Food of some British Birds, The,
reviewed, 315.
—— of Red-breasted Merganser,
311; of Eider Duck, 344,
384.
FORREST, H. E., Notes on Nuthatches breeding at Llan-
dudno, 59; Golden Oriole in
Shropshire, 59; Hoopoe in
Shropshire, 60; Short-eared
Owl Breeding in Pembroke-
shire, 60; An Early Recorded
Waxwing in Wales, 91; Black
Redstarts in Merioneth, 165;
Honey-Buzzard in Shropshire,
204; Smew in Montgomery,
311; Lesser Spotted Wood-
pecker in Merioneth, 343;
Hoopoe in Merioneth, 343;
Velvet Scoter in Shropshire,
345; Red Variety of the
Common Partridge, 345; Chaf-
finch Breeding in Winter, 381,
424.
FOWLER, W. WARDE, Note on
Little Owl in North-west
Oxfordshire, 280.

fulicarius, Phalaropus. See Phalarope, Grey.
fuliginosa, Sterna. See Tern, Sooty.
Fulmar. See Petrel.
fusca, Œdemia. See Scoter, Velvet.
fuscus, Larus. See Gull, Lesser
Black-backed.
—— *Totanus.* See Redshank,
Spotted.

Gadwall (Noble), 20, 94; (additions), 51, 415; in Somerset,
100; in Aberdeenshire, 140;
Probable Nesting in Scotland,
245; in Fifeshire, 348, 426;
in Orkney, 426.
galactodes, Aedon. See Warbler,
Rufous.
galbula, Oriolus. See Oriole,
Golden.
GALE, A. R., Note on Yellow-
browed Warbler in Yorkshire,
201.
gallicus, Cursorius. See Courser,
Cream-coloured.
Gallinule, Allen's (additions), 146.
—— Green-backed, in Norfolk, 134.
gambeli, Anser. See Goose, White-
fronted.
Garganey (Noble), 22; (additions),
54, 415; in Shetland, 245;
Breeding in East Yorkshire,
348.
garrulus, Ampelis. See Waxwing.
—— *Coracias.* See Roller.
GILROY, NORMAN, Notes on Ducks'
Eggs and Down, 94; Nesting
Habits of the Marsh-Warbler,
235; On the Nesting of the
Goosander, 400.
giu, Scops. See Owl, Scops.
glacialis, Colymbus. See Diver,
Great Northern.
—— *Fulmarus.* See Petrel, Ful-
mar.
—— *Harelda.* See Duck, Long-
tailed.
GLADSTONE, HUGH S., Note on Red
Grouse and Black Grouse Hy-
brids, 425.
glandarius, Garrulus. See Jay.
glareola, Totanus. See Sandpiper,
Wood.

LEIGH, A. G., Notes on Redshank
Breeding in Warwickshire, 33 ;
Curious Site for a Robin's
Nest, 90 ; Late Nests of the
Great Crested and Little Grebes,
171 ; Little Owl in Warwickshire, 240.
lentiginosus, Botaurus. See Bittern,
American.
leucoptera, Hydrochelidon. See
Tern, White-winged Black.
leucopterus, Larus. See Gull,
Iceland.
leucorodia, Platalea. See Spoonbill.
leucorrhoa, Procellaria. See Petrel,
Leach's Fork-tailed.
—— *Saxicola œnanthe.* See Wheatear, Greenland.
longicaudata, Bartramia. See Sandpiper, Bartram's.
lugubris, Motacilla. See Wagtail,
Pied.
luscinia, Daulias. See Nightingale.
luscinioides, Locustella. See Warbler, Savi's.
LYNES, COM. H., R.N., Note on
Pebble Nest of a Ringed
Plover, 136.

MACKEITH, T. THORNTON, Note on
the Courting Performance of
the Cuckoo, 239.
macrura, Sterna. See Tern, Arctic.
macularius, Totanus. See Sandpiper, Spotted.
maculata, Tringa. See Sandpiper,
Pectoral.
MAGRATH, MAJOR H. A. F., Notes
on the Common Cuckoo in
India, 197.
major, Dendrocopus. See Woodpecker, Great Spotted.
—— *Gallinago.* See Snipe, Great.
—— *Parus.* See Titmouse, Great.
Mallard (Noble), 20 ; Want of
Down in Nests of, 62 ; Hatching in October, 245.
MAPLETON, H. W., on the Song of
the Wood-Warbler, 226.
Mapping Migratory Birds in their
Nesting Areas, On a Plan of,
322.
marila, Fuligula. See Scaup-Duck.
martius, Picus. See Woodpecker,
Black.

maruetta, Porzana. See Crake,
Spotted.
marinus, Larus. See Gull, Great
Black-backed.
Marked Birds, 35, 171, 245, 246.
Marking Birds, A Plan for, 35.
—— —— Notes on the Work at the
Rossitten Station, 362.
MARRIAGE, A. W., Note on Little
Owl in Hampshire, 310.
Martin, Sand (nestling), 192.
May, Isle of, Rare Birds on the, 346.
MAY, W. NORMAN, Note on Grey
Wagtail Nesting in Berkshire,
90.
MEADE-WALDO, E. G. B., Notes on
Pied Wagtail Rearing Three
Broods, 130 ; Old English
Nesting Bottles, 164.
MEDLICOTT, W. S., Note on Curious
Site of a Wood-Warbler's Nest,
380.
melanocephala, Motacilla. See Wag-
tail, Black-headed.
melanocephalus, Larus. See Gull,
Mediterranean Black-headed.
melanope, Motacilla. See Wagtail,
Grey.
melba, Cypselus. See Swift, Alpine.
merganser, Mergus. See Goosander.
Merganser, Red-breasted (Noble),
40 ; Food of, 311.
merula, Turdus. See Blackbird.
MEYRICK, COL. H., Note on Autumn
and Winter Singing of Buntings,
237.
migrans, Milvus. See Kite, Black.
(Migration) " Report on the Immigrations of Summer Residents
in the Spring of 1907," Reviewed, 247.
—— On a Plan of Mapping Migratory Birds in their Nesting
Areas, 322.
MILLAIS, J. G., Note on Red Grouse
and Black Grouse Hybrids, 384.
minor, Dendrocopus. See Woodpecker, Lesser Spotted.
minuta, Ardetta. See Bittern, Little.
—— *Sterna.* See Tern, Little.
—— *Tringa.* See Stint, Little.
minutilla, Tringa. See Stint,
American.
minutus, Larus. See Gull, Little.
modularis, Accentor. See Sparrow,
Hedge.

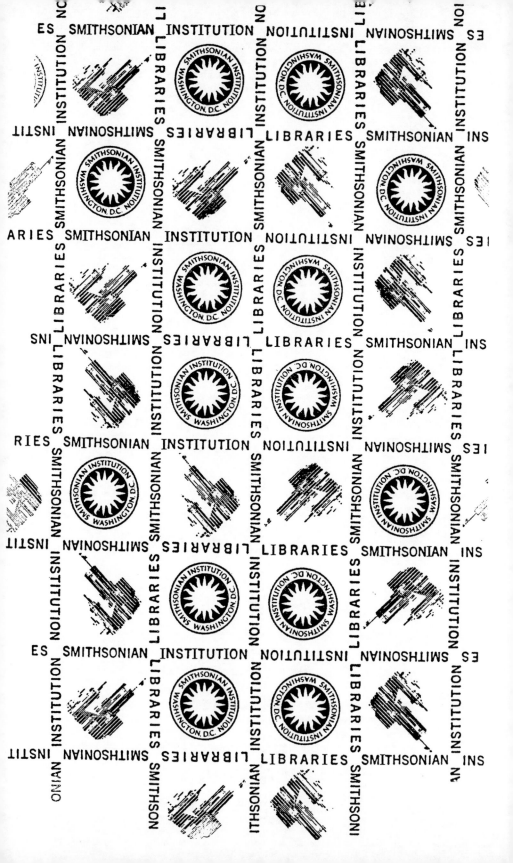